Lake Prespa, Northwestern Greece

Developments in Hydrobiology 122

Series editor
H. J. Dumont

Lake Prespa, Northwestern Greece

A Unique Balkan Wetland

Edited by

Alain J. Crivelli & George Catsadorakis

Reprinted from Hydrobiologia, volume 351 (1997)

Springer Science+Business Media, B.V.

Library of Congress Cataloging-in-Publication Data

A C.I.P. Catalogue record for this book is available from the Library of Congress.

ISBN 978-94-010-6179-7 ISBN 978-94-011-5180-1 (eBook)
DOI 10.1007/978-94-011-5180-1

Printed on acid-free paper

This volume is dedicated to all the people living in the Prespa region in northwestern Greece, Albania and FYROM in the hope that they will succeed in their quest for an environmentally friendly way to stay in the area they love

Contents

Lake Prespa, northwestern Greece.

Hydrobiologia **351:** ix–xi, 1997.
A. J. Crivelli & G. Catsadorakis (eds), Lake Prespa, Northwestern Greece.

Preface

After the Lakes Skadar and Ohrid, the Lakes Megali Prespa and Lake Mikri Prespa, actually forming one wetland, are the largest waterbodies of the Balkans (Figures 1 and 2). They belong to three countries: Albania, FYROM (Former Yugoslavian Republic of Macedonia) and Greece. They are of Tertiary origin and have only underground outlets. The lakes are at 850 metres above sea level amidst mountains rising to over 2500 m asl.

Prespa, as the area of Lakes Prespa area is called, is particularly well-known in Europe for its birdlife, but also for its flora. It is a centre of endemism of noteworthy importance for the biodiversity of Europe. Within the catchment area of Lakes Prespa or its immediate vicinity, three National Parks (Mikri Prespa, Greece, since 1974; Pelister, FYROM since 1948 and Galicia, FYROM since 1958) exist, a fact clearly indicating the high biological, ecological and aesthetic value of this region.

Being in a border area, at least from the beginning of the present century, Prespa area suffered repeately from international political conflicts, and more recently in the Greek civil war (1944–1949). Consequently this remote area remains relatively underdeveloped with a more or less traditional agriculture and fisheries as the main economic activities.

Thus, Prespa is presently facing a difficult problem: how to be developed socio-economically for the welfare of its people, without jeopardizing its tremendous natural and cultural assets. We believe that whatever the solution to this problem will be, it cannot be achieved without a close international collaboration of the three countries sharing this ecosystem. Particularly, a trilateral agreement on the management of the lakes is urgently needed considering it is the key compartment of the whole ecosystem. The fate of Prespa will depend on how successfully this issue will be challenged. We envisage our special issue as a contribution to creating a sturdy reference for the future discussions and negotiations between the various local, governmental and non-governmental bodies for the Prespa area.

The aims of this special issue, which sprung out almost exclusively of studies executed at the Greek parts of Prespa, is threefold:

– to present the available scientific knowledge on the area's ecosystems. Most of this knowledge came out of contributors conducting original studies the last ten to fifteen years.

– to assess, when feasible, changes, threats and trends in the flora and in the fauna (bird, fish) and in the land use and socio-economic activities of the area such as fisheries, agriculture, livestock rearing.

– to try in the light of old and recent knowledges to enhance the understanding about potential management and land use activties in the area which should preserve its exceptional biodiversity and at the same time allow local people increase their standard of living.

The core of this volume concerns unavoidably only the Greek part of the area but comments, ideas and extrapolations are made for the Albanian and FYROM parts whenever available data permit it.

We would like to thank the following referees who have greatly improved the manuscripts: G. E. Hollis, P. Isenmann, H. Källander, J-L. Martin, H. Strijbosch, J. Jalbert, J. Skinner, T. Papayannis, P. S. Maitland, E. Franquet, A. Vaquer, H. J. Dumont. We would like also to thank the inhabitants of Prespa for their collaboration and friendship.

We particularly recognize the contribution of Dr G. E. Hollis who died suddenly in autumn 1996 before the completion of this special issue. In the late eighties he was the dynamic project leader of a WWF-International project whose aim was to produce a Management Plan for the Prespa National Park in Greece. His contribution to the understanding of the function-

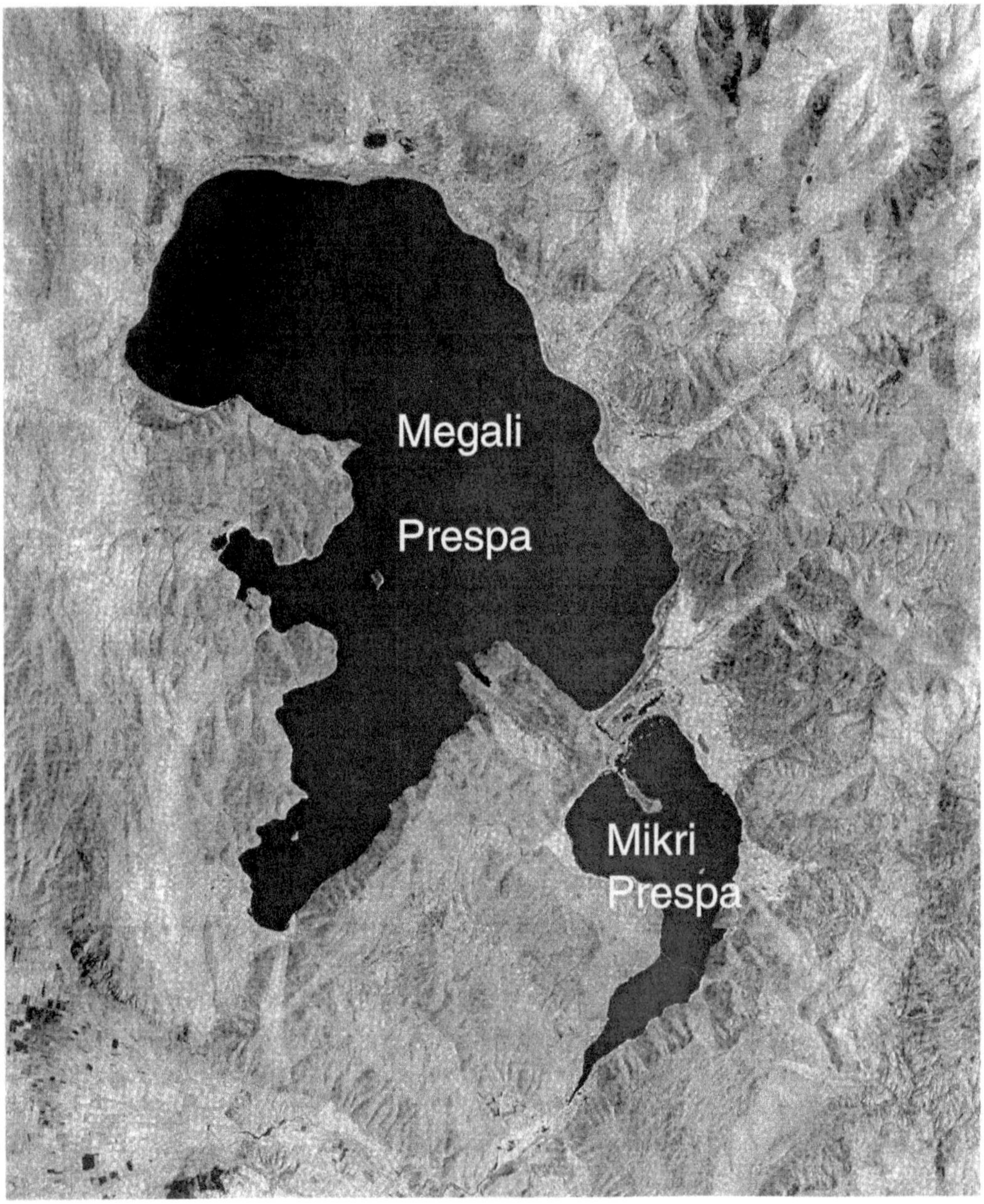

Figure 1. Landsat satellite image (1994) of the Lakes Prespa.

Figure 2. View over Lake Mikri Prespa.

ing of this wetland will long be remembered. Finally we would like to acknowledge R.H. Britton for correcting the English of the numerous non-English speaking authors and C. Heurteaux and D. Poisson for finalizing the manuscript.

The work to prepare this special issue would not have been possible without the financial support of the Tour du Valat Foundation.

ALAIN J. CRIVELLI
Station biologique de la
Tour du Valat,
Le Sambuc,
13200 Arles, France

GIORGOS CATSADORAKIS
Society for the
Protection of Prespa,
530 77 Agios Germanos,
Greece

Hydrobiologia **351:** xiii, 1997.
A. J. Crivelli & G. Catsadorakis (eds), Lake Prespa, Northwestern Greece.

List of contributors

BOUSBOURAS, D., 37 Pergamou Street, 60 100 Katerini, Greece.

CATSADORAKIS, G., Society for the Protection of Prespa, 530 77 Agios Germanos, Greece.

CRIVELLI, A.J., Station Biologique de la Tour du Valat, Le Sambuc, 13200 Arles, France.

DANALATOS, N.G., Agricultural University of Athens, Lab of Soils and Agricultural Chemistry, Iera Odos 75, Botanikos 11855, Athens Greece.

ECONOMIDIS, P. S., Aristotle University of Thessaloniki, Department of Zoology, P.O.Box 134, 54006 Thessaloniki, Greece.

HOLLIS†, G.E., Wetland Research Unit, Department of Geography, University College London, 26 Bedford Way, London WC1H 0AP, United Kingdom. († deceased)

IOANNIDIS, Y., 34 Ch. Trikoupi Street, 185 36 Piraeus, Greece.

KOSMAS, C.S., Agricultural University of Athens, Laboratory of Soils and Agricultural Chemistry, Iera Odos 75, Botanikos 11855, Athens Greece.

MALAKOU, M., Society for the Protection of Prespa, Agios Germanos 530 77, Greece.

MICHALOUDI, E., Aristotle University of Thessaloniki, Department of Zoology, P.O.Box 134, 54006 Thessaloniki, Greece.

MOUSTAKA-GOUNI, M., Institute of Botany, School of Biology, Aristotle University of Thessaloniki, 54006 Thessaloniki, Greece.

MOUSTAKAS, N.K., Agricultural University of Athens, Laboratory of Soils and Agricultural Chemistry, Iera Odos 75, Botanikos 11855, Athens, Greece.

PAVLIDES, G., Akropoleos 36, 54634 Thessaloniki, Greece.

PETRIDIS, D., T.E.I of Thessaloniki, Department of Food Technology, P.O. Box 14561, 54101 Thessaloniki, Greece.

ROSECCHI, E., Station Biologique de la Tour du Valat, Le Sambuc, 13200 Arles, France.

SINIS, A., Department of Zoology, University of Thessaloniki, P.O. Box 134, 54006 Thessaloniki, Greece.

STEVENSON, A.C., Department of Geography, University of Newcastle upon Tyne, Newcastle NE1 7RU, United Kingdom.

TRYFON, E., Institute of Botany, School of Biology, Aristotle University of Thessaloniki, 540 06 Thessaloniki, Greece.

ZARFDJIAN, M., Aristotle University of Thessaloniki, Department of Zoology, P.O.Box 134, 54006 Thessaloniki, Greece.

Hydrobiologia **351:** 1–19, 1997.
A. J. Crivelli & G. Catsadorakis (eds), *Lake Prespa, Northwestern Greece.*

The physical basis of the Lake Mikri Prespa systems: geology, climate, hydrology and water quality

G. E. Hollis[1†] & A. C. Stevenson[2]
[1] *Wetland Research Unit, Department of Geography, University College London, 26 Bedford Way, London WC1H 0AP, UK (†deceased)*
[2] *Department of Geography, University of Newcastle upon Tyne, Newcastle NE1 7RU, UK*

Received March 1997, accepted May 1997

Key words: geology, climate, hydrology, lake level regime, water quality, management, Greece

Abstract

Lake Mikri Prespa is 47 km^2 in area, with a maximum depth of 8.4 m and a normal water level at around 850 m above sea level. It collects water from a granitic and karstic catchment of 189 km^2 which includes snowy mountains rising to over 2000 m. The main flat land is the sedimentary plain to the east and the sandy isthmus that separates Mikri Prespa from the much larger Megali Prespa into which the former normally drains. Rainfall averages around 750 mm because the 600 mm to 700 mm falling on the plains is augmented by heavier falls including much snow on the mountains. There is rainfall each month with an autumnal maximum. River inputs to the lake and the lake level itself peak in the spring with the snowmelt. Mikri Prespa normally rises by about a metre to flood the surrounding wet meadows landward of the reed fringe. The annual cycle of water level fluctuation is superimposed on infrequent upward surges in the level of Mikri and Megali Prespa because of particularly wet and snowy winters and, in recent years, the steady decline of the level of Megali Prespa because of tectonic activity.

There has been considerable human modification of the hydrology of the area with the diversion of the Agios Germanos torrent from Mikri to Megali Prespa, the creation of a canal in Albania which can input or withdraw water from the lake, the culverting of the canal linking Mikri and Megali Prespa, and the creation of an irrigation scheme taking water directly from Mikri Prespa and from the Agios Germanos stream.

The lake water is base rich because of the limestone in the catchment. Whilst concern has been expressed at the eutrophication of the lake, recent studies have proved that there has been no significant change in nutrient status this century and the lake is not eutrophic. There has been a significant increase in turbidity but this may be the result of sediment disturbance by an introduced fish species.

The lake and its supporting hydrological system will need careful monitoring if it is to be effectively managed. A particularly high priority is the development and implementation of a water level management plan for the lake.

Introduction

The lakes, Mikri Prespa and Megali Prespa, are located at the northern end of the Pindus Mountains at the intersection of the frontiers of Greece, Albania and FYROM (Former Yugoslavian Republic of Macedonia). The lakes are at 850 metres above sea level (m asl) amidst mountains rising to over 2000 m asl. There are extensive areas of dolomitic limestone in the area together with granite and hard crystalline metamorphic rocks. Tertiary and Quaternary sediments comprise the relatively small areas of flat land that surround parts of the lakes. The catchment feeding the lakes is a largely closed basin with a hypothesized underground karstic outflow to Lake Ohrid, 180 m lower, and thence via the river Drin to the Adriatic (Figure 1). Canal and sluice construction at the western extremity of Mikri Prespa in Albania has, however, effected a carefully managed link to the River Devoll.

Mikri Prespa (47.4 km^2) and Megali Prespa (253.6 km^2) are linked by a small channel which traverses the alluvial isthmus that separates the lakes. Precipitation varies from around 750 mm in the lowlands to probably well over 1200 mm on the moun-

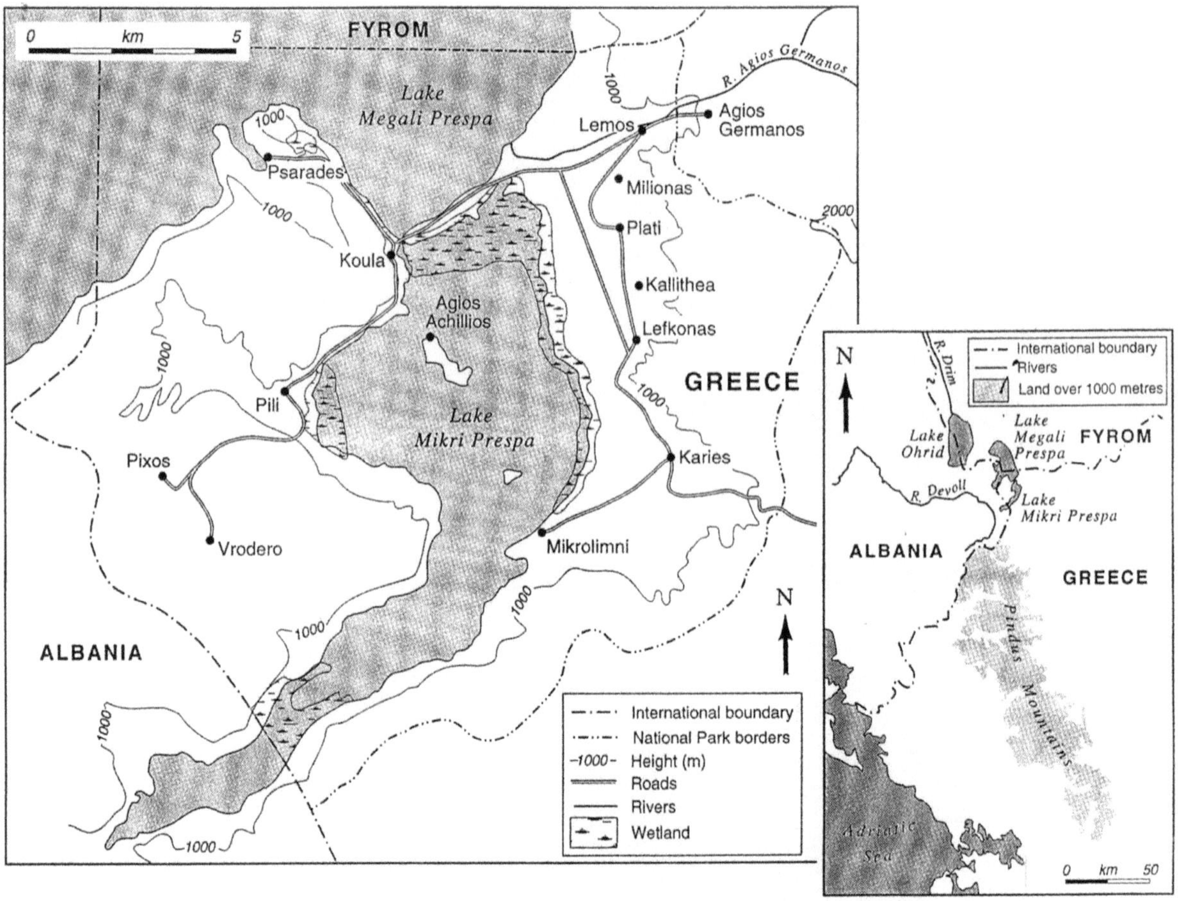

Figure 1. Location map (after Pano 1984).

tains. Annual open water evaporation is of the order of 980 mm. There is a large fall of snow during most winters and the short steep torrents that feed the lakes have their peak flows with snow melt in the late spring. It is likely that the karstic inputs to the lakes follow a similar pattern. The lake levels follow an annual cycle with peak levels in May and June and the low water levels in the autumn. Superimposed on this annual cycle are longer period fluctuations caused by particularly wet or dry periods. The early 1960s were a time when both lakes reached exceptionally high levels whilst the late 1980s saw particularly low lake levels. In the early 1990s, the long established complementarity of levels in Mikri and Megali Prespa has been broken with the former rising to high levels thanks to large winter snow falls and the latter suffering a continual fall in level, possibly because of tectonic changes to its karstic outflow routes.

Lake Mikri Prespa has been affected by human action. In 1936 the Agios Germanos Stream was diverted from Mikri Prespa to its present artificial channel leading into Megali Prespa. In 1953 the Albanians linked Mikri Prespa to the River Devoll so as to allow water to enter the lake in winter and to drain from the lake in summer for irrigation. In 1969, the Albanians added a dam and sluice to the canal leading from Mikri Prespa. In the same year, the Greeks transformed the connection between Mikri and Megali Prespa from a shallow weedy channel into the present deep and narrow concrete flume and road bridge. More recently a sluice gate has been placed at the Mikri Prespa end of this channel. From 1985 onwards, irrigation abstractions gradually took larger amounts of water from Mikri Prespa for the surrounding irrigated fields. The Greek portion of the Prespa region including Mikri Prespa and part of Megali Prespa was declared a National Park in 1974. However, no management

Table 1. Meteorological data for the Prespa Region.

Station	Variable	Data record
Koula	Rainfall	1953–1965
Koula	Meteorology	1965–1990
Agios Germanos	Rain/snow	1965→
Vrondero	Rain/snow	1965→
Kallithea	Rain/snow	1962–1980
Mikrolimno	Rain/snow	1965→

plan was implemented and there was virtually no control of land-use activities within the catchment. The intensification of agriculture and an EU funded project to build a fish farm adjacent to Lake Mikri Prespa (Pyrovetsi et al., 1984), coupled with a major intensification of scientific research (Pyrovetsi & Crivelli, 1988) and conservation activity (Kassioumis et al., 1990), has caused concern about water quality, especially eutrophication, in the lakes.

The hydrology of Lake Mikri Prespa is a major factor in controlling the wetland ecosystems of Prespa with its internationally important avifauna, significant fish populations and fishery. The late spring flooding of the wet meadows behind the reedbeds is crucial for fish spawning and for migratory and breeding water birds. The pelican colonies in the reedbeds are generally kept secure from predators through the maintenance of lake water inundation in the reedbeds.

This paper describes the geology and climate of the Prespa region as an introduction to the hydrology. The hydrology of Lake Mikri Prespa is covered with attention being given to the effects of past human activity and possible future schemes affecting the lake. A review of the available water quality data and related publications for Lake Mikri Prespa is undertaken and shows that there have been minimal changes in lake water quality this century. The paper concludes with recommendations for the hydrological monitoring of this internationally important wetland system.

Geology

A summary description of the geology and geological history of the zone is given in Part I of the Prespa National Park Management Plan (Kassioumis et al., 1990). The Prespa lakes, Lake Malik (drained in 1938) and Lake Ohrid used to be a part of a former Dassaretic Lake in the Jurassic and was linked with the former Adriatic sea. This distinguished Prespa from the other west Macedonian lakes because they were once linked in geological history with the Aegean. It is believed that the Prespa basin used to be a karstic collapse feature before the formation of the lakes took place in the Tertiary (around 35 million years ago) (Jakovljevic, 1935). It is thought that the water level of both Mikri and Megali Prespa were about 80 metres higher during the Pliocene than today.

The catchment geology is dominated by limestones and dolomites on the western and southern sides of Mikri Prespa, while on the eastern side they are predominantly granites and gneisses (Figure 2). The lower strata of these granites and metamorphic rocks are Palaeozoic in age, whilst the upper layers, which include some slightly metamorphosed conglomerates and sandstones, is upper Palaeozoic and Lower Triassic in age (Kassioumis et al., 1990).

Geomorphology

The Prespa region is one of high mountains and steep torrents surrounding the twin lakes of Mikri and Megali Prespa with modest areas of low relief surrounding certain parts of the lakes. Lake Mikri Prespa is an elongated shape with a maximum length of 13 km and a maximum width of 6 km. It has a maximum depth of 8.4 m with several islands within it. A channel, 50 m long and 2 to 10 m wide, connects Mikri Prespa with Megali Prespa. The channel crosses a narrow isthmus of alluvial deposits 4 km long and 50 to 500 m wide. An elderly fisherman in Psarades has reported that during the Second World War the channel linking Mikri and Megali Prespa was very wide and boats could easily sail from one lake to the other.

Field and map evidence suggests that Lakes Megali and Mikri Prespa were once a single lake. It appears that sediments carried into the lake by the Agios Germanos torrent, in particular, have been sorted and moved by the lake's currents and waves in such a way that the Koula – Vromolimni isthmus has been constructed. The Agios Germanos stream was first diverted from Mikri Prespa to Megali Prespa in 1936. Pano (1984), presumably using an old map source, shows the Agios Germanos stream flowing to Lake Mikri Prespa whilst maps from World War II all depict the Agios Germanos discharging to Megali Prespa (Hollis et al., 1989).

There is no high land shown between the Albanian end of Mikri Prespa and the floodplain of the River Devoll. The 1944 British War Office map gives two

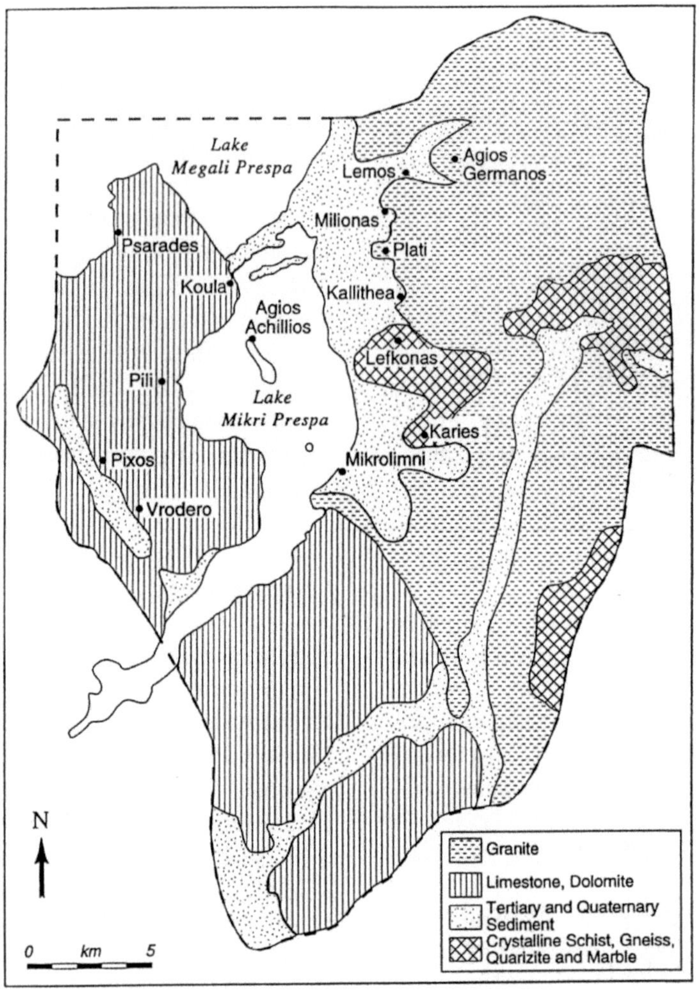

Figure 2. Geology of the land around Lake Mikri Prespa.

Table 2. Long-term meteorological averages for the Prespa Region (after Kassioumis et al., 1990).

Rainfall (mm)	Oct	Nov	Dec	Jan	Feb	Mar	Apr	May	Jun	Jul	Aug	Sep	Year
Koula	59	71	70	50	48	44	44	50	29	26	34	47	572
Kalithea	63	67	67	65	56	47	60	62	35	35	38	52	646
Mikrolimni	58	78	77	80	71	57	56	61	32	33	36	47	685
Vrondero	73	92	98	74	57	49	53	58	34	24	39	44	692
Ag. Germanos	65	78	62	56	46	48	56	60	33	39	39	46	638
Meteorology at Koula													
Mean temp. °C	12	7	3	1	3	6	9	14	20	22	20	17	
Humidity %	65	71	76	75	72	64	62	60	55	51	53	57	
Wind speed m/sec	1.7	1.5	0.9	1.1	1.4	2.0	2.1	2.3	2.3	2.3	2.3	1.8	
Sunshine	184	119	93	105	103	139	192	242	297	324	280	240	

spot heights on the floodplain some 2.5 km west of the end of Mikri Prespa with elevations of 857 and 851 metres. It appears, therefore, that it is possible that there could be a link if the level of the lake or of the river were particularly high (Broikos, 1962). It is indeed possible that the ancient surface outflow from the Megali and Mikri Prespa catchments was into the River Devoll in Albania. Crivelli (pers. comm.) visited the Albanian portion of Lake Mikri Prespa in 1992 and observed a canal, dug between 1951 and 1953, leading from the lake. The canal, with a width of 30 to 80 m drained into the River Devoll and was used to irrigate the regions of Diviaka, Lushnje and Berat. It was reported that in 1969 a dam, with sluices, was constructed across the canal so as to better regulate water flows. Since 1969, the dam has permitted water to enter Mikri Prespa in the winter and spring, and to flow out for irrigation purposes in the summer months.

Megali Prespa is a much larger lake (253 km^2) than Mikri Prespa, being 26 km long, 20 km wide and up to 55 m deep. Megali Prespa appears, from the map evidence, to have no surface outflow (Figure 1). It is widely believed that Megali Prespa flows down to Lake Ohrid in FYROM via underground karstic channels (Pano, 1984).

The land around the lakes is mountainous. The highest point is Kalo Nero (2156 m) to the east. Triklarion mountain (1756 m) stands to the south and Vrondero (1456 m) to the west. Between the two lakes is Devas with an altitude of 1373 m. Hence between the lakes and Kalo Nero there is a elevation change of 1300 m in 7 km. The only flat areas are the alluvial plains of Holocene age around the eastern side of Mikri Prespa, in the Kefalorema valley west of Pili in the west and on the isthmus between the lakes. Much of the shore of Mikri Prespa is low lying wet meadow, marsh and reedbed as is the Mikri Prespa side of the isthmus. The Megali Prespa side of the isthmus has a wide sandy beach but much of the rest of the Greek shore of Megali Prespa is rocky with cliffs and caves in several places.

Soils are formed mainly of alluvia and colluvia on the agricultural land and are characterised by medium pH (6.0–6.5) and low amounts of phosphorus and potassium. The soil types of the forest zone are poorly known but consist of calcareous rendzinas and podsolised brown earth forest soils (Pavlides, 1985).

Climate

The climate shows the characteristics of a hot and dry Mediterranean type during summer (July 23.6 °C) and Mid-European type during the winter with long periods of high rainfall, snow, increased cloudiness and low temperatures (January 0.8 °C). Under an Emberger bioclimatic classification the area is typical of a humid mediterranean type (Emberger, 1963).

There are neither complete nor elaborate meteorological records for the Prespa region. The station at Koula, on the isthmus between Lakes Mikri and Megali Prespa, has operated since 1953 but only gathered a comprehensive set of meteorological data from 1965. Four other sites have recorded rainfall since the 1960s (Table 1). The meteorological data is not formally published in Greece and is not widely available. The fact that there are several governmental agencies responsible, e.g. Public Power Corporation (PPC), Forestry Service, Geological Research Institute etc, causes even greater complications. The death in 1990 of the long serving PPC observer at Koula has seriously degraded the record from this station and caused the loss of certain data.

All of the sites shown in Table 1 are on the plains around the lakes at between 850 and 1100 m asl. There are no meteorological stations or rain gauges in the surrounding mountains and so the meteorological data probably represents conditions as being drier and more equable than is the case in the catchment as a whole.

Table 2 gives the seasonal cycle of weather conditions at Prespa. The overall average rainfall of 647 mm per year is obviously representative of the low-lying agricultural lands. Bonazoundas (1988) gives a catchment average of 742 mm and it seems certain that the precipitation over the highest peaks is greatly in excess of 1000 mm. Kassioumis et al. (1990) report that snowfall averages 300 mm/year. Again this may be representative of the lowlands but the highest land probably receives more than 1 m of snow annually.

The mean air temperature remains above freezing throughout the year but Lake Mikri Prespa is frequently frozen. It seems likely that the deep ravines descending from the mountains, especially that between Triklario and Vrondero, assist the passage of cold air descending from the mountains and so cause the lake to freeze over. Summer temperatures reach a maximum of 22 °C in July with an absolute maximum of 35 °C. Kassioumis et al. (1990) give the open water evaporation rate for Koula as 980 mm per year with a maximum of 204 mm in July.

Table 3. The catchments of the Prespa lakes.

Megali Prespa	km^2
Catchment area in FYROM	1795.5
Catchment area in Albania	162.0
Catchment area in Greece	71.6
Total catchment area	2029.1
Lake area	253.6
Mikri Prespa	
Catchment area in Albania	51.0
Catchment area in Greece to east of lake	94.0
Catchment area in Greece to west of lake	44.0
Total catchment area	189.0
Lake area	47.4
Total area of the combined drainage basins and lakes	2519.1

Table 4. Hydrological data for the Prespa Region.

Station	Parameter	Data record	Data
Megali Prespa at Koula	Lake level	1953–91	Daily
Mikri Prespa at Koula	Lake level	1953→	Daily
Agios Germanos Stream	River flow	1961→	Daily
Oxia Torrent	River flow	1961–68	–
Karies Torrent	River flow	1961–68	–
Kalithea Torrent	River flow	1961–68	–
Levkonas Torrent	River flow	1961–68	–
Plati Torrent	River flow	1961–68	–

Hydrology

The entire internal drainage basin area of the two Prespa lakes, Megali and Mikri, is 2519.1 km^2 (Table 3). Megali Prespa has an area of 253.6 km^2 and a catchment area of 2519.1 km^2 whilst the much smaller Mikri Prespa, 47.4 km^2, has a catchment area of 189 km^2. As such Megali has 8.0 km^2 per km^2 of lake area and Mikri's catchment comprises 4.0 km^2 for each km^2 of lake area.

Table 4 shows the limited hydrological data available at present. As with the meteorological data, this hydrological data is neither published nor centrally archived in Greece. The quality of much of the information, which can be found on hand written sheets, has not been subject to any statistical control procedures. At present the authors have not been able to access any hydrological data deriving from either FYROM or Albania. However, it is believed that the Institut d'Hydrométéorologie in Tirana has certain data pertaining to Prespa.

River flow

The Agios Germanos stream is the best documented of the rivers in the Greek part of Prespa. Its catchment area at Agios Germanos village is 51.6 km^2 with impermeable granites and gneisses being the main lithology. Below the village there is a further 20 km^2 of catchment area contributing to the stream before it flows into Lake Megali Prespa. The only channel of the Agios Germanos torrent flows from Agios Germanos to Lemos and thence to Megali Prespa in a deeply excavated and wide channel between high embankments. River water is diverted from the stream above Agios Germanos to feed the higher levels of the Prespa irrigation scheme.

Bonazoundas (1988) found that the minimum discharge of the Agios Germanos torrent was 210 $l\ s^{-1}$ and its mean discharge was 1115 $l\ s^{-1}$. Table 5 shows the monthly regime of the stream with the annual average flow as 1.17 $m^3\ s^{-1}$ ($37 \times 10^6\ m^3$ annually). The peak flow is in May when snowmelt adds to rainfall inputs. The lowest flows are from July to October when rainfall is slight and evapotranspiration is at its peak. The relatively modest winter runoff is caused by the low temperatures and the maintenance of large snow fields on the mountains in almost all winters.

Figure 3 shows the daily flow hydrograph for October 1978 to September 1984. This reveals the dominance of the snowmelt hydrograph in the late spring. However the highest daily flow recorded was in the autumn of 1979 and was associated with particularly heavy rainfall. Similar late autumn runoff peaks also occurred in 1981 and 1982.

Table 6 shows that in the 1960s the monthly flow in the Agios Germanos torrent was between 5 and 37 times greater than the aggregate inflow to Mikri Prespa along the five torrents which do regularly carry flow. Table 5 shows that, for the more recent data for the Agios Germanos, the months April to October cover 60% of the annual flow. However, the exact location of the gauging stations on the Mikri Prespa torrents is not known and it is possible that there were substantial infiltration losses to the groundwater of the alluvial fans upstream of the gauging stations. Similarly, it must not be assumed that the total inflow to Mikri Prespa is represented by the torrent data in Table 6, because there is direct rainfall input and, almost certainly, underground inflow from the substantial limestone areas of the catchment.

Table 5. Average flow for the Agios Germanos Torrent for 1978–87 (m^3/sec).

Oct	Nov	Dec	Jan	Feb	Mar	Apr	May	Jun	Jul	Aug	Sep	Year
0.36	0.74	1.10	1.16	1.02	1.43	2.87	3.48	1.04	0.48	0.27	0.25	1.17

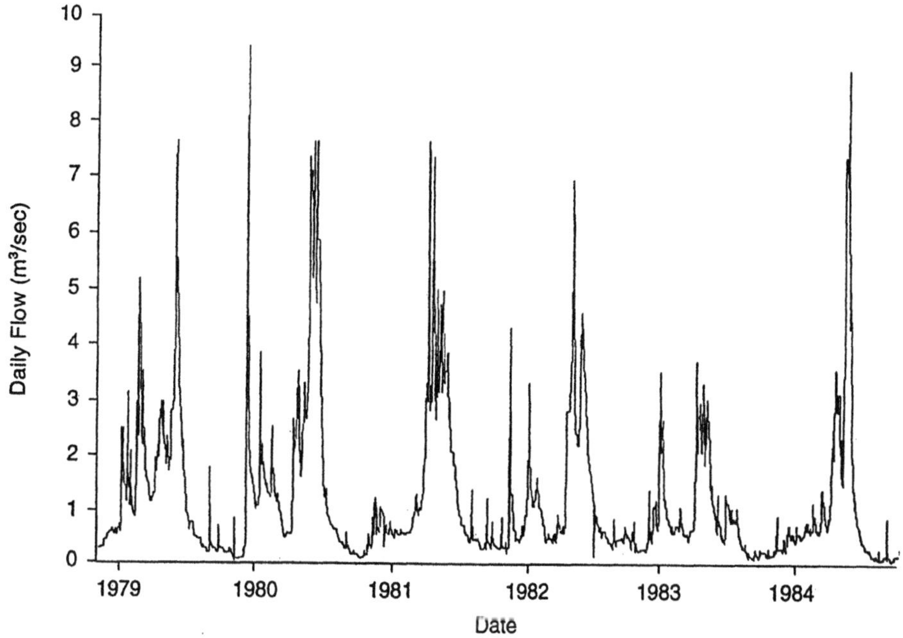

Figure 3. The mean daily flow of the Agios Germanos stream at Agios Germanos.

Water balance of Mikri Prespa

Bonazoundas (1988) calculation of the water balance of the lake at Mikri Prespa (Table 7) used Penman procedures and made informed estimates of such imponderables as groundwater inflow and outflow. The estimate of 4.8 years as the turnover time of the lake is fair in the light of the scant data available for Greece. There must remain large question marks over many parts of the water balance given the karstic inputs and outputs and the large altitudinal variations that are not represented in the instrumental record. In addition, Bounazoundas was not able to make allowance for the transfers of water into and out of the lake which have been made by the Albanians over the years.

Lake level regimes of Mikri and Megali Prespa

Daily records of the level of both Mikri and Megali Prespa at Koula in Greece have been made since October 1953. These manuscript records vary in the manner in which lake level is recorded and various figures are given for the elevation of the zero point on the stage boards (Hollis et al., 1989). However, a consistent and homogeneous record of lake levels has been constructed (Figures 4 and 5). The record for Megali Prespa currently ends in November 1991 because the lake level has fallen steadily since 1986 and it is now below the bottom of the lowest stage board in the lake. Megali Prespa is reported (Catsadorakis, pers. comm.) to have fallen by 5 m since 1990 and in 1995 it was continuing to fall by 10 cm/month.

The record of the level of the two lakes is divisible into four periods. First, from 1953 to 1963, both lakes followed a strong annual cycle within a relatively narrow range and at a relatively low level. Mikri Prespa was usually between 853.4 and 854.1 m asl, whilst Megali Prespa was between 852.7 m and 853.5 m asl on a slightly rising trend. Second, between October 1962 and June 1963, both lakes rose rapidly by between 2.5 and 3 metres. In June 1963, both lakes stood at the same level of 855.7 m asl. Both lakes declined in level steadily from 1963 to 1978 with a clear annual cycle superimposed on the linear trend of decline. The low levels of 852.8 m at Mikri and 852.0 at Megali in

Table 6. Mean discharge for 1961–69 in the Prespa torrents (l/sec) (after PPC mimeo note).

Torrent	Apr	May	June	July	Aug	Sep	Oct
Plati	15	7	4	1	1	0	0
Kalithea	124	25	26	17	14	11	13
Karies	15	9	14	9	6	3	6
Lefkonas	61	18	24	14	13	13	13
Oxia	122	24	16	9	8	6	7
Total direct to Mikri Prespa	337	83	84	50	42	33	41
Agios Germanos	2635	3053	873	480	270	282	210
Ratio of Agios Germanos to direct inflows to Mikri P.	7.8	36.8	10.4	9.6	6.4	8.5	5.1

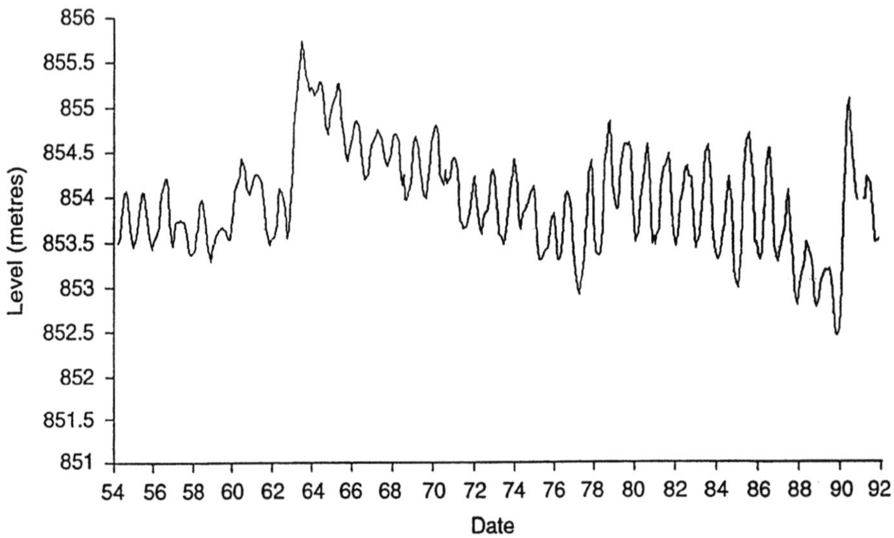

Figure 4. Water level in Lake Mikri Prespa from October 1953 to December 1992.

1978/79 were the lowest in the record to that date. The third period from 1978 and 1988 saw a general rise followed by a general fall in level at Megali Prespa whilst Mikri Prespa oscillated widely each year. The final period from March 1988 onwards saw a marked divergence in the patterns followed by the lakes. Mikri Prespa declined to a record low of 852.4 m asl in November 1990 only to rise by 2.64 m in the winter of 1991–92 to stand at 855.04 m asl in June 1991. A fairly rapid fall in level then ensued to the end of the record (853.47 in December 1992). At Megali Prespa there was an almost continuous fall in level from 853.44 m asl in June 1987 onwards. The record ceases in November 1991 at 851.20 m asl.

When the level of Mikri and Megali Prespa are plotted on the same axes, Figure 6, several additional points become clear:

(i) the data is consistent as the two lakes respond in a similar way at a similar time,

(ii) Megali was generally lower than Mikri during 1953 to 1963 and from 1977 to 1988,

(iii) from 1988 onwards the level of Megali Prespa fell markedly faster than that of Mikri Prespa and Megali seemed to benefit rather little from the apparently wet and snowy winter of 1991,

(iv) after the very high stage of 855.7 m in June 1963, Mikri was generally lower than Megali for much of the following twelve years.

These observations raise a host of questions about the hydrology of Mikri Prespa. First, it seems likely that the Prespa lakes overflowed into Albania and the River Devoll in 1963 and perhaps for some time after that. Broikos (1962) indeed describes the surface out-

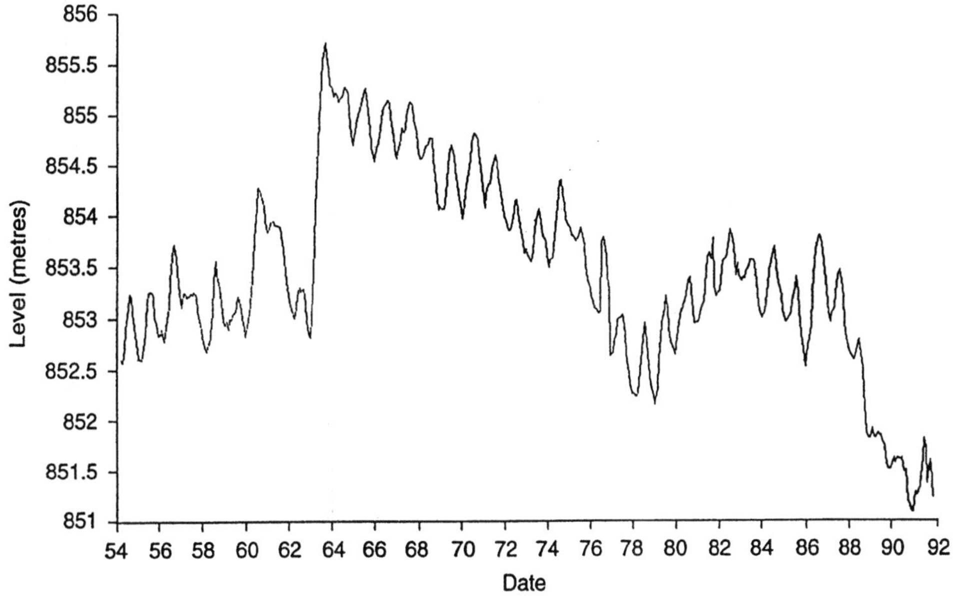

Figure 5. Water level in Lake Megali Prespa from October 1953 to December 1992.

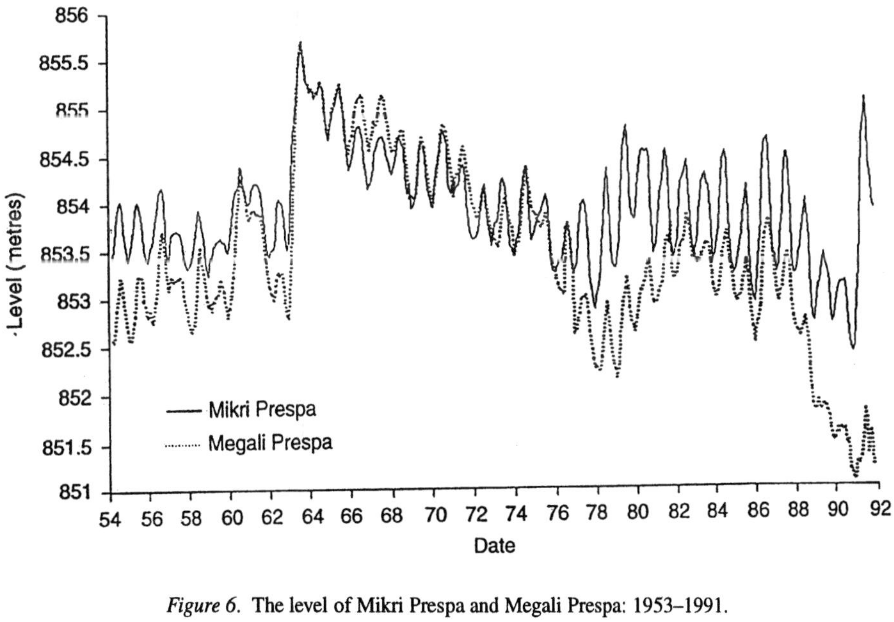

Figure 6. The level of Mikri Prespa and Megali Prespa: 1953–1991.

flow from Mikri Prespa as being to the River Devoll. The canal observed by Crivelli joining the River Devoll and Mikri Prespa must have taken a large part of this hypothesized outflow. Second, since the high water levels of the 1960s are recalled by the local people, are the 'reversed flow' conditions of the 1960s equivalent to the period during World War II when apparently there was a wide channel between the lakes with boats passing in either direction? Whilst this story is firmly fixed in the minds of local fishermen, it is important to note that none of the wartime maps of Prespa, by the British, Americans or Germans (Hollis et al., 1989) gives any indication of such a condition. Third, to what extent has human agency caused the record low levels in Mikri Prespa in 1978 and 1990? Finally, given the ratios of catchment areas to lake areas of Mikri and Megali Prespa, why should the level of Megali Prespa have fallen so rapidly from 1988 to 1995?

The mechanism for the recent fall in level of Megali Prespa can only be speculative without exten-

Table 7. Annual water balance of Lake Mikri Prespa (after Bonazoundas, 1988; Scoulos and Hatzianestis, 1989; Koussouris et al., 1987).

Catchment area	$10^6 m^3$
Precipitation on the land	137.9
Real evapotranspiration	93.0
Surface runoff	27.6
Deep seepage	17.3
Lake area	
Precipitation	35.3
Surface inputs	27.6
Groundwater discharge	10.4
Evaporation	54.2
Irrigation abstraction	8.5
Surface and underground flow to Megali Prespa	10.6
Turnover time	4.8 years
Volume	324 $10^6 m^3$
Maximum depth	8.4 m
Mean depth	4.12 m

sive research in FYROM relating to irrigation practices and hydrogeological conditions. There are informal reports of mild earth tremors in the Lake Ohrid catchment and these could have caused an opening of the underground karstic channels which are believed to carry water from Megali Prespa to Lake Ohrid. More work is needed before a definitive answer can be given.

The range of water level change at Mikri Prespa

One of the keys to the ecological functioning of the Mikri Prespa ecosystem is the flooding of the wet meadows around the lake. Spring flooding of these meadows is important for the breeding water birds and a relatively high and stable water level in May and June is needed for a successful spawning season for the fish.

Figure 7 shows the annual range of water levels in Mikri Prespa. Apart from the exceptionally wet year of 1963, the annual range was under 1 metre until 1977. Annual ranges were in excess of 1 metre from 1978 to 1989 but in 1989–1992 the annual ranges have settled back to under 1 metre. 1969 saw the Albanian's construction of a sluice on the canal linking Mikri Prespa with the River Devoll and the Greek channelisation of the link with Megali Prespa; it is possible to look at water level ranges before and after this date (Table 8).

Table 8 suggests that up to 1969, annual and May–June water level ranges were modest and they varied

Table 8. The water level regime of Lake Mikri Prespa before and after 1969.

	1953–1969	1970–1992
Mean annual range of water level (m)	0.66	1.06
St. deviation of range of annual water level	0.18	0.47
Mean range of water level in May-June (m)	0.16	0.21
St. deviation of range of water level in May-June	0.08	0.12

little from year to year. After 1969, the average annual range of water level increased by 60% to 1.06 metres and the average range of water level in the critical May–June period rose by 31% to 0.21 metres. These simple analyses are only pointers to possible hydrological changes in Lake Mikri Prespa. A much fuller analysis and more complete information is required for definitive conclusions.

Water management schemes affecting Lake Mikri Prespa

Diversion of the Agios Germanos torrent

The proposal to redivert the Agios Germanos torrent back into Mikri Prespa has not been formalised but it is known and 'understood' by local people, scientists, and administrative officials. Pyrovetsi and Karteris (1986) typify assertions as to the benefit of the scheme:

> 'this stream brought large quantities of water to the area, particularly in the snow-melt season, causing an increase in local water volume and in the lake depth at Mikri Prespa, at the same time preventing the spread of reeds on the northern shorelines. The water quality of the lake, and particularly of the wetland, was improved as the highly oxygenated stream entered the lake.'

The diversion of the Agios Germanos torrent to Mikri Prespa has the potential to, at least, double the volume of river inflow to the lake. The increase in the catchment area of the lake would be 27% but its largely impermeable nature would bring a much larger proportionate change in the volume of inflow. The result of an Agios Germanos diversion was calculated by Hollis et al. (1989) using a simple simulation model. The diversion raises water levels by around 30 cm in the winter and spring. The rise in water levels in the summer is of the order of 15cm. This summer rise in levels

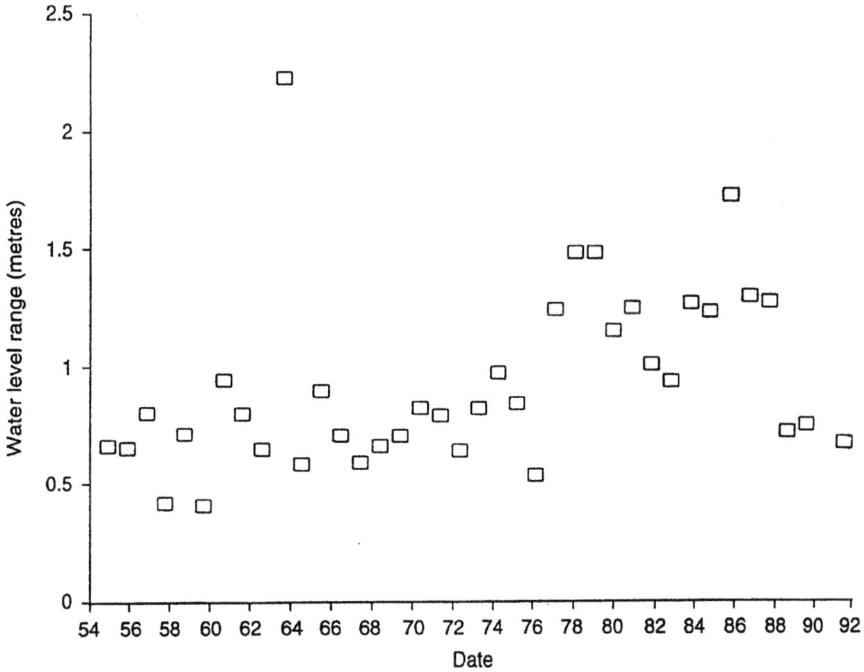

Figure 7. The annual range (water years Oct–Sept) of water levels in Mikri Prespa.

would have been approximately halved if an off take by pumping for irrigation had been simulated. The diversion of the torrent was shown to be mildly beneficial in dry years but in wet years water levels would have to be moderated through the construction of a long overflow weir to Lake Megali Prespa. An alternative would be to build a flow proportioning structure at Lemos so that only a part of major flood flows in the Agios Germanos were transferred to Mikri Prespa. The increase in the input of flood flows from the Agios Germanos and the narrow restricted outflow down the Koula culvert was shown to cause even greater level fluctuations in May and June than in the natural situation and so this regime does not meet the need for a stable water level in May-June when the fish are spawning. In addition, Hollis et al. (1989) recommended against the scheme on the grounds of cost, land disturbance around Lake Mikri Prespa and the poor quality of the Agios Germanos stream at times of low flow because of sewage pollution.

Irrigation scheme

The irrigation scheme at Prespa has been under development for almost thirty years. Fountoukis (1962) said that irrigation was needed between May and September and quoted a projected irrigation area of 2150 ha. Broikos (1962) refers to the contract specification for 1816 ha to be irrigated in a low zone (1161 ha) between 851.5 (lake level) and 876 m, and a high zone (555 ha) between 876 and 921 m. He quoted the water need as 4500 m^3 ha^{-1}. Broikos (1962) said that the irrigation pumping would lower the lake level by 50 mm in summer. With an evaporative loss of 650 mm, he calculated a total fall in water level of 700 mm. Broikos (1962) states that in 1961–62 there was already 190 ha under irrigation. Kemenidis (1971) found that very limited data had been used by previous studies. Kemenidis proposed a total irrigated area of only 1000 ha of which only 900 ha would be cultivated. Zisopoulos (1982) gives the perimeter of the irrigation network as 1800 ha with a net cultivated area of 1424 ha. The works in 610 ha of the low zone were completed in 1970. The full area of 860 ha was completed in 1977. The high zone was not in receipt of irrigation in 1982. During the period 1977–81 the area actually irrigated varied between 175 and 200 ha out of the total available of 860 ha.

The data on the use of the pumping station is said to be unreliable by Bonazoundas (1988). He cites evidence that the station has been allowed to run for 24 hrs per day with the surplus water draining directly back into the lake. Table 9 gives Bonazoundas' estimates of the actual extent of the irrigation.

Table 9. The extent of irrigation of Prespa (after Bonazoundas, 1988).

Year	Total ha	Source	Total ha draining to Mikri Prespa	Source
1977–81	175–200	Zisopoulos, 1982		
1982–83	200–250			
1984	368	Prefect, local knowledge	266	Min. of Ag.,
1985	449	" "	353	Agric. Bank,
1986	512	" "	378	and local
1987	730	Agricultural Bank, Florida	520	knowledge
1988	750	Local knowledge	540	

The pumping station has 4 pumps rated at 250 l s^{-1}. During 1987 the pumps operated for 12 hours per day from 15 May to 26 June and from 27 August to 15 September. They operated for 16 hours per day from 27 June to 26 August. This gives 1732 hours of pumping and a total volume of 6.235 10^6 m^3. This means that 9500 m^3 ha^{-1} was applied which is just over twice the volume of water needed according to Broikos (1962). Bonazoundas (1988) states that in 1988 about three times the irrigation need was pumped. The reasons given are bad management of the network, errors in the studies of the network, unsuitable irrigation methods, and heavy losses from the irrigation canals.

The effects of the 1987 pumping were simulated by deducting the appropriate volume of water from the lake on a daily basis between 15 May and 15 September (Hollis et al., 1989). The results for 1978 to 1984 are that the fall in water level was only a few centimetres during the summer months. The simulation showed that if the sluice at Koula were to be closed at all times then the effect of irrigation water pumping would be slightly magnified. It may be concluded that the volume of water taken by the irrigation scheme has a negligible effect upon the water level of Mikri Prespa.

Psychoudakis et al. (1993) examined the economics of the irrigation project. They found that the 750 ha irrigated area was only 41.3% of the planned area, and this represented only a 450 ha increase in irrigated area since there was some irrigation before the project. Yet, the actual budgetary cost of the project was 4.4 times the initial cost estimate at constant prices. The high cost over-run was due to underestimation of the costs in the preparation stage, lack of coordination during construction, cost increases over the long construction period and lack of government commitment. The project also failed to deliver the planned cropping regime with only beans covering a significantly larger area after the implementation of the scheme (Table 10).

Table 10. Previous, planned and actual composition of main crops in the irrigated area (after Psychoudakis et al., 1993).

Main crops	1960 previous %	1962 planned %	1991 actual %
Maize	3.7	25.0	0.8
Lucerne	0.6	25.0	0.8
Beans	8.8	25.0	55.6
Lentils	2.1	–	–
Peas	4.0	–	–
Potatoes	1.1	–	–
Meadows	11.3	–	–
Winter cereals	68.2	15.0	42.8

Psychoudakis et al. (1993) measured the development values of the project as a reflection of the opportunity cost of the conservation of the area's natural ecosystems. The plan estimates of the development value (PV) were high raising the cost of conserving the environment. However, the actual development values were negative (Table 11), showing that the social opportunity cost of conservation was negative i.e. there would have been a positive benefit from retaining the area in its natural state as measured by the losses incurred by the project. Psychoudakis et al. (1993) found that the benefits to farmers, although less than planned, were substantial, explaining the local pressures for the expansion of the irrigation. The aid-incentive schemes of the recent reform of the Common Agricultural Policy are probably not sufficient to prevent pressures for the expansion of irrigation in similar ecologically sensitive areas, especially where costs are subsidised by the State and users do not pay a realistic charge for water.

Table 11. Project costs and benefits per hectare and EU 'sustainability' subsidies (after Psychoudakis et al., 1993).

		Plan ex-ante	Actual ex-post
1. Costs (Ministry of Agriculture)			
At market prices	ECU/ha	400	1119
At social prices	ECU/ha	372	906
2. Benefits			
At market prices	ECU/ha	1100	473
At social prices	ECU/ha	734	392
3. Available aid under Article 2			
i Prog. for annual crops	ECU/ha	350	350
ii Land set aside	ECU/ha	600	600

Cost and benefits were estimated in terms of annuities (interest 6% for 30 years). For the conversion of Drachmae to ECU, the 1990 rate was used (202Dr/ECU)

HEP scheme

A small scale hydro-electric plant has been proposed for construction on the Agios Germanos torrent above Agios Germanos village. The project was a pilot plant that will be remotely controlled and it was designed for distant frontier areas. The project consisted of a 7 metre high dam 'with no lake' (PPC 1988), a penstock to convey the water down 200 metres, and a power house. The dam has an overflow spillway, a sedimentation by-pass system and a series of fish steps. It was to be 25 metres long with a spillway depth of 1.3 metres. The structure accommodated a downstream release of $200\,l\,s^{-1}$ during the low flow summer months assuming this discharge will be available in the river. The project espoused care for the environment by aesthetic adaptation including re-afforestation; good practice during construction; a minimum downstream discharge from the dam during low flow periods; securing of water supplies for agriculture and livestock; maintenance of the river pools downstream of the dam; construction of a meteorological station; and, where possible, the employment of local people.

The scheme was outside the formal boundaries of the National Park but will impact upon the lower reaches of the river. There has been a strong reaction against the scheme (Friends of Prespa, 1989) on a number of grounds and it seems unlikely to go ahead in the foreseeable future. The revelation that the main driving force for the scheme was the need to absorb EU funds gives it little justification in this sensitive zone.

Modification of the Koula canal linking Lakes Mikri and Megali Prespa

There have been suggestions, emanating from the Greek Forest Service and others (Crivelli & Catsadorakis, pers. comm.), that the canal at Koula linking Mikri Prespa with Megali Prespa requires modification. High water levels cause distress to the farmers in spite of the fact that they extend their fields into low lying lands when lake levels are low. It seems likely that there is pressure to increase the discharge capacity of the canal so as to limit future high water levels. A simple dredging of the canal would be dangerous since this may both limit the ability of the lake to flood the wet meadows and may reduce the stability of the lake level during the critical fish spawning period in May–June. The continuing fall in level of Megali Prespa will also permit enhanced discharges from Mikri Prespa and may facilitate engineering which will destabilise the Mikri Prespa ecosystem. The existence of the water control structure at the Albanian end of the lake adds a further uncertainty to the situation.

A water level management plan, such as those required elsewhere in Europe (Min. of Agriculture et al., 1995), needs to precede any engineering work on the Koula Canal. The management plan needs to involve the Greek and Albanian authorities and all of the relevant organisations concerned with the lake. The plan needs to agree an ideal water level regime and operational rules should actual water levels rise above, or fall below, the agreed optimum levels. Once this plan is finalised, it will then be possible to examine the existing sluice and canal at Koula, and that in Albania, and to decide if further investment needs to be made.

Water level management plan

There is a strong scientific and local consensus (Hollis et al., 1989) that the regime of water levels which occurred in 1986 should be the model for that managed regime (Figure 8). That this is a natural regime in the recent past, is a major advantage in justifying the choice of the regime. The features that make 1986 a model regime are:

- a high maximum water level of over 854.60 m,
- a maximum level of 855 m to avoid flooding fields,
- small variability in water level (16 cm) in May–June,
- a minimum level above the sill of the Koula culvert giving continuous outflow.

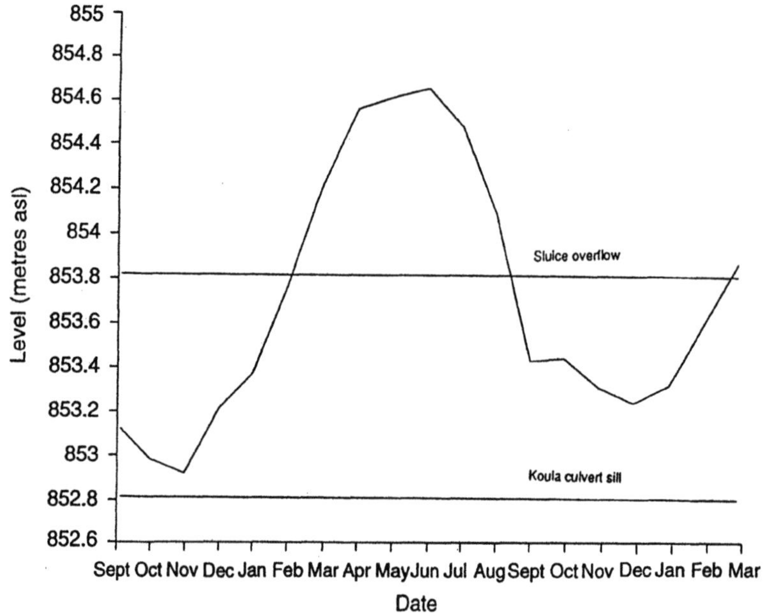

Figure 8. The 1986 water level regime at Mikri Prespa.

Table 12. Water chemistry data for Lake Mikri Prespa 1978–88.

		1977[1]	1978[2]	1978[2]	1984[3]	1985[3]	1986[3]	1987[3]	1988[3]	1987[4]
Conductivity	μs	243	–	–	275	265	277	303	285	175
pH		8.3	–	–	7.8	7.7	7.5	7.3	7.3	8.1
Cl^-	meq l^{-1}	0.1	0.1	0.1	0.1	0.1	0.1	0.2	0.1	0.1
SO_4^{2-}	meq l^{-1}	–	–	–	0.4	0.2	0.2	0.3	0.2	–
HCO_3^-	–	–	–	2.6	2.6	2.7	2.9	2.9	–	
NO_3^{2-}	$\mu g\ l^{-1}$	56.0	16.5	0.06	<440[5]	1067	1043	<440[5]	<440[5]	61.0
NO_2^-	$\mu g\ l^{-1}$	2.0	3.5	1.6	54.0	23.0	18.0	5.0	20.0	1.9
NH_4^+	$\mu g\ l^{-1}$	34.0	32.4	14.1	206.0	146.0	169.0	56.0	164.0	48.4
Total P	$\mu g\ l^{-1}$	28.0	–	–	22.0	19.0	23.0	14.0	24.0	28.0
Orthophosphate as P-PO_4	$\mu g\ l^{-1}$		4.0	4.63	9.3	–	–	–	–	10.0

[1] Data from Mourkides et al. (1978).
[2] Koussouris and Satmadjis (1987).
[3] Land Reclamation Service laboratory results - means of available figures.
[4] Koussouris et al. (1989).
[5] All values below detection limit.

Water quality

Table 12 presents the results of the available tabulated water chemistry for Mikri Prespa. Since it is unclear whether standard methodologies have been used in all analyses, the values contained within this table should be treated with caution. For instance, the results of the Land Reclamation Survey Laboratory are a factor of ten higher than the others, especially for nitrogen compounds. For nitrate it appears an insensitive analytical method is being used since the detection limit is 440 $\mu g\ l^{-1}$. The very high mean nitrate figures for 1985 & 1986 lie outside previous and current measured maxima (ca 340 $\mu g\ l^{-1}$) produced by the other studies. Similar conclusions can be drawn for nitrite and ammonia with current maxima being of the order of ca 2.3 $\mu g\ l^{-1}$ and 30–50 $\mu g\ l^{-1}$ respectively (Koussouris et al., 1989, Mourkides et al., 1978) compared to the results of the Land Reclamation Laboratory of 20–30 $\mu g\ l^{-1}$ and 160 $\mu g\ l^{-1}$ respectively.

The phosphorus results from all workers appear to be in line with one another and show that if total phosphorus levels are used the lake would be classified as meso-eutrophic (Wetzel, 1975).

Tryfon et al. (1994) monitored soluble reactive phosphorus, dissolved inorganic nitrogen and silica concentration monthly from May 1990 to September 1992 and took samples at various depths. They found that 'soluble reactive phosphorus concentration ranged from 0 to 54 $\mu g\ l^{-1}$'. The peak values were in September 1990, March 1991 and July 1992. Values were generally below 10 $\mu g\ l^{-1}$. 'Dissolved inorganic nitrogen varies between 3 and 831 $\mu g\ l^{-1}$, showing the highest values in winter and spring N:P ratio ranged from 6 to 1664 and was lowest at the end of summer and in the autumn. Dissolved silica ranged from 19 to 5535 $\mu g\ l^{-1}$. The vertical distribution of nutrients appeared relatively homogeneous throughout the water column'.

Palaeoecological and phytoplankton studies

There have been claims that Mikri Prespa has become eutrophic by runoff of agricultural fertilizers and untreated raw sewage entering the lake (Pyrovetsi et al., 1984; Pyrovetsi & Gerakis, 1987; Koussouris et al., 1989). These assertions have been supported by the few limnological studies of the lake (Koussouris & Satmadjis, 1987; Koussouris et al., 1987) which report high phosphorus and nitrate loadings suggesting that lake Mikri Prespa is meso-eutrophic/eutrophic. However, the nutrient data presented in these two papers are at variance with one another, the data quoted in Koussouris et al. (1987) are thought to be erroneous with both phosphorus and nitrate values far higher than any eutrophic lake in, for example, the U.K. Data from Koussouris & Satmajdis (1987) have available phosphorus values of 4.53 $\mu g\ l^{-1}$ and nitrate values of 15.7 $\mu g\ l^{-1}$ and would indicate low nutrient availability. These compare with total phosphorus measurements of between 51 $\mu g\ l^{-1}$ and 93 $\mu g\ l^{-1}$ from Pyrovetsi et al. (1984). In addition, Koussouris et al. (1989) use a hydrological desk study, together with an assessment of phosphorus loadings from the catchment, to conclude that Lake Mikri Prespa is meso-eutrophic and is lying very close to a eutrophication threshold. The trophic status of the lake has, therefore, been uncertain and moreover none of these papers present evidence of historical changes in the algal populations within the lake that would corroborate the inferred eutrophication. Two recent pieces of work, Stevenson & Flower (1991) using palaeoecological techniques and that by Tryfon et al. (1994) examining phytoplankton and physical-chemical features of the lake, have shown conclusively that the lake is not eutrophic and that its water quality has changed little during this century.

The application of palaeolimnological techniques (Battarbee, 1986) have provided a convenient way of assessing ecosystem change over time (> 10^1–10^3 years). Such techniques have proved essential in evaluating the role of anthropogenic activity in causing both the eutrophication of many lakes in temperate regions (e.g. Battarbee 1978). Stevenson & Flower (1991) used diatom analysis and related analyses of a short sediment core from Mikri Prespa to assess the impact of catchment disturbance and water enrichment.

Their dating, physical and diatom analyses of the core demonstrated that the early part of the core represents a period of relative ecological stability. Sediment accumulation rates increased steadily from the core base to 15.5 cm (1943) and this rise was related to the continued flow of the Agios Germanos torrent at high flows into Lake Mikri Prespa and the continued cultivation of the easily eroded soils of the upper catchment. The decline in rates was almost certainly related to the diversion of the Agios Germanos torrent into Lake Megali Prespa and the cessation of the cultivation of the upper parts of the catchment associated with a 78% decline in the population of the area from 1940 to 1951 (Pyrovetsi et al., 1984). The more recent increase in sediment accumulation rates from 7 cm (1970) derived from agricultural intensification within the catchment associated with the irrigation-drainage system from 1971 onwards.

The long-term water quality implications for these increased sediment accumulation rates did not appear to be significant since the diatom flora was remarkably stable throughout the time span of the core. The dominance of *Cyclotella ocellata* in the core suggested that the lake is, as yet, little affected by increased nutrient loadings and is still meso-oligotrophic in nature. The relatively high concentration of sedimentary planktonic diatoms indicated that the periods of increased catchment disturbance have had little impact on the lake water quality at least during the growing season of this diatom. Overall, the diatom evidence showed only slight eutrophication of the lake from the 1950's associated with the agricultural development of the alluvial plain to the east of the lake.

These conclusions contradict the work of both Pyrovetsi et al. (1984) and Koussouris et al. (1989)

who report that phosphorus concentrations in Mikri Prespa are relatively high and suggest that the lake may be more eutrophic than that indicated by the diatom results. Moreover, the desk study by Koussouris et al. (1989) calculates phosphorus loadings from point source estimates and shows the lake to be in a critical state with regard to phosphorus loading and eutrophication. By using the Vollenweider critical phosphorus loading model they calculate that the lake receives ca 0.24 g m^{-2} yr^{-1} of phosphorus more than the recommended permissible P loading figure. Stevenson and Flower (1991) showed that Koussouris et al. (1989) overestimated the phosphorus loading because of the simplistic assumption in their calculations that all the phosphorus produced within the catchment is transferred to the lake. In reality, much of the phosphorus produced within the catchment is immediately immobilised within the soil complex and only a small fraction is ultimately transferred to the lake (Wood & Gibson, 1973). The areal phosphorus loading figures calculated by Koussouris et al. (1989) are strongly overestimated and erroneous conclusions have been drawn about the trophic status of Mikri Prespa.

In addition to immobilization of phosphorus within the soil system, Stevenson & Flower (1991) argued that additional mechanisms may be at work in reducing phosphorus in the lake water and its availability to the diatoms. Studies of both phosphorus and nitrate runoff from the agricultural land, entering the lake system through the reed system that fringe the lake, have shown that available phosphorus and nitrate concentrations are reduced markedly as water passes through the reed-bed. Second, phosphorus availability may be reduced by the formation of biologically unavailable detrital phosphorus as phosphate cations react with calcium in calcium-rich soils and hard waters (Boyer & Wheeler, 1989). Third, and related to the second mechanism, is the coprecipitation of phosphorus with calcite formation in the lake (Murphy et al., 1983). The necessary conditions for authigenic calcite formation are known to occur in the lake (Scoullos & Hatzianestis, 1989) and studies of other calcareous lake systems have found this to be a powerful mechanism that can remove up to a third of the total phosphorus (Kleiner, 1988). Calcite formation may be further enhanced by high rates of photosynthesis (Stumm & Morgan, 1970) and in Lake Mikri Prespa this may be associated with blooms of green and blue-green algae (*Microcystis* and *Anabaena*) in the summer.

These considerations when combined with the evidence from the oligotrophic diatom flora, do not support the hypothesis that Mikri Prespa is a seriously enriched lake. This conclusion is further supported by the limited lake chemistry data (Table 12) which demonstrates that available phosphorus values are generally low. Moreover, the values of total-P show little change over the ten years of available chemistry data despite agricultural intensification within the catchment.

Tryfon et al. (1994) examined the structure and dynamics of phytoplankton assemblages in Mikri Prespa. Biomass values varied between 1.2 and 38.0 g m^{-3} and revealed one autumnal and one vernal peak. Cyanophytes, the most important algal group remained dominant throughout the year. Diatoms, the second abundant group, dominated the winter and spring. The periodicity of the phytoplankton and the height of the annual peaks resembled those of many temperate eutrophic lakes but the absence of a summer maximum was opposed to the PEG model statement that eutrophic systems are characterised by increased summer biomass.

Oral history

Another strand of evidence was developed during initial work for a management plan for the National Park. A series of semi-structured interviews were held with eleven of the village chairmen (Hollis et al., 1989). One area of enquiry concerned the evolution of lake water quality during their lifetime. The chairmen were virtually unanimous that no significant changes to the water quality had taken place and several stated that the amount of algae in the lake was largely unchanged too. Whilst almost all of the villagers had used the lake as a source of drinking water in the past, they did not do so today because they now have adequate mains supplies. Almost all of the Chairmen said that they would be quite happy to revert to drinking water from the lake if that were to become necessary for some reason.

Turbidity

Crivelli (1995) reports a statistically significant increase in turbidity in Mikri Prespa as measured by Secchi disc values. This is clearly shown in data from 1984 to 1993 and is even more marked when pre-1984 data is included. Accepting Stevenson and Flower's rebuttal of a eutrophication explanation for this increase in turbidity, Crivelli advances the view that the marked increase in the goldfish population of the lake, following its introduction in the late 1970s, explains the increase in turbidity. It appears that *Carassius auratus* suck up sediment when feeding and so have been observed to increase the turbidity of the waters they inhabit. This exotic species comprised over 30% of the total fish catch from the lake during 1990 and so is now a major feature of the fauna of the lake.

Overall then, whilst the turbidity of the lake appears to have increased, it is still fairly pristine in nature and not subject to serious eutrophication. However, any future management should aim to prevent practices which promote: soil erosion, increased use of fertilizers on catchment soils and direct inputs of nitrogen and phosphorus to the lake.

Hydrological monitoring required

The effective management of the wetland areas of the Prespa National Park, and Lake Mikri Prespa in particular, requires a firm basis of hydrological monitoring data. This is needed for the assessment of natural and human-induced trends affecting the lake and wetlands; for evaluating the effectiveness of management action; for the real-time control of water management structures such as sluices; and for scientific studies.

It is essential that the existing data from the PPC, Forestry Service, Geological Survey, Land Reclamation Service, various universities and others is centralised into a computerised and quality controlled archive. The existing arrangements with material disappearing into various and disparate archives and stores are unsatisfactory.

It is essential that the long established meteorological records for the Koula meteorological site be continued. Daily readings of rainfall; snowfall; mean, maximum and minimum temperature; ground temperature; humidity; wind speed and direction; sunshine hours; and open water evaporation are central to any environmental monitoring scheme for the National Park. The daily read rain gauges at Agios Germanos, Vrondero, and Mikrolimni need to be maintained. If possible at least one rain gauge and snowfall monitoring station, with weekly readings, needs to be established in the mountains at an elevation of 1500 m or more.

The level of Lake Mikri Prespa needs to be read daily at Koula on the established system of stageboards. The stageboards in Megali Prespa need to be expanded and extended so that the continuing decline in the lake level can be monitored and better understood. The status of the sluices at Koula and in Albania need to be recorded so that environmental data can be understood against the background of management activities. The flow down the canal at Koula towards Megali Prespa needs to be current metered approximately every month so that there is a good understanding of the relationship between water level in Mikri Prespa and outflow. This relationship will change with vegetation growth and die back, sedimentation or erosion in the channel and any changes engineered in it.

The existing river gauging station at Agios Germanos needs to be rehabilitated so that continuous flow records are produced that will allow the calculation of a full range of hydrological parameters for this major stream in the area. The other torrents, which were gauged for a brief period in the 1960s, have heavy sediment loads and would probably not repay the expense and difficulty of further gauging station construction. Their flow can, in any case, be estimated by reference to the flow in the Agios Germanos stream.

The volume of water pumped from the lake, and that diverted from the Agios Germanos stream, for the irrigation scheme needs to be recorded, as does any artificial input or output in Albania.

The water level needs to be recorded in about half a dozen wells that are not routinely used for water supply. This will give an indication of the levels of groundwater storage. The wells should be selected from around the basin and need to include the limestone, recent sediments and granitic rocks.

There needs to be routine monitoring of water quality using internationally acceptable laboratory methods. Rainfall, collected from Koula with inert polythene funnels and collecting vessels needs to be measured on a monthly basis for pH, conductivity, dissolved inorganic nitrogen, soluble reactive phosphorous, Pb, Fe, Zn and Cu. The Agios Germanos torrent, selected ditches and torrents within the irrigation scheme and agricultural area, outputs via Koula and inputs or outputs in Albania require monthly monitoring to determine BOD, dissolved oxygen, pH, conductivity, NH_4-N, NO_3-N, NO_2-N, PO_4, and suspended load.

The quality of the lake water needs to be determined at around six sites chosen to replicate as far as possible the sites used in previous surveys. Water needs to be taken from the surface and at about 1 metre intervals with depth. The parameters to be recorded should cover temperature, dissolved oxygen, turbidity with a Secchi disc, NO_3-N, NO_2-N, NH_4-N, and reactive and total phosphate.

Conclusion

The location of the Prespa lakes on the frontiers of three countries inhibits intensive integrated studies. This paper has focused on the physical basis, and especially the hydrology, of Lake Mikri Prespa but within the context of the wider basin encompassing Megali Prespa, Lake Ohrid and the managed linkage with the River Devoll. The region is mountainous with peaks rising 1300 m above the lakes within 7 km. There is relatively little climatological or hydrological data available for the region and the total absence of data for the mountain areas makes detailed computations hazardous. The large component of karstic limestones in the region makes total inflows and outflows of water from the lakes rather uncertain whilst the torrents draining the crystalline rocks have a strong peak of flow during the snow melt of the spring. Megali and Mikri Prespa have 8 km^2 and 4 km^2 respectively of mountainous catchment area for each km^2 of lake area. Their water levels have changed in unison until the late 1980s and 1990s. Today, Lake Megali Prespa appears to have settled to a long period of continual decline occasioned by tectonic opening of its karstic outflow channels to Lake Ohrid.

The annual water balance of Mikri Prespa is dominated by direct precipitation of 35 10^6 m^3, river inputs of 28 10^6 m^3 and evaporation of 54 10^6 m^3. Groundwater discharge, flow to Megali Prespa and abstraction for the Prespa Irrigation Scheme are each around 10 10^6 m^3

The lakes have a strong seasonal cycle of level with a longer term fluctuation caused by particularly wet periods. The early 1960s saw especially high water levels and Lake Mikri Prespa has been similarly high in the early 1990s thanks to extensive snowfall. The Agios Germanos stream, the largest in the Greek portion of the catchment, was diverted from Mikri to Megali Prespa in the 1930s. The Albanians have linked Mikri Prespa with the River Devoll and the channel linking Mikri Prespa and Megali Prespa was substantially deepened and straightened in 1969. There is some evidence that these changes have contributed to a larger range of water levels and a more rapidly changing water level regime, especially in the ecologically important months of May and June.

It has been shown that it is undesirable to pursue the diversion of the Agios Germanos stream back into Lake Mikri Prespa, even though it would effectively double the river inputs to the lake. The existing irrigation scheme has never achieved the command area discussed at the outset. Overall it has been estimated that it lowers the water level in Mikri Prespa in the summer by around 5 cm. The run of the river hydro-power scheme that has been proposed for the headwaters of the Agios Germanos stream will deliver few local benefits and adversely affect the stream's ecology. It has a poor rationale both hydrologically and in terms of the local economy. The dredging or re-engineering of the Koula canal which connects Lakes Mikri and Megali Prespa would threaten the ecological stability of the wetlands of the Prespa National Park if it were undertaken without the prior completion of a comprehensive water level management plan for Mikri Prespa.

The palaeoecological study of lake sediment from Mikri Prespa and the recent phytoplankton research have contradicted the assertions in the literature that the lake is, or is fast approaching, a eutrophic condition. The dated core showed that sediment inputs to the lake have risen somewhat since the 1940s but that overall water quality in the lake has not changed significantly. There is only the slightest of evidence for nutrient enrichment from agricultural fertilizers and local oral knowledge confirms the conclusion that Mikri Prespa's water quality is stable and meso-oligotrophic in nature. The lack of a summer peak of phytoplankton indicates that Mikri Prespa does not follow the pattern of classical eutrophic lakes.

Overall, the hydrology of the Prespa Lakes is dominated by the winter surfeit of rain and snowfall with summer droughts, the periodic occurrence of particularly wet years which raise water levels substantially, and special natural phenomena such as the earth quake which appears to have precipitated the steady drainage of Lake Megali Prespa and the formation of the sedimentary bar which now separates Mikri and Megali Prespa. Onto this natural pattern of short and long term fluctuation the effects of human activity have been superimposed although these seem to have been relatively modest compared to the natural changes. In particular, there seems to be no evidence that the quality of the lake water in Mikri Prespa has been greatly affected by human activity this century.

References

Battarbee, R. W., 1978. Observations on the recent history of Lough Neagh and its drainage basin. Phil. Trans. r. Soc., Lond., Series B. 281: 303–345.

Battarbee, R. W., 1986. The eutrophication of Lough Erne inferred from changes in the diatom assemblages of ^{210}Pb and ^{137}Cs dated sediment cores. Proc. r. Irish Acad., B. 86: 141–168.

Bonazoundas, M., 1988. Environmental Impact from the irrigation network under construction at Lake Mikri Prespa in the Prefecture of Florina: Phase A Preliminary Evaluation of the Impact of the Network. Mimeo.

Boyer, M. L. H. & B. D. Wheeler, 1989. Vegetation patterns in spring-fed calcareous fens: calcite precipitation and constraints on fertility. J. Ecol. 77: 597–609.

Broikos, A., 1962. Final Study for the Development of Areas of Prespes, Prefecture of Florina: Brief Introductory Report. Ministry of Agriculture, 18 pp. (In Greek). Mimeo.

Crivelli, A. J., 1995. Are fish introductions a threat to endemic freshwater fishes in the Northern Mediterranean Region? Biol. Conserv. 72: 311–320.

Emberger, L., 1963. Carte Bioclimatique de la Région Méditeranéene. FAO/UNESCO, Rome/Paris, 2 sheets (1:5 000 000).

Fountoukis, T., 1962. Supplementary soil report on the areas of Prespes, Prefecture of Florina. Land Reclamation Service, 8 pp. (In Greek). Mimeo.

The Friends of Prespa. 1988. Remarks on the small hydroelectric project of Agios Germanos, Prespa, Florina Prefecture, 20 pp. Mimeo.

Hollis, G. E., H. Athanassiou, S. Bradford, G. Catsadorakis, A. J. Crivelli, R. Flower, A. P. Gerakis, B. Goldsmith, N. Rose, N. Stedman, A. C. Stevenson, D. Thomas, & Y. Troumbis, 1989. A Management Plan for the Prespa National Park, Greece. Volume 1: Five Research Studies. Report to WWF (International) on Project 3535, Department of Geography, University College London: 108 pp.

Jakovljevic, S., 1935. Recherches limnologiques sur le lac de Prespa. Verh. int. Ver. Limnol. 7: 187–226.

Kassioumis, K., 1991. Prespa National Park Management Plan. Forestry Service, Ministry of Agriculture, Athens, Greece. Mimeo. 3 volumes. (In Greek).

Kemenides, E., 1971. Revision of the final study of the irrigation network for Prespes area: Technical Report. Ministry of Agriculture, Land Reclamation Service, 99 pp. (in Greek).

Kleiner, J., 1988. Coprecipitation of phosphate with calcite in lake water: A laboratory experiment modelling phosphorus removal with calcite in Lake Constance. Wat. Res. 22: 1259–1265.

Koussouris, Th., A. Diapoulis & E. Balopoulos, 1987. Limnological situations in two shallow Greek lakes (Kastoria and Mikri Prespa Lakes). Geojournal 14: 377–379.

Koussouris, Th., A. Diapoulis & E. Balopoulos, 1989. Assessing the trophic status of Lake Mikri Prespa, Greece. Ann. Limnol. 25: 17–24.

Koussouris, Th. & J. Satmadjis, 1987. Changes in plankton assemblages from spring to summer in a Greek lake. Rev. Int. Oceanogr. Med. 87/88: 51–66.

Ministry of Agriculture, Fisheries and Food, Welsh Office, Association of Drainage Authorities, English Nature, National Rivers Authority. 1995. Water level management plans: A procedural guide for operating authorities. Ministry of Agriculture, Fisheries and Food, London, 25 pp.

Mourkides, G. A., G. E. Tsikritsis, S. E. Tsiouris & U. Menkisoglou, 1978. The lakes of Northern Greece. I. Trophic Status, 1977. Sci. Ann. School of Agriculture and Forestry, Agriculture Section, Aristotelian University of Thessaloniki, 21: 94–119. (In Greek).

Murphy, T. P., K. J. Hall & I. Yesaki, 1983. Biogenic regulation of iron availability in a eutrophic hardwater lake. Sci. Total Envir. 28: 37–50.

Pano, N., 1984. Hidrologjia E Shqiperise. Akademia E Shkencave E Rps Te Shqiperise, Insitituti Hidrometeorologjik, Tirane: 43 pp.

Pavlides, G., 1985. Geobotanical study of the National Park of Lakes Prespa (N.W. Greece) Part A: Ecology, Flora, Phytogeography, Vegetation. Aristotelian University of Thessaloniki, Greece: 308 pp. (in Greek)

Psychoudakis, A., S. Papoutsi-Psychoudakis & A. M. M. McFarquhar, 1993. An assessment of the irrigation project affecting the wetlands of Prespa, Greece. Greek Biotope/Wetland Centre and Department of Agricultural Economics, School of Agriculture, University of Thessaloniki, Greece: 44 pp.

Pyrovetsi, M. D., A. J. Crivelli, P. A. Gerakis, M. A. Karteris, E. P. Kastro & N. Komninos, 1984. Integrated environmental study of Prespa National Park. Final Report to the Commission of the European Communities DGXI, Thessaloniki, Greece. ix + 205 pp + maps.

Pyrovetsi, M. D. & M. A. Karteris, 1986. Forty year Land Cover/Use Changes in Prespa National Park, Greece. J. envir. Mgmt 23: 173–183.

Pyrovetsi, M. D. & P. A. Gerakis, 1987. Environmental Problems from Practicing Agriculture in Prespa National Park, Greece. Environmentalist 7: 35–42.

Pyrovetsi, M. D. & A. J. Crivelli, 1988. Habitat Use by water-birds in Prespa National Park, Greece. Biol. Conserv. 45: 135–153.

Scoullos, M. & J. Hatzianestis, 1989. Dissolved and particulate trace metals in a wetland of international importance: Lake Mikri Prespa, Greece. Wat. Air Soil Pollut. 44: 307–320.

Stevenson, A. C. & R. J. Flower, 1991. A palaeoecological evaluation of environmental degradation in Lake Mikri Prespa, N.W. Greece. Biol. Conserv. 57: 89–109.

Stumm, W. & J. J. Morgan, 1970. Aquatic Chemistry, Wiley, Chichester.

Tryfon, E., M. Moustaka-Gouni, G. Nikolaidis & I. Tsekos, 1994. Phytoplankton and physical-chemical features of the shallow Lake Mikri Prespa, Macedonia, Greece. Arch. Hydrobiol. 131: 477–494.

Wetzel, R. G., 1975. Limnology, Saunders, Philadelphia.

Wood, R. B. & C. E. Gibson, 1973. Eutrophication and Lough Neagh. Wat. Res. 1: 173–187.

Zisopoulos, A., 1982. A brief report on the land reclamation development project of the Prespa area. Land Reclamation Service, Western Macedonia Office, Kozani. (Greek, gov. report).

Hydrobiologia **351:** 21–33, 1997.
A. J. Crivelli & G. Catsadorakis (eds), Lake Prespa, Northwestern Greece.

The soils

C. S. Kosmas, N. G. Danalatos & N. K. Moustakas
Agricultural University of Athens, Lab of Soils and Agricultural Chemistry, Iera Odos 75, Botanikos 11855 Athens, Greece

Key words: soil units, physical and chemical characteristics, land suitability assessment, Greece

Abstract

The soils of Lake Prespa can be distinguished according to their degree of development, drainage condition, texture, content of coarse fragments, topography, degree of erosion and the presence of impermeable layers. A generalized land evaluation was carried out according to which land suitability was assessed for major uses including arable farming, pasture, forestry, wildlife, and recreation and tourism development. Furthermore, the soils were classified in different groups for environmental protection by means of both the protection of areas with high ecological value against any agricultural activity and the delineation of the lands susceptible to degradation such as soil erosion, through intensive arable farming, and groundwater contamination, through intensive irrigation and the application of high amounts of fertilizers and pesticides.

Introduction

The following discussion concerns the soils of Prespa basin and is mainly based on the results obtained in the project 'Evaluation of Soil Resources of the Prespa Region, Greece', a joint research effort of the Agricultural University of Athens, Lab. of Soils and Agricultural Chemistry, and the EEC, D.G. Environment, Nuclear Safety and Civil Protection, Proj. no. B6617-25-89.

The Prespa area, located in western Macedonia, Greece, is an area internationally known for its ecological importance. Part of Prespa's lowland is occupied by a National Park, a unique genetic reservoir for endangered waterbirds, which has been listed in the Ramsar Convention as an internationally protected wetland. However, the natural ecosystem in the Park is under continuous threat close to the Lake Mikri Prespa from grazing and clearance of the natural vegetation in order to extend agricultural land. Livestock production and farming are the most important sources of income in the Prespa lowland. The most important crops are beans and rye, covering 93% of the agricultural land.

The entire landscape of the lowland has been remodeled after the construction of the drainage-irrigation network. About 700 hectares of agricultural land are irrigated by surface irrigation methods with water supplied from the Lake Mikri Prespa.

Intensive cultivation with high amounts of irrigation water supplied by surface methods and the application of ammoniacal fertilizers (ammonium phosphate at rates of 140–200 kg N ha^{-1} and 70–100 kg P ha^{-1} per year) has resulted in the drastic acidification of agricultural soils. Under very acid soil conditions, manganese and aluminum are known to be liberated, often causing toxicities to the growing plants.

Many studies have already been conducted on various ecological aspects of the area, and different agricultural development programs have been presented by both authorities and other organizations. A detailed soil study would be useful for a proper land evaluation for agricultural development, respecting the environmental conditions of the area. Apart from the great scientific interest, the knowledge of the soil sources and their potential would complete the existing studies so that an overall approach can be used for further development, taking into consideration environmental requirements.

Study area and methods

An area of 3360 hectares including the Prespa basin and parts of its surrounding hills and mountains has been studied in detail by the Laboratory of Soil Science and Agricultural Chemistry of the Agricultural University of Athens (EEC-Project: Evaluation of soil resources of the Prespa region, Proj. N° B6617-25-89). The study area (Figure 1) lies between the latitudes 40°43′ and 40°51′ North and the longitudes 20°00′ and 20°10′ East; it has an average elevation of 853 m and the highest peak at 1020 m. The study area is defined to the north by the Greek-FYROM (Former Yugoslavian Republic of Macedonia) borders and the Megali Prespa, and to the west by the Devas, the Vrondero mountains and the village of Pili. It extends as far as to Mikrolimni and Oxia villages to the south and to Karies, Lefkonas, Plati, Milionas and Lemos villages to the east.

The main soil boundaries on the soil map were delineated on black and white air-photos (scale 1:20 000) supplied by the Greek Army Geographical Service. The complexity of the soils made it necessary to frequently correct these boundaries during an intensive field survey in which 850 routine augerings to a depth of 1.5 m were described. The soils were described and mapped according to the mapping system developed by Yassoglou et al. (1982) and currently used in the National Soil Survey Project. This system includes information on geomorphology, direction of soil development, drainage, texture, gravel and stone content, presence of carbonates, soil erosion, and the kind and depth of limiting layers or special properties. The soils are classified according to the Soil Taxonomy (Soil Survey Staff, 1975).

Soil samples were collected from 46 representative soil profiles and analyzed for physical and chemical properties (at the laboratory of Soils and Agricultural Chemistry of the Agricultural University of Athens), such as particle size distribution, organic matter content, pH, cation exchange capacity, available phosphorus and total nitrogen, field capacity,wilting point and bulk density. Infiltration measurements were carried out at the major soil mapping units using the double-ring infiltrometer.

Climatological data were collected from the Koula Meteorological Station and represent the typical climate of the Prespa basin. Data for the surrounding mountainous areas are not available.

Description of land characteristics

Climate

The climate of the Prespa lowland is Mediterranean, with hot dry summers and cold humid winters. More precisely it is 'Continental Mediterranean' as defined by Papadakis (1970), as this area and more generally the Macedonean plains are situated in a climatic transition zone between the Mediterranean climatic zone to the south and the Continental climatic zone to the north.

The mean annual temperature is 11.3 °C, fluctuating between 1.4 °C in January and 21.6 °C in July. The annual rainfall is 572 mm, with 60% falling in the period September–February, and 24% in the spring. The mean annual air humidity is 63%. The soil water balance tentatively calculated according to the Thornthwaite-Mather method (1964), assuming a deep groundwater table, is presented in Figure 2. From this approximation, it appears that potential evapotranspiration exceeds rainfall in an average year. From October till April, however, precipitation is much greater and particularly effective for leaching. In the period May–September, precipitation lags behind potential evapotranspiration leading to a pronounced drought at the end of summer.

Based on the average meteorological data, the Prespa area has a mild climate with warm summers, designated as Csb according to Koeppen (1931). However, the marginal character of this climate towards the Cfb humid climates of northern latitudes should be stressed. According to the ombrothermic diagram of Bagnouls-Gaussen and the calculated value of the xerothermic index (x = 50 days), the area has a submediterranean climate (FAO/UNESCO, 1963). According to the soil Taxonomy (Soil Survey Staff, 1975) and based on the values of average winter soil temperature (6.8 °C), average summer soil temperature (12.3 °C), the soil temperature regime of the Prespa soils is characterized as mesic. The 'xeric' moisture regime applies to all soils having a very deep groundwater table that do not receive run-off from adjacent areas. The rest of soils are saturated with water for some time in the year and have an 'aquic' moisture regime.

Parent material

The substrata of the hills and mountains surrounding the Prespa basin to the north, east, and south are mainly gneisses and granito-gneisses. The soils developed

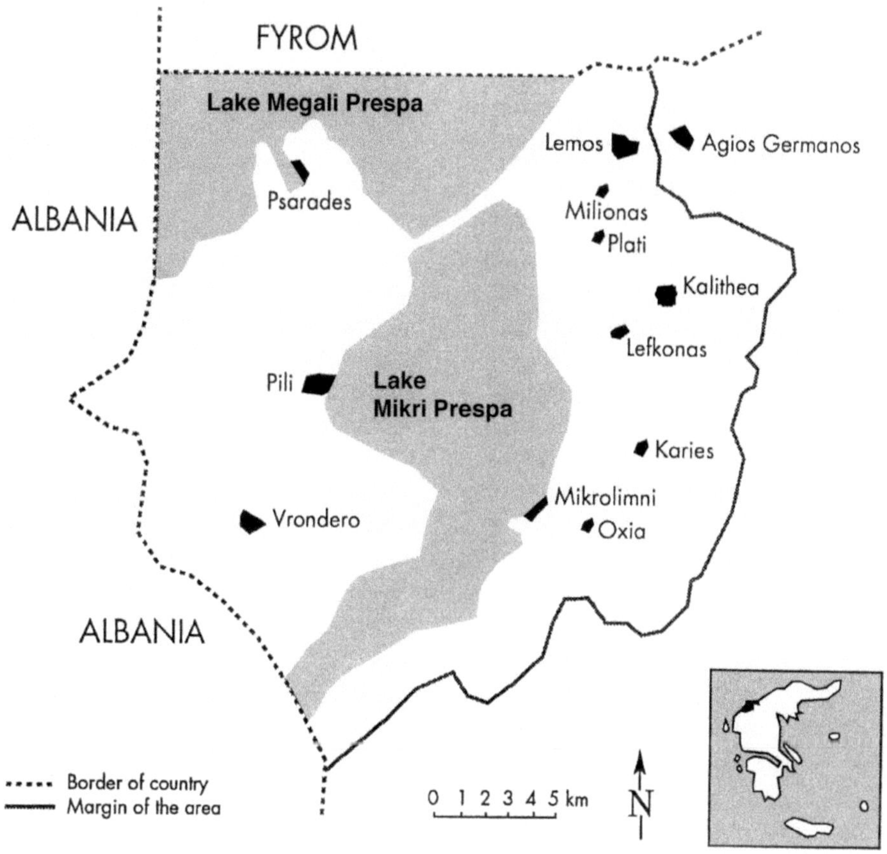

Figure 1. The location of the Prespa study area.

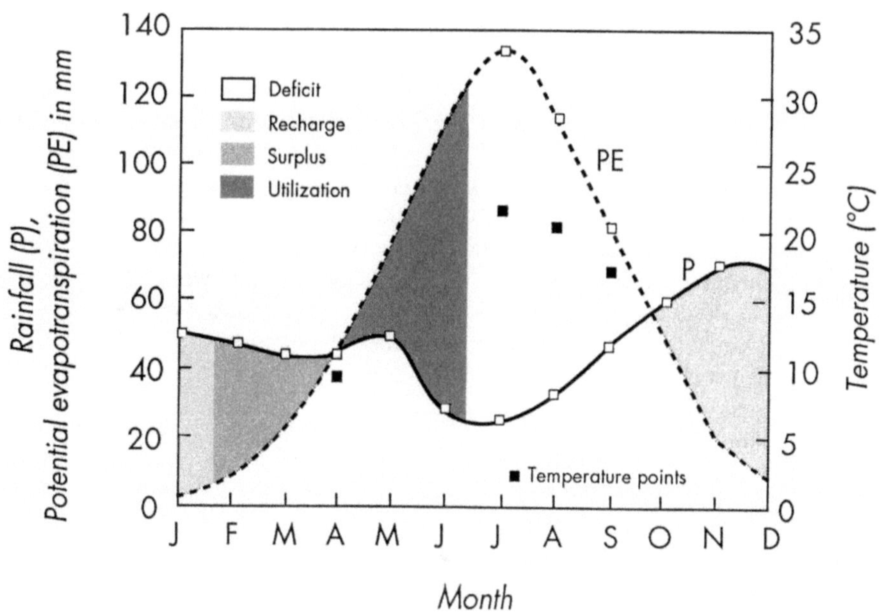

Figure 2. Climatic data and soil water balance of the soils of the Prespa region (Source: Kosmas et al., 1990)

from these parent materials are friable, rather permeable, generally acid and of moderately low base status. In contrast, soils developed on limestone deposits occurring in the western part of the basin are clayey, very slowly permeable, and are characterized by a high pH and base saturation.

In the basin, the materials are mainly recent alluvia and, to a lesser extend, old alluvial sediments. Some lacustrine deposits mixed with alluvial materials and few organic deposits are also present (Figure 3). These materials are the erosion products of the soils formed on the surrounding mountains, transported by streams and channels and deposited at lower levels. Their high textural variation is attributed to the speed of water upon deposition. Figure 3 shows that Prespa basin is crossed by seven major streams. These streams originate from granitic and metamorphic mountains to the east and flow westwards to the lowland, heavily loaded with erosion products. The marked decrease in flow gradient has resulted in the deposition of alluvium, peaking at the point of emergence onto the plain and spreading out in a fan-like form into the Prespa lowland.

The fan deposits vary in texture from coarse boulders and gravel near the fan apex where the Lemos, Plati, Milionas, Kalithea, Lefkonas villages are situated, to finer materials deposited at lower positions (Figure 4). The Agios Germanos stream has the greatest gradient and suspended sediment load in the area resulting in considerable deposition of very coarse materials (sand and gravel) along its course towards the lakes where it enters. The lack of vegetation in the watershed of this stream and the occurrence of rather steep slopes are responsible for the high quantities of materials transported to the lowland. In contrast, the gradients of the streams are much lower in the Mikrolimni plain; the watersheds are forested so that the soil materials transported and deposited in the alluvial plain are mainly medium to fine-textured.

Between the Mikri and the Megali Prespa lakes the materials are mainly sandy. The sand has been deposited and has filled the bottom of the shallow lake or beach ridges that existed in the past. Two main beach ridges, the first running in a curve and the second parallel to the shoreline of the Mikri Prespa lake, were mapped, while in the area along the shoreline of the Megali Prespa, the beach ridges have collapsed by wind erosion and human activities. Some of the beach ridges are seperated from each other by 'fingers' of finer deposits of silt and clay.

The area along the shoreline of Mikri Prespa lake near Pili village (Figure 3) has received less quantities of alluvial material, so that sedimentation in the lake is restricted allowing the deposition of organic materials.

Relief and soil characteristics

A considerable variation in topography exists in the area, with slopes ranging from 0–10 percent in the alluvial plain to more than 35 percent in the surrounding hills. Such variation markedly influences the soils of the Prespa through its effects on natural drainage, plant cover and soil erosion.

Natural soil drainage ranges from well drained on the beach ridges and the upper parts of the alluvial fans, to very poorly drained in the depressions and the flat bottomland (Figure 5). Normally, the level of the groundwater table fluctuates seasonally resulting in a distinct mottling especially in the lowlands in the central plains of the Prespa basin. The lowest parts of the region are temporarily ponded, and the groundwater table rises close to the soil surface resulting in the formation of reducing, gray horizons known as 'gley horizons'. In many cases, however, the groundwater table has been lowered by the construction of the drainage network.

Relief is closely related to the degree of soil development in the basin. Soil erosion is severe, especially on the hilly soils around Agios Germanos, Milionas and Plati villages. Intensive cultivation of the sloping agricultural land and clearance and/or over-grazing of the uplands under natural vegetation has resulted in accelerated erosion and deposition of new materials in the lowland. In the lowland, however, soil erosion coexists with deposition of new materials. This is clearly observable in the vicinity of Lefkonas and Karies villages where old alluvial soils (Alfisols) have been buried below very recent alluvial materials. Moreover, in the depressions of the lowland (west of Plati village), recent fine-textured, very gravelly soil materials have filled the bottoms of previous depressions or ponds. Comparison of aerial photographs taken in 1945 and 1982 demonstrated that the land has extended into the Mikri Prespa lake by an average of 43 meters, through deposition of alluvial materials carried by streams.

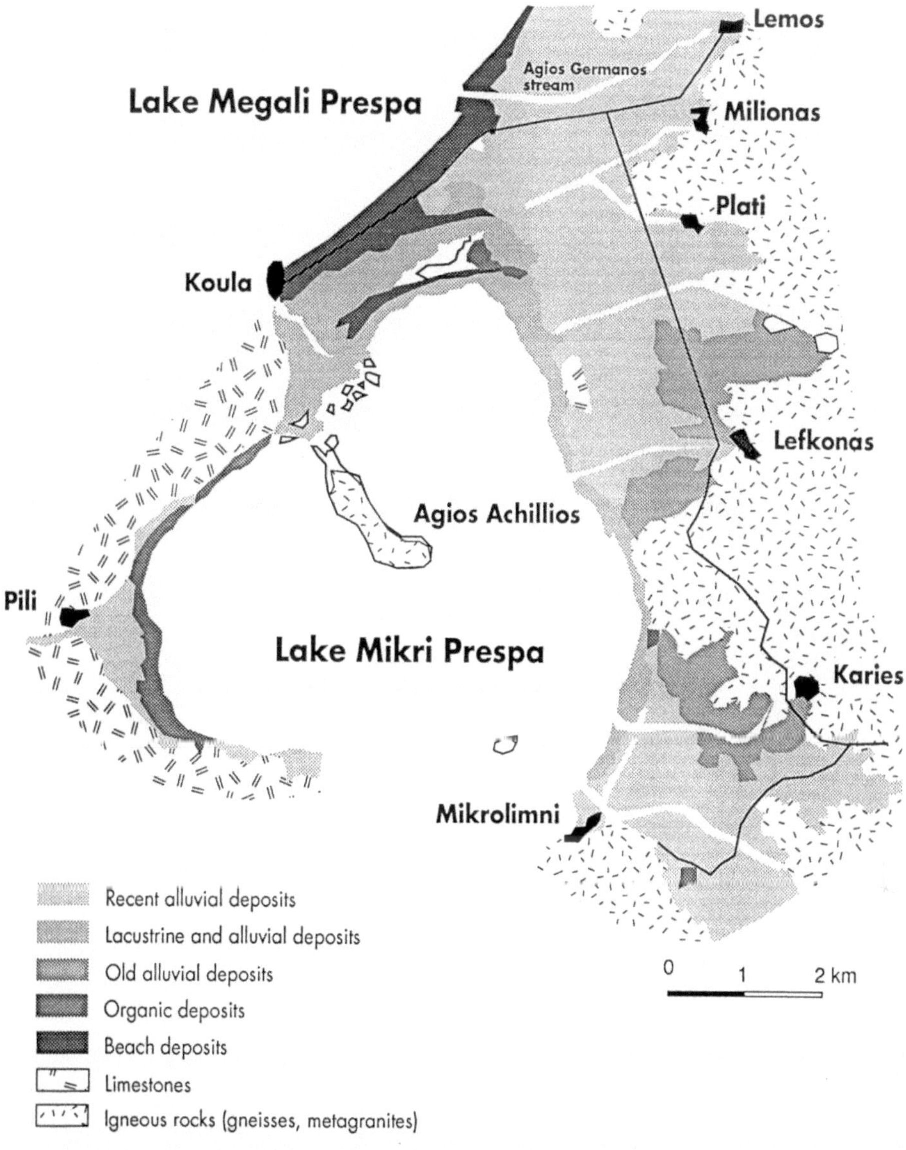

Figure 3. Distribution of parent materials of the soils formed in Prespa basin (Source: Kosmas et al., 1990).

Description of the soils

Major soil units

The soils of the Prespa basin can be grouped into the following major units:

– Entisols: these are recent alluvial soils lacking diagnostic horizons other than an ochric epipedon. Deposition of new materials by streams and channels on the surface of these soils was faster than development of pedogenic horizons, so that these soils show hardly any profile development. This soil unit occupies 2229 hectares (66.3% of the mapped area) in the bottom lands and on some alluvial fans.

– Inceptisols: these soils occur on the alluvial fans but only in the west part of the area where the deposits are calcareous. The presence of carbonates delays clay translocation and soil development, so that these soils show only moderate horizon differentiation such as color and structure formation (cambic horizon). This

Figure 4. Distribution of textural classes of the soils in the Prespa basin (Source: Kosmas et al., 1990).

soil unit covers only 42.3 hectares (1.3% of mapped area).

– Alfisols: these are the oldest soils of the area, formed on old alluvial fans and terraces occurring at higher elevations. They show distinct horizon differentiation such as clay transformation and/or translocation (formation of an argillic horizon), and cover 376.2 hectares or 11.2% of the mapped area.

– Histosols: these are soils rich in organic materials formed along the shorelines of the Mikri Prespa lake and of some ponds in the lowland formed from the accumulation and decomposition of bushes, reeds and sedge-type vegetation with a thickness of 10 cm to

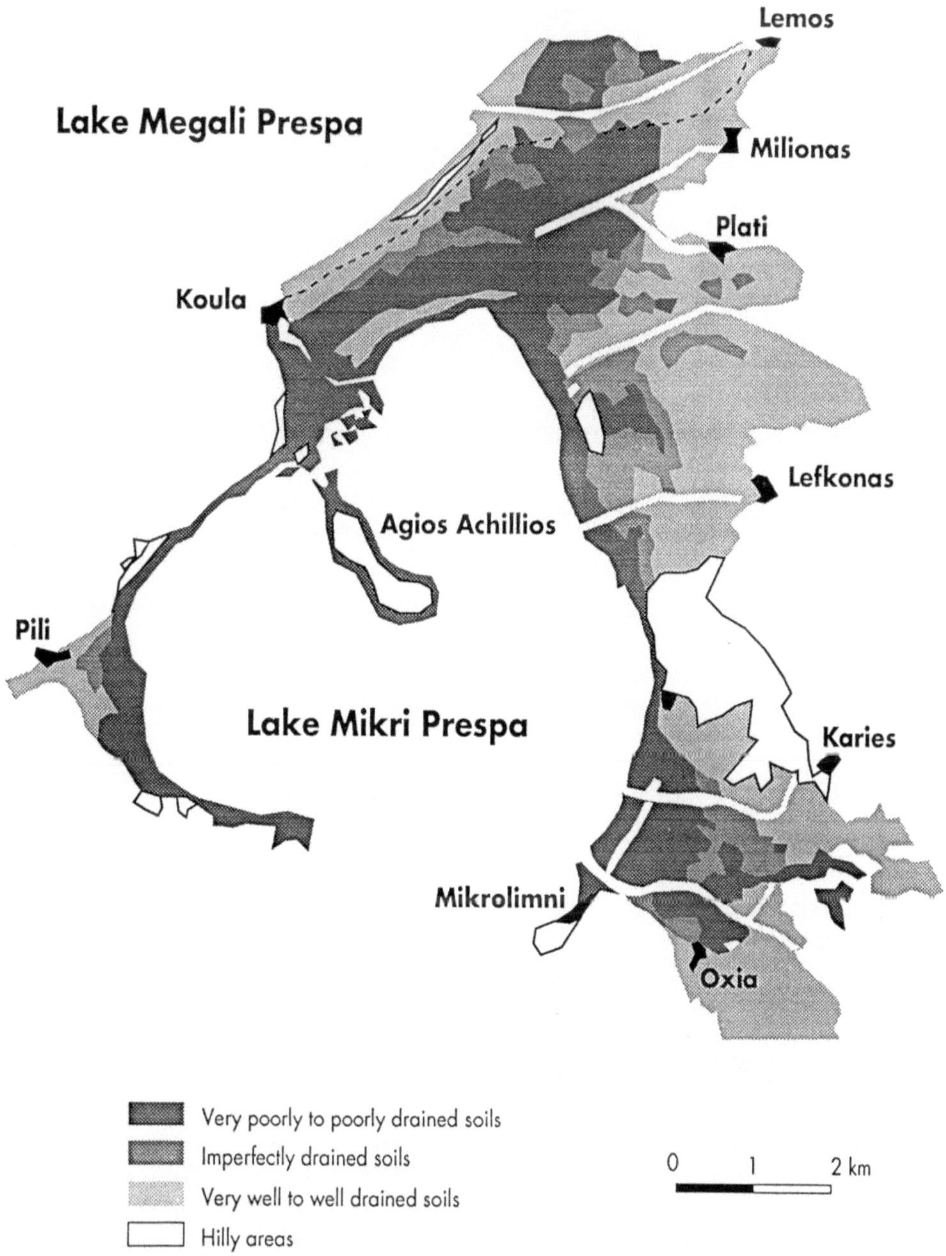

Figure 5. Distribution of drainage classes in the fluvial deposits of the Prespa basin (Source: Kosmas et al., 1990).

more than 100 cm. This soil unit occupies only a small part of the mapped area (47.5 hectares or 1.4%).

The soils of the uplands are distinguished according to the degree of erosion and profile development. They are classified as Entisols if the parent material has been exposed on the soil surface through severe erosion, and Inceptisols or Alfisols if they are moderately to slightly eroded and have a cambic or argillic horizon, respectively. The main characteristics of the upland soils under grass or shrubs is their advanced degree of erosion. In contrast, soils under dense forest are adequately protected and relatively deep. The parent material of these soils is confined to igneous (gneiss and granito-gneiss) or limestone formations. Due to

Table 1. Distribution of drainage classes among recent alluvial soils (Entisols)(Source: Kosmas et al., 1990).

Drainage class	Area (ha)	Percentage
Very well drained	678.8	30.4
Well drained	114.0	5.1
Moderately well drained	121.9	5.5
Somewhat poorly drained	145.2	6.5
Poorly drained	42.4	1.9
Very poorly drained	1126.9	50.6
Total	2229.6	100.0

the steep topography, they show erosion and restricted plant root penetrability. There is limited information available for these soils.

Physical and chemical characteristics

Entisols

The recent alluvial soils are mainly nearly flat. Thus, 1738 hectares (or 78% of the area) have a gradient of less than 2%, whereas slightly sloping soils (gradient 2–6%) cover an area of 432.6 ha (19.4%). Moderately sloping soils (gradient 12–18%) cover 53.4 ha and 4.6 ha, respectively. Due to the moderately high rainfall, the geomorphological characteristics of the area and the vicinity of the lakes, these soils are usually poorly drained and contain reduced 'gley' horizons over an area of 1127 ha (Table 1). Well drained soils are found in the uppert part of the basin, on the alluvial fans or along the streams or channels on coarse soil materials and cover an area of 792.8 ha.

As Table 2 shows, the dominant textural classes in the surface layer (depth 0–35 cm) of these soils are sandy clay loam and clay loam (44.3% of the area), followed by sand and loamy sand (23.5% of the area). The classes loam and sandy loam have about the same distribution (15.4% and 13.9%, respectively) in the area, whereas fine textural classes are rather limited (2.9%). In the subsoil (35–75 cm) and substratum (75–150 cm), the dominant textural classes are also clay loam and sandy clay loam (Table 2). Limiting layers of coarse sand are present at different depths and thickness as following: 123.6 ha with layers between 25 and 50 cm depth, 63.1 ha with layers 50 and 100 cm depth, and 30.0 ha with layers between 100 and 150 cm. Additionally, soils with a sandy texture throughout the soil profile cover a total area of about 112.2 ha.

The soils are mainly acid (pH usually ranges from 4.4 to 6.5), and of medium base status (base saturation usually ranges from 30 to 83%). The fertility status is generally moderate, characterized by low organic matter contents (viz. 1.0–2.5%) and total nitrogen content (viz. 0.06–0.12%) in the surface horizon. Both organic matter and nitrogen contents are higher in the top-soil than in the subsoil and substratum. Their distribution with depth, however, is erratic due to the alluvial origin of the soils. The poorly drained soils have a slightly higher amounts of organic matter ranging from 1.7–2.5% if cultivated, or have a thin layer of organic materials if they are under natural vegetation.

The total phosphorous content ranges from 234–2858 mg P kg^{-1} of soil. The amount of available phosphorus (P-Bray) is high ranging from 3–374 mg P kg^{-1} of soil (average 37.2 mg P kg^{-1}), with the highest concentration in the surface layer. The cation exchange capacity (CEC) ranges from 3.2 to 38.7 cmol kg^{-1} soil, with an average of 13.7 cmol kg^{-1} soil. The extractable aluminum and manganese range from traces to 0.3 cmol kg^{-1} soil and from 0.008–0.16 cmol kg^{-1} soil, respectively.

In terms of their physical properties, Entisols are characterized by a friable to firm consistency, a weak fine to medium sized subangular blocky structure in the topsoil with a tendency to become weaker and coarser in the lower sections of the profile. The bulk density ranges from 1.45 to 1.55 g cm^{-3} in the topsoil to 1.55–1.60 g cm^{-3} in the deeper layers. The available water holding capacity ranges from very low to high depending on soil texture.

Water infiltration data show a considerable scattering. In most soils, however, the basic (final) infiltration rate is low to moderate with an average value of 2.4 cm h^{-1}.

The soils of this group are classified into the following subgroups according to Soil Taxonomy:

– Typic Xeropsamments: sandy soils mainly along the lakes, on the beach ridges, or along the streams or channels.

– Aquic, Typic Xerofluvents: depending on the presence or absence of hydromorphic characteristics within 150 cm of the soil surface. Aquic are Xerofluvents that are saturated with water during any period in most years. The rest of the Xerofluvents belong to the Typic subgroup.

– Aeric, Typic Fluvaquents: very poorly drained soils. They are characterized by a shallow water table (but usually deeper than 45 cm) and are gray colored throughout most of the soil profile (Typic Fluvaquents); or they contain a perched water table and have iron mottling (Aeric Fluvaquents).

Table 2. Distribution of textural classes among recent alluvial soils (Entisols) throughout the soil profile (Source: Kosmas et al., 1990).

	Distribution of textural class at depth					
	0–35 cm		35–37 cm		75–150 cm	
Textural class	area (ha)	%	area (ha)	%	area (ha)	%
Sand, loamy sand	524.7	23.5	401.1	18.0	275.9	12.4
Sandy loam	344.0	15.4	591.2	26.6	771.4	34.8
Loam, silty loam	310.5	13.9	266.6	11.9	140.8	6.3
Sandy clay loam						
Silty clay loam	986.4	44.3	766.6	34.4	10411	46.7
Clay loam						
Clay, silty clay	63.6	2.9	203.6	9.1	–	–
Total	2292.2	100.0	2292.2	100.0	2292.2	100.0

Inceptisols
These soils are slightly gravelly sandy clay loam or clay loam. In a few cases the surface layer is stony (stones 15–35%). They are mainly well drained. In some areas, the soils are characterized as somewhat poorly drained, remaining wet for short periods in the year.

The organic matter content of the topsoil is low (<2.2%) and decreases with depth. The total nitrogen content ranges from 0.03 to 0.1%. The soils are rich in extractable calcium (2000–7500 mg Ca kg^{-1} soil), magnesium (350–780 mg Mg kg^1 soil) and potassium (140–360 mg K kg^{-1} soil).

In terms of their physical properties, these soils are characterized by a firm consistency, a moderate medium to coarse-sized subangular blocky structure in the surface layer with the tendency to become weaker in the lower sections of the soil profile. The bulk density ranges from 1.4 to 1.7 g cm^{-3} in the topsoil and increases slightly in the deeper sections. Water infiltration is slow to moderate (0.6 to 4.2 cm h^{-1}).

At the subgroup level, these soils are classified as Typic Xerochrepts because they have high base saturation and they do not have iron or manganese mottling within 75 cm of the soil surface.

Alfisols
These soils are characterized by a clay loamy texture. Due to geomorphic characteristics, they are mainly very well drained (168.6 ha or 44.8% of Alfisols), while well drained and imperfectly drained variants cover 144.2 ha and 63.6 ha, respectively (38.3% and 16.9% of Alfisols, respectively). Gravels are present in most soils with contents ranging from 5 to 35 percent. Additionally, the surface layer of most soils is slightly stony to stony and rarely bouldery. The substratum is occasionally bouldery too.

Most soils are slightly sloping (gradient 2–6%) covering an area of 218.2 ha (58.0%) to moderately sloping (6–12%) covering an area of 98.6 ha (26.2%). Flat soils (gradient 0–2%) cover 34.5 ha (9.2%), and steep soils (gradient 12–18%) cover an area of 24.7 ha (6.6%).

The surface soil is strongly acid (pH = 4.5–5.5), and the subsoil is medium to slightly acid (pH – 5.1–6.8). Base saturation ranges normally from 40 to 70 percent.

The organic matter content is 1.7 to 2.5% in the topsoil and decreases regularly with depth to 0.5–0.7% in the deeper layers. The total nitrogen content of the surface layer is also low, ranging from 0.06 to 0.14%.

The total phosphorous content ranges from 387 to 1589 mg P kg^{-1} soil (average 930 mg P kg^{-1}). The available phosphorous content ranges from 2.8 to 138.5 mg P kg^{-1} with an average of 49.8 mg P kg^{-1} soil. However, in heavily fertilized soils, extremely high values of available P may be found. In the subsoil and substratum, the available phosphorous decreases rapidly ranging normally from 0.3 to 135.6 mg P kg^{-1} soil.

The extractable manganese ranges from 20 to 50 mg Mn kg^{-1} soil with the highest values measured in the topsoil. Extractable aluminum ranges from trace to 0.1 cmol kg^{-1} soil.

The structure is typically weak to medium subangular blocky in the topsoil to moderately coarse angular blocky in the deeper horizons. Basic infiltration rates are usually low to moderately low (0.8 to 1.5 cm h^{-1}).

The soils of these groups are further classified into:

– Ultic, Aquultic Haploxeralfs: soils with a base saturation less than 75 percent are classified as Ultic Haploxeralfs, while the soils with base saturation less than 75 percent and with iron-manganese mottling within 75 cm of the soil surface are classified as Aquultic Haploxeralfs.

– Ultic Rhodoxeralfs. Soils having red colors (redder than 5YR) in the argillic horizon and a base saturation of less than 75 percent are classified in this subgroup.

Histosols

These soils consist mainly of well decomposed organic materials (sapric) with a thickness of at least 70 to 150 cm. In some areas, the soils also contain inorganic materials, which have been deposited by the surrounding streams and subsequently enriched with small amounts of carbonates ($CaCO_3$ <4%). They are normally saturated with water and have a very high water-holding capacity. All these soils are classified as Typic Medisaprists. They are under natural vegetation inside the National Park and should be protected against any agricultural use in order to preserve the unique wildlife of the area.

Land suitability assessment

Land use planners need guidance for taking decisions on land use change so that the environmental resources be put to the most beneficial use, taking additionally into account the conservation of these resources for the future (FAO, 1976). The following major kinds of land use can be defined in the Prespa region: annual crops, annual and/or perennial crops, pasture, forestry, wildlife, recreation and tourism development. The suitability of land was assessed for these uses taking into consideration the following characteristics: drainage, effective soil depth, presence of organic layers, particle size distribution, topography, the presence of impermeable layers, and erosion hazard.

For this purpose, the following categories of soil units were defined:

cultivated soils

– deep, somewhat poorly to poorly drained, coarse to moderately fine-textured soils with or without a slight profile development (including a number of Entisols and Inceptisols of the Prespa basin). These soils are more suitable for annual summer crops (Figure 6),

– deep, very well to moderately well drained, nearly flat to moderately sloping, medium to fine-textured variants of Entisols, Inceptisols and Alfisols occurring in the basin. These soils are more suitable for annual or perennial crops.

From the above soil groups, 384 ha of the recent alluvial soils and 224 ha of the well developed soils show no significant limitations for the development of intensive agriculture other than the need for proper management (fertilization, irrigation, etc.). Very coarse-textured soils occupy 266 hectares. They present severe limitations for agricultural development because of their marginal fertility status. Fertilization and irrigation of these soils involve severe hazards for groundwater pollution. Coarse-textured soils occupy 259 ha in the mapped area. These soils have a low fertility status and show moderate limitations for agricultural development. Moreover, the risk of groundwater pollution by the use of fertilizers and pesticides is high.

soils for wildlife and tourism development

– shallow, very well drained, very coarse to coarse-textured soils, without any profile development, occurring along the shoreline of the Megali Prespa lake (including some Entisols of the basin). They are more suitable for recreation and tourism development,

– deep to shallow, very poorly drained, coarse to fine-textured soils occurring in the depressions. These soils contain organic layers on their surface; they are frequently ponded and are most suitable for wildlife protection.

The soils along the shoreline of the Mikri Prespa lake occupy 587 ha. These soils are frequently ponded, they are covered by natural vegetation and constitute the primary wildlife value of the National Park. The great ecological importance of the Prespa region and particularly the conservation of the existing birds populations dictates the protection of these soils against any agricultural development.

upland soils

– shallow, moderately steep to very steep soils of the uplands, formed on residual parent materials. These soils are more suitable for pastures,

– moderately deep soils of the uplands formed on residual parent materials. These soils are more suitable for forestry.

Soils with slopes greater than 6% occupy 210 hectares of the mapped area. Such soils are very susceptible to erosion if cultivated. Water run-off and soil erosion should then be controlled by means of crop rotation, conservation tillage, stream diversions and contour farming. If these soils are grazed, protection

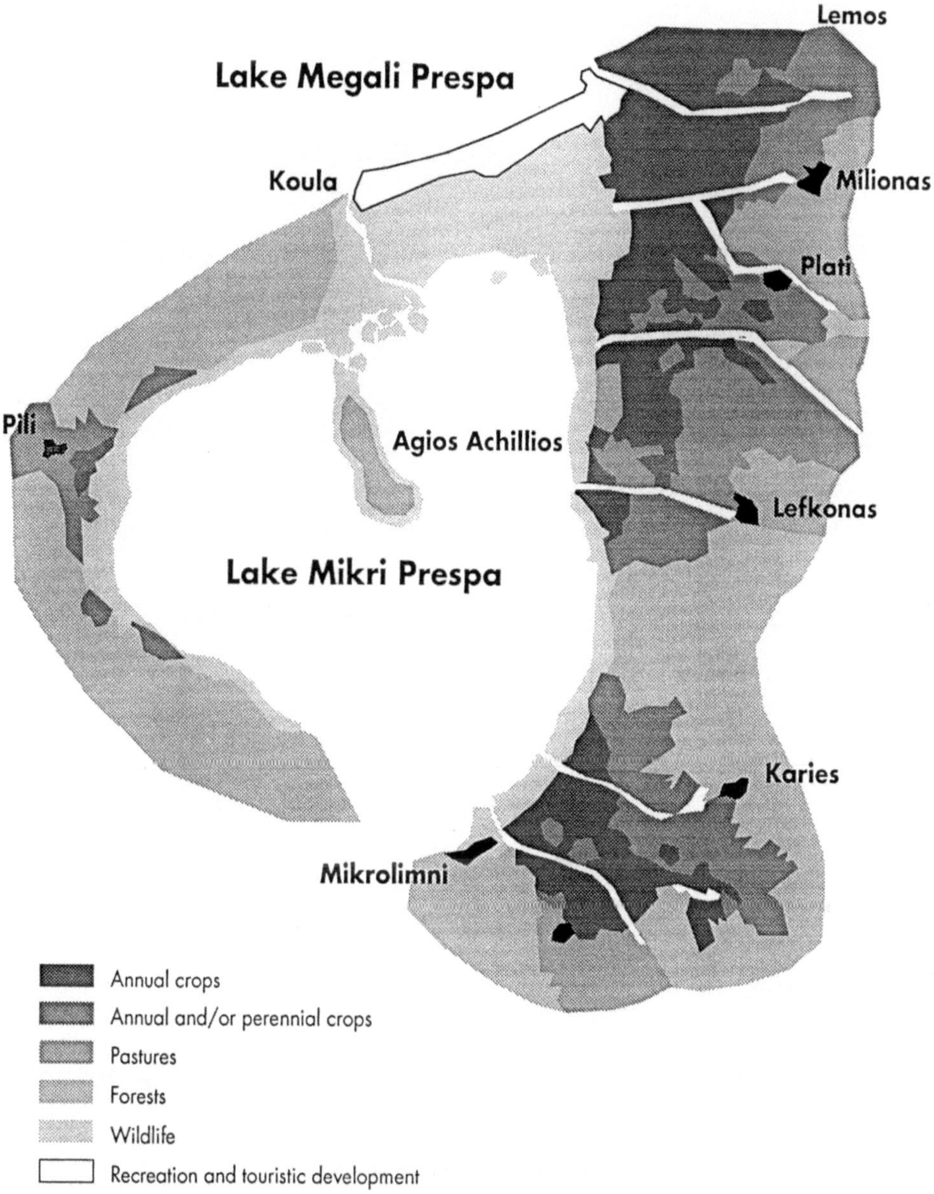

Figure 6. Land suitability map (after Kosmas et al., 1990).

from overgrazing comprises an additional soil conservation measure.

Soils on the uplands with slopes greater than 12% present very serious erosion hazards, despite their natural vegetation cover. In the majority, these soils are not suitable for exploitation of profitable arable agriculture, but they should be managed as forest or pastures under the condition that water run-off and soil erosion are controlled.

Farming and groundwater protection
Most of the land in the Prespa basin is intensively cultivated with annual crops. Large amounts of fertilizers and pesticides are applied annually. Such management may ensure high yields, but in the long term it is associated with adverse consequences for the environment, viz. environmental pollution and land degradation. The soils of the Prespa basin are classified according to the required inputs for maximum environmental protec-

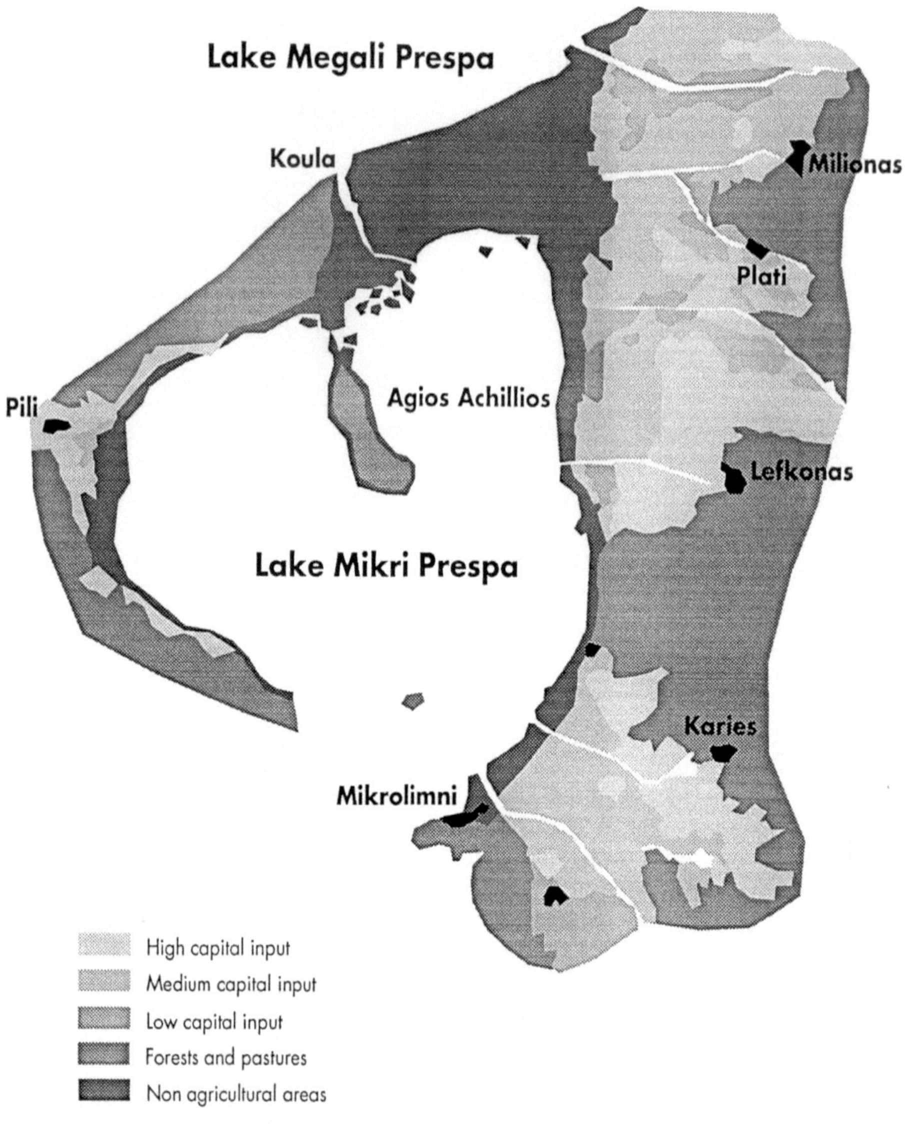

Figure 7. Map of proposed capital investment (after Kosmas et al., 1990).

tion. Three levels of capital inputs are distinguished for the soils under cultivation:

Low capital input. Local varieties are used. Fertilizers and pesticides are not applied. Yields depend primarily on soil fertility and environmental conditions.

Medium capital input. Improved varieties are used. Insufficient amounts of fertilizers are applied and inadequate disease control is undertaken. Some measurements are taken to control soil erosion.

High capital input. Improved varieties are used. Use of fertilizers and disease control are adequate. Management is as recommended by the local Agricultural Extension Service, so that yields are maximum for the prevailing climatic conditions. Measurements are taken to control soil erosion.

According to the above classification, the cultivated soils in the Prespa basin are placed into the following broad groups (see also Figure 7):

– nearly flat to slightly undulating, medium to moderately fine-textured, well drained soils with very deep groundwater table. These soils cover 951 hectares and should be used for high capital input agriculture,

– moderately well to somewhat poorly drained, medium to moderately fine-textured soils; and poorly drained, moderately fine to fine-textured soils having a relatively deep groundwater table. These soils cover about 730 hectares and should be used for medium capital input agriculture,

– coarse to very coarse-textured soils with highly variable drainage conditions. These soils cover 924 hectares and should be used for low capital input agriculture.

Intensive cultivation associated by the application of large amounts of fertilizers and pesticides may ensure high yields on the one hand, but on the other it will create adverse consequences for the environment, viz. groundwater pollution and land degradation.

The soils along the shoreline of the Mikri Prespa lake are covered by natural vegetation and constitute the primary wildlife value of the National Park. The great ecological importance of the Prespa region and particularly the conservation of the existing bird populations dictates the protection of these soils against any agricultural development.

Conclusion

The soils of the Prespa basin were derived mainly from gneisses and granito-gneisses and are coarse-textured, friable, rather permeable, generally acid and of moderately low base status.

A considerable variation in the topography of the area, with slopes ranging from 0–10 percent in the alluvial plain to more than 35 percent in the surrounding hills, markedly influences natural drainage, plant cover and soil erosion.

Intensive cultivation and application of ammoniacal fertilizers has drastically acidified the agricultural soils. Under low acid soil conditions, manganese and aluminum are liberated often causing toxicity to the plants. Artificial drainage has removed the excess of water from the upper soil layers. The effectiveness of the drainage network was higher in the alluvial plain of the Agios Germanos stream and in the alluvial plain west of Plati village. Summer crops such as beans – one of the most popular crop – grow normally and produce high yields without significant problems. A detailed environmental and economic study is necessary before any decision is taken about further reclaiming these soils.

References

FAO/UNESCO, 1963. Carte Bioclimatique de la zone Mediterranee. Notice explicative. O.N.M. Paris.

FAO, 1976. A framework for land evaluation. FAO Soil Bulletin 32, Rome: 79 pp.

Kosmas, C., N. Moustakas, B. Tsatiris & N. Danalatos, 1990. Evaluation of soil resources of the Prespa region, Greece. EEC project. Directorate General, Environment, Nuclear safety and Civil Protection No. B6617-25-89, 145 pp.

Koppen, W. S., 1931. Die Klimate der Erde. 2nd Auflege Walter der Gruyter, Berlin: 369 pp.

Papadakis, J., 1970. Climates of the world. Their classification similarities, differences and geographic distribution. Buenos Aires: 47 pp.

Soil Survey Staff, 1975. Soil Taxonomy: a basic system of soil classification for making and interpreting soil surveys. USDA Agric. Handbook 436 U.S. Government Printing Office, Washington, D.C. 754 pp.

Thornthwaite, C. W. & G. R. Mather, 1957. Instruction and tables for computing potential evapotranspiration and the water balance. Drexel, Inst. of Techn. Lab. of Climatology. Publ. Climatol. Vol. X; No. 3, fifth print. Cenerton, N.J, 311 pp.

Yassoglou, N., A. Nychas & C. Kosmas, 1982. Parametric designation of mapping units for soil survey and evaluation in Greece base d on Soil Taxonomy. Am. Soc. of Agr., Annual Meetings, Anacheim, California.

Hydrobiologia **351:** 35–40, 1997.
A. J. Crivelli & G. Catsadorakis (eds), Lake Prespa, Northwestern Greece.

The flora of Prespa National Park with emphasis on species of conservation interest

G. Pavlides
Akropoleos 36, 54634 Thessaloniki, Greece.

Key words: flora, endemic plant, biogeography, conservation, Greece

Abstract

The flora of Prespa National Park is composed of 1326 plant species. Among them are 17 Pteridophytes and 1309 Spermatophytes, represented by 11 Gymnosperms, 1057 Dicotyledons and 241 Monocotyledons. The flora is evaluated and categorised into species which can be considered as new for the Greek flora, species under international protection, and rare local species threatened with extinction. Plant species are also categorised according to their ecological requirements, with special reference to the endemic species and their centres of origin in respect to their total percentage (15.4%).

Introduction

The Prespa National Park constitutes a geomorphologically diverse system, with a distinctly mountainous relief. In the lower part of the drainage basin lie the lakes and the major role can be ascribed to Lake Mikri Prespa, which is the primary nucleus of the biochore. The environment of the National Park is mainly defined by the existence of both the lacustrine and terrestrial ecosystems. The altitudinal range of the area is between the level of the lake (853 m asl) and the summit of Mt Varnous (peak Dikorpho, 2118 m), an altitudinal difference of more than 1250 m. The diverse micro-habitats with particular hydrological and climatic parameters, in combination with the long-standing geopolitical isolation of the area on the one hand, and with the eventful history and the general human impact of recent years, on the other, have contributed to the creation of a rich and diverse flora as well as to an impressively diverse vegetation. The geological substrate, almost equally divided into calcareous and granitic, has had a major role in the floristic diversity. Granitic soils have facilitated the migration of Central European species via the Dinaric Alps to Mt. Varnous and some reach the calcareous Pindus mountain tops. On the contrary, plant species abounding on the limestone mountains of continental Greece, reach the limestone part of the National Park which for many of them constitutes their northern distribution limit.

The flora

The flora of the Park is composed of 1326 plant species (Pavlides, 1985 and unpubl. data). Their identification and nomenclature is based almost exclusively on the Flora Europaea (Tutin et al., 1964–1980) which should be consulted for authors' names. Among the 1326 plant species of the National Park, 17 are Pteridophytes and 1309 Spermatophytes. Of the latter, 11 are Gymnosperms, 1057 Dicotyledons and the remaining 241 species are Monocotyledons (Table 1). The flora is subsequently classified qualitatively into various plant categories.

Species new for the flora of Greece

Some plant species found in the Park are new for the Greek flora. In this category belong 21 species which are endemic to a wider area of the Balkans or even a wider area in Eastern Europe. In this respect, we stress the presence of these endemics in the Hellenic area for the first time in the Park, which thus seems to be

Figure 1. Diphelypaea (Phelypaea) boissieri (Reuter) Stapf, an endemic of the Central Balkans. (Photo: G. Catsadorakis)

within the range of distribution of the above-mentioned species. From the phytogeographical point of view the following species are of great interest: *Vinca balcanica, Dianthus perviformis, Centaurea rufidula, Cirsium brachycephalum, Tragopogon pterodes, Alyssoides sinuata, Barbarea balcana, Corydalis ochroleuca, Ranunculus aduncus, Alchemilla remiformis, Jovibarba hirta ssp. hirta, Rhinanthus ovifugus, Eryngium serbicum* and *Lilium jankae.*

To the above list must be added *Diphelypaea (Phelypaea) boissieri* (Reuter) Stapf (Figure 1), an endemic of the Central Balkans found in Prespa in 1990 (F. Thorning-Lund & A. J. Crivelli, pers.comm.)

In addition, the following 17 non-endemic species were recognised as new for the Greek flora: *Aster bellidiastrum, Cirsium palustre, Senecio fluviatilis, Thalictrum foetidum, Potentilla emilli-popii, Peucedanum palustre, Carex ericetorum, C. pseudocyperus, Juncus tenuis, Luzula subetica, Sparganium emersum, Thlaspi alliaceum, Scabiosa columbaria ssp. portae, Aldrovand vesiculusa, Stachys alpina, Agrostis rupestris* and *Festuca ovina.*

Table 1. The taxonomic composition of the flora of Prespa National Park.

Taxon	Family	Genus	Species	%
Pteridophyta	10	12	17	1.28
Gymnospermae	3	4	11	0.83
Dicotyledones	84	365	1057	79.71
Monocotydelones	16	96	241	18.18
Total	113	477	1326	100.00

For certain plant species, the region of the National Park constitutes their second recorded site of occurrence in Greece. This second find will contribute to the removal of any potential reservations about their presence in Greece and consequently will extend the corresponding geographic ranges of the species. Among these species the following 11 are endemics of a wider area of the Balkans or of Eastern-Central Europe: *Alkanna pulmonaria, Taraxacum palustre, Barbarea longirostris, Trifolium velenovskyi, Galium cinereum, Onosma echioides, Pulmonaria rubra, Sempervivum zeleborii, Potentilla hirta, Festuca varia* and *Globularia cordifolia.* There are also the following 19 non-endemic species: *Adoxa moschatellina, Crepis tectorum ssp. tectorum, Hieracium echioides ssp. echioides, Petasites albus, Senecio nemorensis, Scabiosa ochroleuca ssp. rhodopaea, Vaccinium uliginosum, Fragaria moschata, Sorbus aria ssp. aria, Eriophorum latifolium, Avenula pubescens, Calamagrostis pseudophragmites, Koeleria pyramidata, Sesleria coerulea, Luzula spicata ssp. spicata, Potamogeton gramineus, Polygonatum verticillatum, Vaccinium myrtillus* and *Siphonostegia (Lesquereuxia) syriaca.*

The flora of the National Park comprises also various other species, the presence of which in Greece is according to the Flora Europaea questionable (Gr?) or eventually species for which inadvertently no indication of their presence in Greece was registered (Gr), in spite of the fact that these are well-known members of the Greek flora. Among these species, 22 have been found in the National Park, including: *Carex sempervirens, Bonbycilaena erecta, Euphorbia segetalis, Doronicum hungaricum, Ornithogalum pyrenaicum, Satureja montana, Bupleurum commumatum* and *Saxifraga pedemontana ssp. cymosa.*

Endemic plant species of the Park

In this paper the word endemic is used to indicate a plant species known to exist only in a restricted area in the Balkans or in Central-Eastern Europe.

The flora of the National Park comprises various plants of endemic character which having as starting point the centre of their geographical range inside or outside Greece, penetrate into the environment of the National Park. The percentage of these endemics in the local flora amounts to 15.4%. 163 endemic species were recorded which were already known to occur in Greece from previous studies. Besides these, 41 other endemics have been recorded for the first time or their occurrence in the Greece was simply confirmed once more. *Centaurea prespana* is mentioned as a local endemic (Figure2). A mention was already made above, of the 21 endemic species of different origin, for which this is the first recorded occurrence in Greece. A further 11 other endemic species were also recorded in the flora of the Park for the second time in Greece.

The origins of the remaining endemic plants is in detail as follows:

Exclusively endemic to Greece (25): *Centaurea prespana, Anchusa variegata, Anthemis pindicola, Rhinanthus sintenisii, R. pindicus, Fibigia eriocarpa, Knautia magnifica, Hypericum apollinis, H. rumeliacum, Teucrium divaricatum ssp. graecum, Astragalus parnassi, Trifolium parnassi, Scrophularia pindicola, Verbascum epixanthinum, Trinia frigida, Iris pumilla ssp. attica, Fritillaria graeca ssp. graeca, Scorzonera crocifolia, Astragalus lacteus, Armeria undulata, Ranunculus ficarioides, Helianthemum hymettium, Achillea absinthoides, Galium pisoderium* and *Festuca graeca.*

Endemics to Greece also occurring in other countries (23): (a) Greece-Albania: *Alkanna corcyrensis, Minuartia stellata, Centaurea spruneri ssp. guiciardi, Sedum serpentini, Sempervivum reginae-amaliae, Nepeta spruneri, Scutellaria rudicunda ssp. rudicunda, Galium intricatum, Saxifraga taygetea, Pedicularis graeca, Lilium chalcedonicum, Euphorbia heldreichii, Crataegus heldreichii*; (b) Greece-Bulgaria: *Asperula aristata ssp. nestia, Minuartia glomerata ssp. velutina, Hesperis macedonica*; (c) Greece-Italy: *Verbascum macrurum.* (d) Greece-FYROM: *Dianthus myrtinervius, Veronica kindii, Viola eximia, Anchusa serpentinicola, Hieracium naegelianiformae* and *Festuca rechingeri.*

The Balkan endemics are 74 and include: *Campanula foliosa, Silene asterias, Ramonda serbica, Viola orphanidis, Eryngium palmatum, Alyssum corymbosum, Ranunculus serbicus, Lotus stenodon, Jasione heldreichii, Ballota macedonica* and many others.

Figure 2. Centaurea prespana, a local endemic. (Photo: G. Catsadorakis)

There are 13 Balkan endemics also shared with other phytogeographic sub-areas: Balkan-Pannonian: *Achillea crithmifolia, Doronicum hungaricum, Juncus thomasii, Oenanthe banatica, Sedum cepaea, Thlaspi goesingense.* Balkan-Mid-European: *Hieracium alpinum, Oenanthe peucedanifolia, Peucedanum austriacum.* Balkan-Circumalpine: *Geranium macrorrhizum, Plantago argentea, Knautia drymeia.* Balkan-sub-Mid-European: *Asperula purpurea.*

There are 7 Balkan endemics shared with Italy: *Pinus leucodermis, Centaurea alba ssp. deusta, Ptilostemon strictus, Arabis pseudoturilis, Stachys tymphaea, Verbascum longifolium, Trifolium patulum.*

Endemics of various origins are 22: West Mediterranean: *Potentilla hirta.* Mediterranean – Mid-European: *Taraxacum apenninum.* Mediterranean – sub-Mediterranean: *Carlina acanthifo-*

lia. Mid-European: *Sedum sexangulare, Epilobium collinum, Chamaespartium sagittale*. Mid-European – Pontic: *Galium anisophyllon*. Sub-Mid-European: *Chamaecytisus polytrichus, Cirsium eriophorum*. Sub-Mid-European-Pontic: *Veronica urticifolia*. Sub-Mediterranean – sub- Atlantic: *Heracleum sphondylium ssp. sphondylium*. Sub-Mediterranean: *Narcissus poeticus ssp. poeticus, Silene paradoxa, Alchemilla xanthochlora*, Arctic-alpine: *Alchemilla alpina*. Alpine-Circumalpine: *Gentiana lutea*. Alpine-Pannonian: *Gentiana punctata, Tozzia alpina ssp. carpathica*, Alpine: *Ranunculus saxatilis, Vicia oroboides, Luzula luzulina, Arabis allionii*.

Endemic species imperfectly known in Greece or species the occurrence of which in Greece was up to now questionable and were found in the Park are 8: *Pinus peuce, Senecio subalpinus, Knautia macedonica, K. midzorensis, Goniolimon dalmaticum, Saxifraga pedemontana, Carex sempervirens, Genista sericea*.

Plants with a status of international protection (Red Data Book, IUCN 1982)

The flora of the Park comprises also several species which are have been given a status of international protection of various grades (IUCN, 1982). Among the protected endemics in the Park, some belong to the Vulnerable (V) category, like for instance *Ramonda serbica, Trapa natans*, others are rare (R): *Dianthus myrtinervius, Anthemis pindicola, Sedum serpentini, Sempervivum reginae - amaliae, Astragalus lacteus, Ranunculus ficarioides, Rhinanthus pindicus, Centaurea prespana* (a local endemic), *Erodium guicciardii, Ranunculus fontanus, Viola eximi* and *Achillea absinthoides*. There are also various other species, which are not threatened (nt): *Helianthemum hymettium, Scorzonera crocifolia, Knautia magnifica, Trifolium parnassi, Verbascum epixanthinum, Sideritis syriaca*, as well as those placed generally under protection without any gradation of the imminent threat: *Salvinia natans, Juniperus excelsa, Pinus peuce, Acer heldreichii, Corylus colurna, Gentiana lutea, G. punctata, Convallaria majalis, Prunus prostrata, Salix aurita, Atropa bella-donna, Narcissus poeticus, Lilium jankae, L. albanicum, L. chalcedonicum, L. martagon* and *Goniolimon dalmaticum*.

Rare, interesting and threatened species of the National Park

The conservation of the habitats of certain plant species in the environment of the National Park has become gradually problematic, either on account of their restricted distribution and the small size of their populations, or because of the impact of human activities and pressure, and consequently they are threatened with extermination. According to our estimation the conditions mentioned above concern certain species, the future of which is considered as precarious and as a consequence, they are included in the category of locally rare or in the category of probably endangered species, such as the herbaceous: *Botrychium lunaria, Isoetes histrix, Adoxa moschatellina, Ramonda serbica, Gentiana lutea, G. punctata, Parnassia palustris, Trollius europaeus, Smyrnium rotundifolium, Polycnemum arvense, Salsola kali, Rumex maritimus, Sagittaria sagittifolia*, the woody hydrophilous plants *Salix silesiaca, Tamarix parviflora, Salix alba ssp. micans*, as well as the shrubby *Pistacia terebinthus*, the presence of which is uncertain.

The case of *Gentiana lutea* is very characteristic. Makris (1979) mentions that 'the plant is common on the alpine meadows of north continental Greece and particularly in the region of Prespa, where it is known with the local name 'wild tobacco'... Kavadas Botanic Dictionary (in Makris 1979) refers to the plundering overcollection of *Gentiana lutea* in the winter of 1956 by the locals to almost complete eradication for acquiring their fleshy roots'. This happened not only in the Greek part of Prespa but also in the Yugoslavian part, the roots being sold to German medical companies.

Several other plant species have been also observed, which formerly thrived in abundance, but nowadays their conservation has become more and more problematic, among which certain have been brought to the verge of their definite extermination. Among them we can mention for instance the psychrophile conifers *Pinus heldreichii* and *P. leucodermis*, nowadays represented by stunted and usually creeping individuals, the Balkan endemic *Pinus peuce*, the presence of which is accepted with reservations, the aquatic therophyte *Trapa natans*, formerly covering the surface of shallow waters in small bogs and nowadays surviving in loose patches in restricted parts of the lake, the aquatic pteridophyte *Salvinia natans*, which is becoming increasingly rare, the hydrophilous bulbous geophyte of quite remarkable beauty *Narcissus poeticus ssp. poeticus (var. hellenicum)*, forming in the past pure meadow formations near the lake and nowadays found sporadically mainly amidst thorny hedges, while the occurrence of the benthic hydrophytes *Vallisneria spiralis* and *Najas marina* has recently been called into question, furthermore the scattered populations of the ornamental plant *Lilium sp.* are becoming more and more scarce.

Certain plant species of the Park have a very restricted distribution, growing in isolated vegetation stands or amidst sparsely populated patches. The danger that they might be exterminated on account of accidental damage is evident and this is the reason that they must be regarded as locally protected species, like for instance the herbaceous plants: *Vinca balcanica, Silene bupleroides, Saponaria bellidifolia, Chrysanthemum coronarium, Senecio fluviatilis, Xanthium spinosum, Sedum cepaea, Teucrium scordium ssp. scordium, Lupinus luteus, Oxalis acetosella, Corydalis ochroleuca ssp. ochroleuca, Goniolimon dalmaticum, Rumex maritimum, Thalictrum foetidum, Umbilicus erectus, Arabis sagittata, Linaria annua ssp. pachyrrhiza, Teesdalia nudicaulis, Leonurus cardiaca, L. marrubiastrum, Salvia rigens, Filipendula ulmaria, ssp. ulmaria, Potentilla emilli-popii, Rubus schleicheri, Dictamnus albus, Atropa bella-donna, Hyoscyamus niger* and *Tulipa sylvestris ssp. australis*, the rare (for this area) tree species: *Betula pendula* and *Quercus robur*, the semi-arboreous *Euonymus latifolia* and finally the shrubby species: *Lonicera alpigena ssp. formanekiana* and *Rhamnus lycioides ssp. oleoides.*

There is also a specific category of plants whose occurrence in Greece is very restricted, in spite of the fact that they are very interesting from the phytogeographical point of view, such as the herbaceous plants: *Asplenium septendrionale, Onosma echioides, Dianthus pinifolius, Vaccinium myrtillus, Polygonatum verticillatum, Cirsium eriophorum, Valeriana officinalis ssp. officinalis, Deschampsia flexuosa, Silene lerchenfeldiana, Barbarea longirostris, Antennaria dioica, Symphytum tuberosum* and *Myosotis sylvatica ssp. subarvensis.*

There are, however, some plant species in the Park, which have not been included in the flora of the wild plants because they have been considered as semi-wild, ornamental or ruderal species. Among them can be mentioned the semi-wild or cultivated tree species *Castanea sativa, Prunus armeniaca, P. domestica, Populus nigra* cv. *italica*, P. X. *canadensis, Platanus orientalis, Salix babylonica, Morus alba* and *M. nigra*, the shrubby species *Jasminum fruticans, Sophora japonica* and *Pyracantha coccinea*, the species planted for the protection of the soil against erosion, like *Spartium junceum* and *Robinia pseudacacia*, as well as various cultivated horticultural species and cereals.

Table 2. Phytogeographic composition of the flora of Prespa National Park.

Phytogeographic origin	Number species	%
European	341	25.71
Eurasian	290	21.87
Balkan	251	18.93
Mediterranean	243	18.32
Sub-mediterranean	157	11.84
Cosmopolit./American/Asiatic/Tropical	39	2.94
Undetermined	5	0.38

The phytogeographic origin of Prespa National Park flora

The phytogeographic analysis of the flora of the National Park concluded that Prespa National Park can be classified to the Balkan mountains type according to the division of the Balkans by Regel (1947) and the Balkan sub-zone of the Sub-Mediterranean vegetation zone according to the division of Europe by Meusel et al. (1965).

Species of European origin are the most numerous followed by species of Eurasian origin and the Balkan elements (Table 2), their majority having an endemic status. Mediterranean elements are represented mainly by cold-resistant species, while truly Mediterranean thermophilous species are absent. Among the Sub-Mediterranean species are found several interesting forest trees. There are 39 Cosmopolitan, American, Asiatic and Tropical species. Five species, namely *Epipactis persica, Himantoglossum caprinum, Ophrys oestrifera var. grandiflora, Ophrys pseudomammosa* and *Orchis pinetorum* are chorologically undetermined.

Among the life-forms, according to the Raunkiaer system of classification, Hemicryptophytes dominate with 649 species (48.94%), and Therophytes (annuals) follow with 300 species (22.63%). Phanerophytes (120, 9.05%), geophytes (110, 8.3%), chamaephytes (70, 5.28%) complement the spectrum of life-forms along with a sizeable proportion of 77 species (5.8%) of hydrophytes-helophytes. According to the participation of each life form type the National Park flora is classified in the bioclimatic zone of Hemicryptophytes, typical of the Central-European area. By comparison, the life form spectrum of the Prespa National Park is similar to those of Voras, Lailias and Vertiskos mountains of northern Greece.

Regarding the growth forms of the National Park flora, herbaceous perennials dominate with 807 species (60.86%), annuals follow with 309 species (23.3%), woody perennials with 120 species (9.05%) and finally herbaceous biennials with 90 species and 6.79%. From a combination of the ecological forms it can be concluded that the main bulk of 1206 species (90.95%) corresponds to various herbs and chamaephytes while the remaining 120 species are woody plants, i.e. 55 shrubs (4.15%), 21 tree-like species (1.58%) and 44 tree species (3.32%).

Conclusion

The present day visitor of the National Park will have many difficulties in discovering its rare and beautiful wild flowers. Their beauty in combination with their therapeutic properties tempt people to collect them and this results in gradually decreasing populations. Thus, many rare species seek refuge in naturally protected, remote, unapproachable areas, where human presence and intervention is negligible or absent.

The present day condition of the environment of the National Park as regards its flora can be evaluated as only slightly threatened. There are however, encouraging signs that no further impoverishment of the flora will take place in the future, given that certain limitations of land uses will be implemented by the National Park management authorities.

References

Babalonas, D. & G. Pavlides, 1989. A contribution to the aquatic macrophyte vegetation of Lake Mikri Prespa. BIOS 1: 19–29. Thessaloniki (in Greek).

Braun-Blanquet, J., 1964. Pflanzensoziologie. Grundzuge der Vegetationskunde. (1. Aufl. 1928) 3. Aufl. Wien, New York, 865 pp.

Greuter, W., 1987. Some notes on *Lesquereuxia* (Scrophulariaceae). Bot. Jahr.Syst. 108: 251–257.

Hopp, I. & J. Petrides, 1987. Prespa. A continuity of its diversity. Nature. Bull. Hellenic Soc. Protect. Nature 39: 51–55.

Horvat, I., V. Glavac & H. Ellenberg, 1974. Vegetation Sudosteuropas, Stuttgart, 767 pp.

IUCN, 1982. Threatened plants Committee secretariat. Rare, threatened and endemic plants of Greece. Ann. Mus. Goulandris 5: 69–105.

Lavrentiades, G. & G. Pavlides, 1985. Contribution to the research of hydrophytic and helophytic associations of the lake Mikri Prespa. Proc. 4th Symp. Greek bot. Soc., Thessaloniki (in Greek): 144–155.

Lavrentiades, G., 1956. *On the hydrophytic and helophytic flora of embryophytes in Greek Macedonia.* Ph.D. Thesis. Aristotelian University of Thessaloniki, 88 pp. (in Greek).

Makris, K., 1979. The trade and need for protection of the aromatic and medicinal plant species. Proceedings of conference on the protection of the flora, fauna and biotopes in Greece, Athens, October 1979. Hellenic Soc. Protect. Nature, Athens: 87–90.

Meusel, H., E. Jager & E. Weinert, 1965. Vergleichende Chorologieder Zentral-europaischen Flora. Vols 1–2. VEB G. Fischer Verlag. Jena.

Pavlides, G., 1985. Geobotanical Study of the National Park of lakes Prespa (NW Greece). Part A. Ecology, Flora, Phytogeography, Vegetation. I–XI, 308 p. + map. Thessaloniki (in Greek with English summary).

Pavlides, G., 1988. Study of the grazing lands of Kalo Nero (Bella Voda), Varnous mt, Agios Germanos, Florina (NW Greece). Ministry of Agriculture. Thessaloniki (in Greek), 128 pp.

Regel, C., 1947. Zur Frage der Grenze dem Mittel meergebiet und Mitteleuropa auf der Balkanhalbinsel (Dritte Folge). Ber. Geobot. Forschungsinst. Rubel Zurich.

Strid, A. & K. Tan (eds), 1991. Mountain Flora of Greece 2. xxv + 974 pp. Edinburgh University Press, Edinburgh.

Tutin, T., V. Heywood, N. Burges & O. Valentine, 1964–1980. Flora Europaea. Vols 1–5. Cambridge University Press, Cambridge.

Hydrobiologia **351:** 41–60, 1997.
A. J. Crivelli & G. Catsadorakis (eds), Lake Prespa, Northwestern Greece.

Aquatic and terrestrial vegetation of the Prespa area

G. Pavlides
Akropoleos 36, 54634 Thessaloniki, Greece

Key words: aquatic vegetation, terrestrial vegetation, phytosociology, plant biodiversity, Greece

Abstract

We studied the vegetation of the aquatic ecosystems of Lake Mikri Prespa. The lacustrine vegetation comprises three distinct forms: floating plants, benthic hydrophytes and helophytes, which are described and classified from the phytosociological point of view, as follows: (a) the vegetation of the floating plants belong to the Lemnetea class and is represented by two plant communities; (b) the vegetation of the benthic hydrophytes, belongs to the Potametea class consisting of two different categories, namely the submersed formations and the emergent formations of the hydrophytes. Various plant communities were recognised in this type of vegetation and three among them are considered as the most representative; (c) the vegetation of helophytes, the prevailing life form in this wetland, belongs to the Phragmitetea class and is represented mainly by seven well organised plant communities. The respective vegetation of two of the five more important wetland sites is described.

The terrestrial vegetation is composed of forest and meadow vegetation. The forest vegetation of the National Park belongs to the class Querco-Fagetea and shows the following zonation: (a) in the vicinity of the lake, at the elevation of 860–1000 m, two associations have been found: the mixed deciduous and evergreen forests of Ostryo-Carpinion orientalis and the evergreen forests of Ostryo-Carpinion adriaticum; (b) the deciduous oak forests surround the previous zone at the altitude of 900–1300 m with two principal associations, namely the Quercetum frainetto and the Quercetum petraeae; (c) in the upper forest layer between 1200–1800 m asl, dominate beech forests of the association Fagion moesiacum and the less extensive occurrence of the mixed beech-fir stands (Ass. Abieti-Fagetum moesiacum).

The zone above the tree limit is distinguished by its subalpine character semi-shrub vegetation extending higher than the forest (1800–2000 m), whereas alpine meadows cover the vegetation at higher altitudes. On the plains and in the forest clearings exist herbaceous meadow formations of variable structure, in parallel with the vegetation of specific habitats, such as nitrophilous and ammophilous plants.

Introduction

The structure and physiognomy of the Park's vegetation as well as the qualitative and quantitative features of its floristic elements show a distinct submediterranean-continental character and seem to react effectively to any mediterranean influence. Two principal types of ecosystems predominate in this region the aquatic and the terrestrial. The particular floristic and physiognomic features of the various plant assemblages of both ecosystems, with special emphasis on the former which comprises the nucleus of the Park, i.e. the lake of Mikri Prespa, are described below (Figure 1). The methodology which we followed in order to recognise certain plant assemblages comprised in the above mentioned vegetation was based mainly upon the system of Braun-Blanquet (1964) in connection with the phytosociological system of Horvat et al. (1974).

The aquatic vegetation

The lacustrine ecosystem of the Prespa complex is very interesting from the scientific and economic point of view. The lacustrine vegetation as a whole constitutes

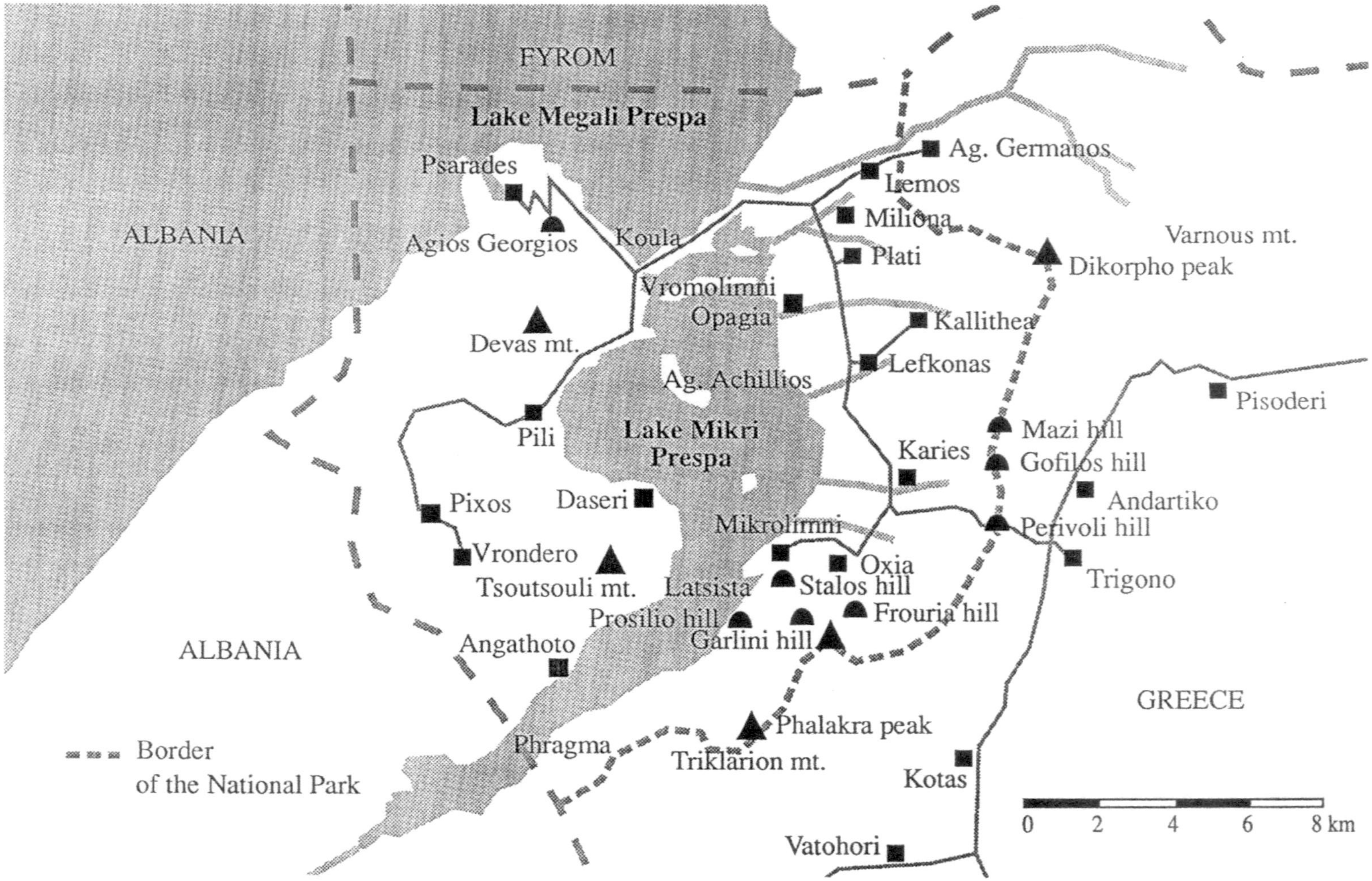

Figure 1. The study area.

the main productive ecosystem of the wetlands and a basic factor for the existence of the avifauna. The structure and layout of the lacustrine plant formations, especially those of Lake Mikri Prespa, is also paramount for the spawning of fish and the development of eggs.

Three basic formations of plant assemblages can be distinguished, each of them being ecologically independent within the particular biochore: the formation of the floating plants, which is made up of loose vegetation stands of small size, floating on the water surface, the formation of the benthic hydrophytes, consisting of stands with loose submerged or semi-submerged roots fixed in the bottom of the lake and with their stems emerging out of water, and the formation of helophytes, comprising solidly rooted half-sunk in the water, high-raised vegetation stands growing mainly in the shallows of the lakeshore.

On the basis of the floristic analysis of the vegetation certain plant entities have been recognised, which make up organised plant communities classified under hierarchically higher entities (Lavrentiades, 1956; Pavlides, 1985; Lavrentiades & Pavlides, 1985; Babalonas & Pavlides, 1989).

The vegetation of floating plants

The vegetation of floating plants (freely floating macrophytes) is composed of small sized aquatic plants, i.e. a few millimetres or centimetres in diameter, floating on the water surface, forming loose and easily movable vegetation stands. They are mostly found in the shallow and limpid waters of the littoral zone and in the ditches surrounding the lake as well as in water ponds among reedbeds.

The floating plants are represented by the Lemnetea class, in which the main and indicator species are *Lemna minor* and *Spirodella polyrrhiza*, both a few millimetres in diameter and lens-shaped, and by the comparatively bigger in size, 1–2 cm in diameter, pteridophyte *Salvinia natans*, which displays an incomplete shallow root system.The above mentioned typical floating plants are often accompanied by certain bigger hydrophytes, like *Trapa natans* and *Hydrocharis morsus-ranae* and constitute this peculiar vegetation identical in character with the associations of the Lemnion minoris alliance.

From the floristic analysis of the floating vegetation stands, two representative associations with a characteristic physiognomy have been recognised:

Ass. Lemnetum minoris. Almost complete dominance over the structure of the vegetation is exerted by the indicator species *Lemna minor.* It is accompanied steadily, although with smaller participation by the similar species *Spirodela polyrrhiza,* as well as by the indicator species *Salvinia natans* and *Trapa natans*, with the participation of various accompanying species, among which the commonest are *Ranunculus aquatilis, Myriophyllum verticillatum, Sagittaria sagittifolia, Utricularia vulgaris*, and *Lycopus europaeus*. The association is often found in ponds, in the middle of clearings among reedbeds in the region of Vromolimni and Koula, as well as on the surface of the slowly flowing waters in the drainage ditches near the same region, where they form small sized patches of loose structure.

Ass. Hydrocharitetum morsus-ranae. It is easily recognised by the predominance and co-dominance of two species mainly, namely *Salvinia natans* and *Hydrocharis morsus-ranae*, as well as due to the participation to a lesser degree of *Nymphoides peltata* and *Trapa natans.* Accompanying species of this association are the benthic hydrophytes *Ceratophyllum demersum, Myriophyllum verticillatum, Ranunculus aquatilis, Sagittaria sagittifolia* and the helophytes *Utricularia vulgaris, Lycopus europaeus, Sparganium erectum, Glyceria fluitans, Rorippa amphibia.* This association often occurs on the eastern coastal zone and especially between Mikrolimni and Opagia, within helophytic formations as well as in marshes surrounding the lake.

The vegetation of the benthic hydrophytes (submersed macrophytes)

The vegetation of the benthic hydrophytes is made up of aquatic plants rooted in the bottom of the shallow parts of the lake up to a depth of 0.2–2.5 m according to the substrate and the contour of the lake-shore. The vegetation stands of the benthic hydrophytes form submersed as well as surface pastures. They often occur in the innerside of reedbeds, where they grow in zones parallel to the shoreline and furthermore in the open water clearings of the helophytic formations of Lake Mikri Prespa.

According to the particular portion of the lake, where they grow and according to their life history traits, two distinct categories of submersed macrophytes are distinguished: those thoroughly submersed, represented by the submersed benthic hydrophytes and

Figure 2. Floating-leaved vegetation of *Nymphaea alba* in an opening of the Phragma area, near the Greek-Albanian borderline in Lake Mikri Prespa.

Figure 3. A floating cluster of *Nuphar lutea* inside a reedbed, in a channel of the Phragma area.

stands growing under water as well as on the water surface, represented by the emergent benthic hydrophytes.

Vegetation of the submersed benthic hydrophytes

The vegetation of the submersed benthic hydrophytes is composed of hydrophytes growing permanently under water and forming the characteristic underwa-

Figure 4. Rich aquatic vegetation of helophytic and benthic combinations: in the background a strip of the Phragmitetum association, in the centre a dense growth of the Myriophyllo–Nupharetum association, and at left and right pure clusters of the herbaceous helophyte *Oenanthe aquatica*, near Vromolimni.

ter pastures. At short intervals and only seasonally certain stems as flowers and inflorescences or even leaves, emerge from the water and float on the surface. Their growth is favoured in the deepest waters of the sublittoral zone and in particular in areas of smooth and flat bottom of sandy and clayey composition, where they are rather loosely rooted. They preferably grow in the lees, bights and outlets of streams surrounding the lake, their presence being more frequent in regions near Mikrolimni, Pili, Angathoto, Latsista, Koula and Phragma. The cohesiveness in the structure of the submersed vegetation stands is differentiated locally, revealing a gradation in the way the stems of the benthic hydrophytes are interlinked with each other, from extremely loose to very dense. They usually comprise few members and are imperfectly organised, so that these stands are not identifiable as typical, well established in the literature, vegetation formations.

Among the locally more interesting vegetation stands of the class Potametea we have identified primarily the following representative species: *Potamogeton perfoliatus, Ceratophyllum submersum, Najas marina,* secondarily the species *Ceratophyllum demersum, Myriophyllum verticillatum, M. spicatum, Potamogeton crispus, P. pusillus, Ranunculus trichophyllus*, the rarely occuring *Vallisneria spiralis* and finally the Chlorophytes of the genera *Cladophora* and *Chara*, which often accompany the benthic hydrophytes.

The vegetation of the emergent benthic hydrophytes (floating leaved macrophytes)

The vegetation of the emergent benthic hydrophytes normally includes aquatic forms solidly rooted in the bottom of the lake and other water bodies, extending their rope-like elongated petioles interwoven with each other up to the water surface. In addition, their most vital stems, like for instance long-stalked usually lanceolate leaves as well as flowers and inflorescences are floating or aerial. Extensive floating communities of striking physiognomy have been observed, among which the most characteristic and widely-spread benthic species such as *Nymphaea alba* (Figure 2), *Nuphar lutea* (Figure 3) and *Nymphoides peltata.*

The floristic analysis of this kind vegetation enabled us to recognise certain well organised plant communities of the Nymphaeion alliance, like for instance:

Ass. Myriophyllo-Nupharetum. It is the most prominent plant community of the wetland among the emergent benthic hydrophytes, forming particularly dense patches with impressive combinations. These stands seem to be coherently attached in the bottom of the sublittoral zone of the lake in water-holes and 40-100 cm deep ponds found at the periphery of the lake, as well as in greater depths in the lake itself. The surface covered by this plant community varies locally from 25 to 40 m^2. Representative species of this association are *Nuphar lutea, Myriophyllum verticillatum, Ceratophyllum demersum* and *Salvinia natans*, which give a particular character to the vegetation. This type of stands are more frequent in the Vromolimni-Agios Achillios-Koula area (Figure 4), however, it occurs sporadically in other parts of the wetland, for example in the Phragma area at the Greek-Albanian border.

Ass. Trapetum natantis. The main indicator species of this association is the hydrophyte *Trapa natans*. It is regularly although sparcely accompanied, by the benthic hydrophytes *Nymphoides peltata* and *Ranunculus aquatilis*. This type of vegetation was formerly met in the shallow 30–40 cm marshes in an area close to Mikrolimni meadows (Figure 5) and Opagia. The structure of the vegetation covering the surface of the water in a layer of some centimetres is dense and cohesive. It often covered the whole surface of the marshes, 12–20 m^2. Recent observations showed that this association as *Trapa natans* itself become scarcer.

Ass. Trapo-Nymphoidetum. It is a variation of the preceding community, where both *Nymphoides peltata* and *Ranunculus aquatilis* co-dominate with *Trapa natans*. We have recognised the community in shallow coastal marshes, close to the outlet of Mikrolimni stream as well as in the area of Vromolimni (Figure 6), where it formed floating patches of 12–25 m^2.

Plant communities of different structure and physiognomy often occur in the littoral zone of the lake, up to a depth of 2 m. More often they comprise species with long-stalked lanceolate floating leaves and with the emerging spike inflorescences of species like *Potamogeton natans, P. lucens, P. nodosus*, and *Polygonum amphibium.* Various benthic hydrophytes participate in the above community as accompanying species, such as *Potamogeton crispus, P. perfoliatum, Nymphaea alba, Ranunculus flaccidus, Utricularia vulgaris,* as well as helophytes such as *Phragmites australis, Eleocharis palustris, Mentha aquatica* and *Rorippa amphibia.*

The helophytic vegetation (emergent macrophytes)

The helophytic marshy vegetation is the most widespread and ecologically important vegetation type in

Figure 5. A pond covered with floating vegetation of the *Trapetum natantis* association, in the Mikrolimni area (photographed in 1974).

this wetland. Helophytes grow around the perimeter of the lake penetrating into its interior up to a depth of 2.5 m and occasionally deeper. The structure of the helophytes consists of partly submersed hydrophytes, which are solidly rooted preferably in clayey sandy sediments, with the largest part of stems emerging above the water's surface, hence their second name superaquatic hydrophytes.

Among the most representative and important helophytes are the gramminaceous species *Phragmites australis, Typha angustifolia, Scirpus lacustris, Carex pseudocyperus,* which render the particular physiognomy to the reedbed vegetation as widely known.

The characteristic feature of these communities is the fact that they are laid out in strips from the shore line lakewards and that one community can encircle inside it another type, while in the space between them, usually clearings, dense patches of benthic hydrophytes grow, forming a complex vegetational mosaic, like for instance in Vromolimni area where this characteristic feature is fully visible (Figure 7).

The characteristic helophytic communities recognised in the littoral zone of the lake are the following:

Ass. Phragmitetum. It constitutes the dominant helophytic community growing around the lake in dense and occasionally impenetrable stands. They can be easily recognised on account of the complete plant coverage by the helophyte *Phragmites australis* (Figures 4, 7). Extensive and coherent reedbeds exist at the eastern and northern shallow parts Lake Mikri Prespa

Figure 6. An aspect of the aquatic benthic vegetation of the Trapo–Nymphoidetum association, with the dominant *Nymphoides peltata*, near Vromolimni.

Figure 7. A general view of the Vromolimni-Koula-Agios Achillios wetland. The zonal arrangement of the vegetation is obvious: in the foreground a meadow of herbs, bordered by elongated patches of helophytes of the Caricetum pseudocyperi association, followed by the aquatic associations of the Nymphaeion alliance, and in the background formations of the Scirpo-Phragmitetum, Typhetum angustifoliae and Phragmitetum associations.

and especially in smoothly sloping sandy-clayey substrate. On the rocky shores of the lake, where water deepens abruptly the helophytic vegetation is absent or interrupted. In the basic composition of the community participate mainly *Scirpus lacustris, Rorippa amphibia, Lythrum salicaria, Myosotis palustris, Stachys palustris,* as well as various other accompanying helophytes of less importance.

Ass. Scirpo-Phragmitetum. This association is a variation of the preceding and it consists of *Phragmites australis* and the following helophytes, all of them participating in equivalent proportions: *Scirpus lacustris* ssp. *tabernaemontani, Typha angustifolia, T. latifolia, Sparganium emersum* ssp. *erectum*. Various other helophytes participate less commonly such as *Iris pseudacorus, Alisma plantago-aquatica, Eleocharis palustris*. This community often grows in zonal formations parallel and externally landwards to those of Phragmitetum (Figure 7), covering comparatively smaller areas. It usually occurs in the vicinity of lakes and in a water depth of 45–100 cm, mainly in the areas of Koula, Mikrolimni and Agios Achillios-Pili.

Ass. Typhetum angustifoliae. The main feature of this association is the dominance of *Typha angustifolia*, the remaining structure being analogous to the preceding association and in a way it is a variation of it (Figure 8). It always grows in stations, where the bottom presents depressions, in such a way that it is surrounded by two communities, the higher located Phragmitetum and the lower located benthic Myriophyllo-Nupharetum. It occurs in the region of Koula, Pili and Opagia-Vromolimni.

Ass. Scirpetum lacustris. This community differs in physiognomy and structure from the remaining helophytic ones. Its dominating species is the reed-like *Scirpus lacustris* with an essential participation of other species, such as *Phragmites australis, Carex pseudocyperus, Lythrum salicaria, Myosotis palustris. Solanum dulcamara, Sparganium emersum, Polygonum lapathifolium* and *Alisma plantago-aquatica* differentiate the community in places. It is always found in protected spots, in the interior of the lake, near reedbeds and usually in shallow waters around the littoral zone (Figure 9) in the eastern and northern parts of Lake Mikri Prespa.

Ass. Caricetum pseudocyperi. The dominance of *Carex pseudocyperus* in particularly dense formations is the main feature of this community. Its composition is differentiated by the participation of herbaceous helophytes such as *Lythrum salicaria, Myosotis palustris, Scutellaria galericulata* and *Nasturtium officinale*. It covers relatively limited flooded or not areas at Koula (Figure 7) and the eastern periphery of the lake in general.

Ass. Oenantho-Rorippetum amphibiae. This community is characterised by the absolute predominance and co-dominance of the two herbaceous hydrophytes *Rorippa amphibia* and *Oenanthe aquatica*. Its composition is accomplished with the frequent presence of *Lycopus europaeus, Lythrum salicaria, Galium palustre, Myosotis scorpioides,* and the participation of various herbaceous hydrophytes as accompanying species. During efflorescence a relatively low, 60–80 cm in height, half-submersed and multicoloured vegetation is formed (Figure 10). It grows in shallow marshes, in water depths of 35–50 cm in Koula (Figure 4) and Agios Achillios areas. Among the numerous accompanying species the most important are: *Alisma plantago-aquatica, Butomus umbellatus, Veronica anagallis-aquatica, Rumex aquaticus, Cyperus fuscus, Catabrosa aquatica, Scirpus holoschoenus, Polygonum amphibium.*

Ass. Apietum nodiflori. This meadow-like community usually borders reed-beds and is more extensive in the eastern perimeter of Lake Mikri Prespa. It consists of relatively low stands, 40–60 cm in height, of loose structure, where the dominant species is *Apium nodiflorum*. It grows in the littoral zone in places where drainage ditches and streams discharge, as well as in 10–20 cm deep hollows with stagnant and flowing water in sandy and clayey soils. The community entails numerous species and among its most abundant representative species are *Veronica anagallis-aquatica* and *Nasturtium officinale*. The participation of the following helophytes is considered essential: *Iris pseudacorus, Eleocharis palustris, E. uniglumis, Lythrum salicaria, Carex pseudocyperus, Stachys palustris.* Among the accompanying species the most important are *Scirpus holoschoenus, Polygonum amphibium, Juncus articulatus* and *Rumex aquaticus*.

The most important wetland habitats of Lake Mikri Prespa

Certain habitats within the National Park are of great interest for their biotic communities. These areas have been established as main nuclei of the Park. As far as Lake Mikri Prespa is concerned these sites are: Phragma, the area around the Biological Station of Mikrolimni and the mouth of Mikrolimni stream, the entire eastern coastal zone, the Vromolimni-Koula, and the Agios Achillios-Pili-Daseri coastline (Figure 1). Two of the most important sites for vegetation are described below.

Figure 8. Clusters of the Typhetum angustifoliae association and in the background the Phragmitetum association in Opagia area.

Figure 9. Evergreen woodland of the Ostryo-Carpinion adriaticum alliance, and below clusters of helophytes of the Scirpetum lacustris association, between Daseri and Angathoto.

The Phragma area at the Greek-Albanian border

From the geopolitical point of view, the main feature of this isolated area is a state of environmental relaxation and calmness with evident absence of any human activities, in particular during the last decades, especially for the area located within the boundaries of the Greek state. This had had a favourable impact on the development of the vegetation. As a result certain portions of this vegetation ended gradually in the final 'climax' stage of evolution. The structure and vitality of the various patches of hydrophytes is particularly impressive in the extreme lateral borders as well as across the lake among the border line, about 1.5 km in length and 1.5 m maximum depth. On both sides of the lake rise densely

Figure 10. Clearly separated helophytic communities. In the background, tall grass-like clusters of the Scirpo-Phragmitetum association, and in the foreground, the herbaceous Oenantho-Rorippetum amphibiae association, around Vromolimni.

vegetated slopes, often with rocky drenched borders ending in the crest of the overhanging mountains. The striking beauty, the ideal conditions of natural protection of the area, its variable aquatic and terrestrial vegetation are the main physiognomic features of this site (Figure 11).

A combination of aquatic and terrestrial vegetation follows the coastline with a succession of shrubby and herbaceous patches, while in the lake thrive reedbeds of impressive vitality among which the Ass. Phragmitetum predominates. The particular characteristic of this vegetation is created, however, by the luxuriant development of the shrubby hydrophyte *Salix cinerea* in pure stands but at the same time also mixed formations with reedbeds. It covers the coastal zones from the shallow portions of the lake towards the interior up to a depth of 1.2 m. The trees are solidly rooted in the bottom emerging with its drenched or half-sunk many branched trunk on the surface and forming a dense supra-aquatic canopy up to heights of 3 m. Among the shrubby patches occur various helophytes usually the species *Phragmites australis, Lythrum salicaria, Lycopus europaeus, Rumex hydrolapathum, R. aquaticus, Galium palustre, Scutellaria galericulata, Thelipteris phaegopteris, Eupatorium cannabinum* and occasionally in some spots the nitrophilous species *Urtica dioica.* The general floristic composition, of this peculiar vegetation leads to the shrubby helophytic association Salicetum cinereae (Figure 11).

The patchy distribution of *Salix cinerea* stands is polymorphous and often labyrinth like with channels and openings. Open water areas are suitable for benthic hydrophytes occuring in various combinations. More conspicuous are the representative formations of Ass. Myriophyllo-Nupharetum. Occasionally are found pure formations of the emergent benthic hydrophytes, like for instance *Nymphaea alba* (Figure 2), *Nuphar lutea* (Figure 3), *Potamogetum lucens, Polygonum amphibium* and *Nymphoides peltata.* Among the submersed benthic hydrophytes the most frequently occurring are *Potamogeton nodosus, P. crispus, P. gramineus, Polygonum rurivagum,* some species of the genera *Myriophyllum* and *Ceratophyllum* and rarely *Trapa natans, Hydrocharis morsus-ranae* and *Najas marina.*

The Vromolimni-Koula area

The extensive and diverse wetland of the northern part of the lake is recognised as the most important nucleus of the Park. Reedbeds are its main feature and among them several open water areas are covered by aquatic vegetation (Figure 4). Slight elevations in the form of islets occur sporadically, on which dense wet meadow vegetation grows. The general morphology and struc-

Figure 11. An aspect of the vegetation of the Phragma area: in the background at left clusters of reeds of the Scirpo-Phragmitetum association, at right clusters of shrubs of the Salicetum cinereae association, and down at right floating patches of *Nymphaea alba.*

ture of Vromolimni area is favourable to several species of waterbirds. The ecological importance of this site is great (Figure 7).

The helophytic vegetation finds here its full expression with the Ass. Phragmitetum prevailing. The associations Scirpo-Phragmitetum, Typhetum angustifoliae and Scirpetum lacustris occur sporadically. On the whole, the participation of helophytic species appears in a sequence of decreasing plant cover as follows: *Phragmites australis, Scirpus lacustris ssp. tabernaemontani, Typha angustifolia, Sparganium emersum, Carex pseudocyperus, Iris pseudacorus, Alisma plantago-aquatica* and others.

In marshy habitats, formed usually in the lakeward borders of the reedbeds grow herbaceous vegetation stands, mainly of the associations Oenantho-Rorippetum amphibiae (Figure 10) and *Apietum nodiflori,* while in peripheral hillocks grows the meadow community of *Eleocharis palustris-Carex hirta,* which have been found in codominance, but usually with the independent predominance of these species locally, where in both stands occur in common the following helophytes mainly: *Lythrum salicaria, Scutellaria galericulata, Scirpus lacustris.* As accompanying species of *Eleocharis palustris* we have usually found *Mentha aquatica, M. microphylla, Juncus inflexus, J. bufonius, Scirpus maritimus, Plantago major, Rorippa sylvestris,* and with *Carex hirta* the species *Galium palustre, Solanum dulcamara, Agrostis capillaris, Sparganium emersum, Ranunculus repens, Rumex sanguineus* and *Scrophularia umbrosa.*

On the borders of the helophytic vegetation, adjacent to cultivated plants, in seasonally flooded sites grows the nitrophilous wet meadow community of *Cirsium arvense-Cirsium vulgare,* in which the participation of the species *Galega officinalis, Trifolium resupinatum* and *Lotus corniculatus* is considerable. Among the accompanying species participating in this community the following species worth mentioning: *Scutellaria galericulata, Agrostis capillaris, Lycopus europaeus, Epilobium angustifolium, Lythrum salicaria* and others.

The vegetation of the submersed benthic hydrophytes occurs usually in the interspace between the small inner lakes and the marshes (Figure 7). Among the aquatic plant communities prevails Myriophyllo-Nupharetum, in the basic composition of which participate as accompanying species: *Nymphaea alba, Nymphoides peltata, Hydrocharis morsus-ranae, Myriophyllum* sp. *Polygonum amphibium.* The extraordinary presence of the loose, free-floating vegetation on the water surface of ditches and marshes constitutes a characteristic feature of this wetland, where the community of the minute floating plants of association Lemnetum minoris reveals its full composition.

The terrestrial vegetation

The terrestrial vegetation of the National Park is variable and is differentiated in herbaceous and – the more widesread – woody formations. The structure and physiognomy of the vegetation corresponds fully to the Continental-Central European character with a slight participation of a few Meditteranean elements. The distribution and structure of each vegetation type is the result of environmental, orographic and edapho-climatic impacts on the one hand, and geopolitical, anthropogenic impacts on the other. A brief description of the terrestrial vegetation follows below.

Vegetation of special biotopes

On the lower plains and hills of the National Park, particularly at the east of the lakes, the various human activities have affected greatly the natural vegetation. The various herbaceous formations of these peculiar biotopes are described below.

Weed vegetation of cultivations
The weed vegetation makes its appearance sporadically in the cultivations. It consists exclusively of therophytes. The majority of natural weeds of winter cultivations appear in the spring. The species composition is rather constant. The most typical species are *Agrostemma githago, Centaurea cyanus, Ranunculus arvensis, Papaver rhoeas, Buglossoides arvensis, Anthemis arvensis, Scandix australis, Legousia speculum-veneris, Caucalis platycarpos, Vicia cracca, Nigella arvensis, Fumaria officinalis, Lactuca viminea, Consolida regalis, Sinapis arvensis,* including some less important accompanying species, such as *Valerianella carinata, Orlaya grandiflora, Vicia villosa, Bromus secalinus, Delphinium halteratum, Daucus carota, Centaurea salonitana* etc. in combinations, corresponding to the plant associations of the Secalinetea class.

In the summer cultivations, as for example those of beans and maize, the composition of weeds seems to be of a nitrophilous type. During summer and before harvest the commonest species are *Polygonum arenaria, Amaranthus hybridus, Atriplex patula, Solanum nigrum, Chamomilla recutita, Chenopodium vulvaria, Cichorium intybus, Vicia tenuifolia* and *Portulaca oleracea.*

Nitrophilous vegetation
Nitrophilous plants occur around human settlements and installations for livestock. Among the most frequent species which often constitute pure patches, are *Urtica dioica, Sambucus ebulus, Cirsium vulgare, Onopordum illyricum, Datura stramonium, Portulaca oleracea*, and others of a secondary importance such as *Chenopodium album, Sonchus oleraceus, Marrubium vulgare, Carduus pycnocephalus, Dipsacus fullonum, Xanthium strumarium, Urtica urens, Amaranthus hybridus, Xanthium spinosum, Saponaria officinalis, Malva sylvestris.*

Generally nitrophilous formations are grouped together in associations belonging to Chenopodietea class.

Vegetation of road sides
It is differentiated according to the habitats fringing the roads. Along the automobile roads of the plain zone, for example, where soil is rich in coarse materials, the vegetation presents a xerophytic character. The most frequent species are *Chenopodium botrys, Melilotus officinalis, Verbascum purverulentum, Linaria genistifolia, Bromus sterilis, Echium italicum, Jasione heldreichii, Tragopogon porrifolium, Achillea millefolium, Artemisia campestris, Sisymbrium officinale, Chondrilla juncea, Dasypyrum villosum, Capsella bursa-pastoris, Cynoglossum creticum.* The most widespread species of the agricultural roads among the croplands, where soils are rich in nitrogen are *Cirsium vulgare, Sisymbrium irio, Vicia pannonica, Melica ciliata, Parentucellia latifolia, Anchusa arvensis, Heliotropium europaeum.*

Along the forest roads, where soils are particularly rich in fine-grained materials and humid, a characteristic high herbaceous vegetation appears with *Epilobium angustifolium, Verbascum phlomoides, Artemisia vulgaris, Tanacetum corymbosum, Campanula patula, Sambucus ebulus, Chaerophyllum aureum, Arabis turrita, Silene noctiflora, Arctium lappa, Cardamine buldifera*, as well as a variety of species of the forest herb layer.

Ammophilous vegetation
Along the coastal strip of Lake Megali Prespa, sand is accumulated in sand-dune-like formations. The most important species which grow on these peculiar biotopes are the rhizomatous geophytes. Owing to these, the sand-dunes may be stabilised. As ammophilous species can be recognised there the

grammineous *Juncus conglomeratus, J. articulatus, Cyperus fuscus, Scirpus holoschoenus, Carex hirta*, and secondarily the herbaceous species *Onosma heterophylla, Anchusa officinalis, Asphodeline lutea, Euphorbia myrsinites, Centaurea cuneifolia, Jasione heldreichii.* Close to the tourist pavilion small patches of *Myosotis refracta, Hordeum murinum, Clypeola jonthlaspi, Alyssum minutum*, and *Veronica triphyllos* appear, while along the sandy banks of the neighbouring channel joining the two lakes, the hydrophilous species *Polygonum hydropiper, P. persicaria, Rorippa prolifera, Myosoton aquaticum, Ranunculus paludosus* and *Carex divisa*, grow in abundance.

Meadow vegetation of plains and hills

According to the soil moisture of each habitat, two types of meadow vegetation are distinguished: the wet meadows, on fertile and deep soils of the littoral zone, and the meadows, which develop on the shallow and dry soils of the surrounding sloping hill-sides. In addition, a variety of intermediate forms of meadow vegetation may be also seen.

Meadow vegetation of moist biotopes

This is the well known wet-meadow vegetation of the littoral zone of the lakes and running waters. In their main combination of plants take part also several pastoral, quick growing herbs, such as *Trifolium repens, T. pratense, Vicia cracca, Lotus corniculatus, Lathyrus pratensis, Potentilla reptans, Plantago lanceolata, Prunella vulgaris*, along with the grasses *Alopecurus pratensis, Festuca rubra, Poa trivialis, Arrhenatherum elatius, Trisetum flavescens, Cynosurus echinatus*. All these plants form together a compact and grassy mat, ca 0.30 m high. Variations are usual along the littoral zone. For example, in the Oxia-Karies area the vegetation is dominated by *Trifolium resupinatum, T. incarnatum* ssp. *molinerii, Plantago altissima, Cladium mariscus, Alopecurus pratensis, Anthoxanthum odoratum, Draba muralis, Taraxacum officinale, Ajuga genevensis, Petrorhagia velutina*, and the scattered bulbous *Narcissus poeticus* ssp. *poeticus* and *Leucojum aestivum*. These two last species, which in the past covered large areas and formed pure patches, decline year after year.

Meadow vegetation of woodlands

Forest edges and clearings are colonised by a type of meadow vegetation, varying in its composition and physiognomy, according to the local ecological conditions. Plant species growing on grazing pastures are divided into the following groups according to their life forms: rosette plants such as *Hieracium prealtum* ssp. *bauchini, H. cymosum, Potentilla micrantha, Plantago lanceolata, Taraxacum serotinum*; tufty herbs: such as *Potentilla pindicola, Achillea coarctata, Anthemis arvensis, Onobrychis arenaria* ssp. *laciniata, Trifolium cherleri*; tufty grasses such as *Koeleria splendens, Melica ciliata, Molinia caerulea, Stipa capillata;* shooty plants such as *Marrubium vulgare, Mentha longifolia, Anchusa arvensis*, and chamaephytic phrygana such as *Clinopodium vulgare, Origanum vulgare, Genista tinctoria, Acinos alpinus, Thymus sibthorpii.*

The meadow vegetation covers a large area, particularly in the submountainous zone of the National Park. Its floristic composition varies according to the local ecological conditions. The majority of low and medium altitude meadow formations of the National Park, belong to the Molinio-Arrenatheretea class.

Forest formations (Forest Vegetation zones)

The forest vegetation of the National Park is divided in successive zones from the lakeshores to the watershed line on the mountains.

The brief description of woodland vegetation and the shrubby formations is based on Horvat et al. (1974) system of classification.

Mixed forests of low altitudes

The woodland vegetation either starts immediately from the perimetre of the lake, as is the case in the west and south-west parts of the National Park, or its lower limits extend down to an intermediate zone next to human settlements and cropland as it happens in the eastern part.

According to their composition and physiognomy, the forests of a first perimetric strip around Lake Mikri Prespa may be distinguished in two different types of forests. However, at many places of this zone the mixed deciduous-evergreen forest is absent, and is replaced by the higher lying vegetation of deciduous oak forests.

Deciduous-evergreen forest of Ostryo-Carpinion orientalis

The mixed deciduous-evergreen forests of shrubs and small trees of the Ostryo-Carpinion orientalis Alliance have a Mediterranean character, and cover a small coastal area at the south-west edge of the National

Park. They appear on the lower slopes of the Mt. Triklarion facing the lake and again on the sides of Oxia hill, near Mikrolimni (860–1000 m).

The most important constituent species are the deciduous small-sized trees *Carpinus orientalis* and *Ostrya carpinifolia*. In addition to these are found *Quercus pubescens, Acer campestre, Fraxinus ornus, Sorbus domestica*, the shrubs *Juniperus oxycedrus, Prunus spinosa, Cornus sanguinea, Coronilla emerus, Colutea arborescens, Rosa canina* and the shrubby climbing *Lonicera etrusca* and *Vitis vinifera*. In the forest understorey grow the chamaephytes *Ruscus aculeatus* and *Asparagus acutifolius*, as well as a rich flora, consisting especially of shade-preferring species, such as *Helleborus cyclophyllus, Lathyrus laxiflorus, Teucrium chamaedrys, Viola odorata, Brachypodium pinnatum*, and many others.

Evergreen forests of Ostryo-Carpinion adriaticum

This vegetation zone, also known as Juniper-Hornbeam-Macedonian Oak zone, extends on the coastal area of the west portion of the National Park on the feet of Mts Devas and Tsoutsouli (Figure 9), and at two small areas of the hillocks Stalos and Prosilio, west of Mikrolimni (860–1000 m, 1100 m respectively). The physiognomy of this type of vegetation is given by its main evergreen tree species *Juniperus foetidissima* (Stinking Juniper) and *Quercus trojana* (Macedonian oak), and the deciduous small tree *Carpinus orientalis* (Hornbeam). Other trees and shrubs taking part in the structure of this kind of forest, are the evergreen *Juniperus oxycedrus, Buxus sempervirens, Phillyrea latifolia*, and the deciduous *Ulmus minor, Rubus canescens, Prunus mahaleb, Pr. webbii* and others, which are constantly accompanied by herbaceous species, as *Euphoria myrsinites, Dorycnium pentaphyllum* ssp. *herbaceum, Potentilla micrantha, Orlaya grandiflora, Arabis hirsuta, Silene italica, Psoralea bituminosa, Luzula forsteri, Digitalis lanata, Iris reichenbachii* and many others.

Evergreen Box-Juniper shrubland

This vegetation type occurs only on limited areas of the sub-mountainous zone west of Vrondero, up to 1050–1120 m asl (Figure 12). The area is dominated exclusively, by the evergreen shrubs *Buxus sempervirens* (Box) and *Juniperus oxycedrus*, which form dense and low clusters of 1.20–1.80 m in height. In the floristic composition of forest also take part sickly individuals of the trees *Quercus pubescens, Pyrus amygdaliformis, Fraxinus ornus*, the shrub *Prunus spinosa*, and the most usual herbs are *Hypericum olympicum, Sedum athoum, Acinos alpinus, Poa bulbosa, Trifolium cherleri, Lathyrus cinerea, Alyssum minutum, Stachys plumosa.*

This kind of vegetation is classified in the *Buxus sempervirens-Juniperus oxycedrus* association. Although it appears far from the coastal belt, according to its structure and physiognomy, it may be classified in the first, lower, coastal forest zone.

Agios Georgios of Psarades forest nucleus

The evergreen conifer forest around Agios Georgios of Psarades (1000–1100 m) on the northeast slopes of mountain Devas is the only absolute protection area (nucleus) of the National Park consisting of forest vegetation.

This impressive forest consist of tall (12 m high) and straight trees of *Juniperus foetidissima* and *J. excelsa* (Figure 13). In the forest understorey can be found scattered small trees of the species *Juniperus oxycedrus, Carpinus orientalis, Prunus spinosa, Crataegus monogyna, Quercus trojana, Fraxinus ornus*, the chamaephytes *Ruscus aculeatus, Asparagus acutifolius, Sideritis montana, Teucrium polium, Thymus leucotrichus* and on its loose herb layer, a variety of species, as *Cardamine graeca, Euphorbia myrsinites, Campanula sparsa, Erophila verna, Viola arvensis, Potentilla recta, Anchusa variegata, Astragalus parnassi, Salvia rigens, Goniolimon dalmaticum, Draba athoa, Campanula albanica, Saponaria bellidifolia* with many geophytes such as *Gagea arvensis, Muscari neglectum, Crocus pulchellus, Ornithogalum refracta, Iris pumilla, Asphodeline lutea, Fritillaria orientalis, Cyclamen hederifolium.*

The whole composition of this forest gives the impression of the Juniperetum excelsae association.

Deciduous oak forests

Deciduous oak forests constitute the dominant vegetation type in Prespa National Park. They form a zone, extending up to an elevation of 900–1200 m on the slopes of the surrounding the lakes hills and mountains. The lower expansion of the forest, well known as thermal-limits of zone, do not follow the rules of theoretical succession. Therefore, the oak forest boundaries on the east portion of the area come near settlements and croplands, as it happens in Lefkonas and Plati, without the interposition of the Ostryo-Carpinion orientalis zone, which is absent in this area.

Figure 12. A view of evergreen scrub of the Buxus sempervirens-Juniperus oxycedrus association, grazed in the foreground, at Pixos-Vrondero area.

Figure 13. An evergreen stand of junipers at Agios Georgios of Psarades, consisting of *Juniperus foetidissima* and *J. excelsa* (Juniperetum excelsae association) old trees.

Similar phenomenon occurs in Latsista area, towards the south, where the oak forests come close to the lake shore. The deciduous oak forests of Prespa can be classified in the Continental or Balkan thermophile zone or Quercion frainetto and some portions with a psychrophile character in the Balkan zone of Quercion frainetto-cerris.

Thermophile oak forests of Quercion frainetto

These forests extend mostly in the eastern part of the area, where they occur as well organised and dense forests of an average height of ca 8 m. But its most typical stands appear in certain southern parts, where they reach 10 m or more in height. Predominant participant is *Quercus frainetto* accompanied by *Quercus pubescens* and numerous other members, as the small trees *Fraxinus ornus, Acer platanoides,* the shrubs *Lembotropis nigricans, Rubus canescens, Cornus mas, Colutea arborescens, Rosa corymbifera,* the chamaephytes *Chamaecytisus eriocarpus, Ch. heuffeli, Acinos suaveolens, Thymus sibthorpii,* the herbaceous species *Lathyrus niger, Astragalus glycyphyllus, Lychnis coronaria, Scleranthus perennis, Dorycnium pentaphyllum* ssp. *herbaceum, Lotus uliginosus, Aster bellidiastrum, Alliaria petiolata, Trifolium aureum* and the grasses *Phleum pratense, Brachypodium sylvaticum, Festuca heterophylla, Anthoxanthum odoratum, Koeleria pyramidata, Arrhenatherum elatius.*

The oak forests of Garlini and Frouria hills in the south, constitute a distinct forest patch enriched with species such as the small trees *Acer obtusatum, Sorbus aria, Tilia cordata,* the shrubs *Rosa pendulina, Cotoneaster nebrodensis,* as well as the typical herbaceous species *Lilium martagon, L. albanicum, Tragopogon balcanicum, Artemisia alba, Geranium aristatum, Lathyrus venetus, Trifolium pignantii, Achillea chrysocoma, Dianthus carthusianorum, Knautia drymeia, Conium maculatum, Polygonatum odoratum, Iris germanica, Convallaria majalis,* and many others.

The composition of these forests is identified as belonging to the Quercetum frainetto association (Quercion frainetto alliance).

Psychrophile oak forests of Quercetum petraeae

The structure and physiognomy of this forest differs greatly from the previously described. The most important role is given by the psychrophile deciduous species *Quercus petraea, Q. cerris, Q. pedunculiflora,* to a certain extent *Quercus frainetto,* and the evergreens *Quercus trojana, Juniperus foetidissima, Buxus sempervirens,* etc. There are also, sporadically, small trees as *Pyrus communis, Crataegus laciniata, Carpinus betulus, Prunus webbii, Pr. mahaleb, Acer pseudoplatanus, A. monspessulanum, Sorbus torminalis, Malus dasyphylla, Tilia cordata,* and more rarely specimens of *Quercus robur, Euonymus latifolius, Ulmus minor, Cornus sanguinea* and *Quercus pubescens.* On the shrub layer of the forest grow specimens of *Juniperus oxycedrus, Crataegus monogyna, Corylus avellana, Ligustrum vulgare, Euonymus europaeus, Colutea arborescens,* and the herb layer make their appearance species such as *Geum urbanum, Mycelis muralis, Helleborus cyclophyllus, Euphorbia amygdaloides, Fragaria vesca, Geranium lucidum, Lathyrus laxiflorus,* as well as sparse individuals of *Dictamnus albus, Eryngium palmatum, Lilium albanicum, L. chalcedonicum, Saponaria glutinosa, Alcea pallida, Ornithogalum refractum* and *Euphorbia seguierana* ssp. *niciciana.*

This forest type is dominant on the western part of the National Park, especially on the mountains Devas (1380 m) and Tsoutsouli (1351 m), where it reaches its maximum vitality and height (12 m). The thermal limits of these forests are very often in touch with the lower lying forest zone (1000–1100 m) or ends up at the coastline of Lake Megali Prespa (860 m). In some parts of these psychrophile forests, particularly towards the west, the composition of forests changes to include various thermophile elements, corresponding to the Quercetum frainetto association.

The psychrophile forests of *Quercus petraea* can be found also at the eastern side of the National Park, on the hillsides of Mazi, Gofilos, Perivoli (1100–1300 m) and elsewhere. These parts can be classified into the Quercion frainetto zone. However, the cool environmental conditions imposed the domination of *Quercus petraea,* either in pure stands or mixed with *Quercus cerris.* The structural form of this psychrophile vegetation differs in many respects from that of the corresponding vegetation of the western area, by the penetration of species with very important role, such as the trees *Acer obtusatum, Fagus sylvatica, Pyrus amygdaliformis, Ostrya carpinifolia, Sorbus umbellata, Prunus cocomilla, Fraxinus excelsior* and the herbaceous species *Trifolium alpestre, T. vesiculosum, Potentilla inglinata, Filipendula vulgaris* and *Hypericum barbatum.*

The structural features of the psychrophile oak forests classify them in the Quercetum petraeae association of the Quercion frainetto-cerris alliance.

Beech forests and beech-fir forests

Generally, the deciduous and humid beech forests form zones, which extend on the mountains of Prespa between 1200 to 1800 m asl, constituting the higher lying forest belt of the National Park. Their distribution epicentre is in the eastern and secondarily the southern part, while they are absent in the western sector.

Beech forests present Central-European features, and in the Balkan Peninsula are represented by the Fagion moesiacum alliance and vegetation zone. According to their floristic composition, the following two types can be distinguished.

Deciduous beech forests of Fagetum moesiacum

Due to anthropogenic impacts, this kind of vegetation is highly variable in structural terms represented by a spectrum of forms from low monolayer coppice vegetation (5–8 m high) to more vigorous and multi-layered high forests (15–20 m high). The dominant elements of this kind of vegetation are *Fagus sylvatica* and *F. moesiaca (fa. taeniolepis* and *fa. spatulolepis*). The most frequent tree species found in beech forests are *Quercus petraea, Acer obtusatum, Carpinus betulus, Betula pendula, Acer heldreichii, Prunus avium, Sorbus aria, Populus tremula, Corylus colurna*, with regular occurence of shrubs such as *Juniperus communis, Crataegus heldreichii, Sambucus nigra*, and less frequently *Lonicera alpigena* ssp. *formanekiana, Cotoneaster integerimus, Rubus idaeus, R. schleicheri*, and *Rosa arvensis*. There is also a great variety of herbaceous flora on the forest understory. The majority of herbs blossom at the end of spring and the summer. The most representative among these are *Convallaria majalis, Geranium sanguineum, G. sylvaticum, Cardamine buldifera, Polygonatum odoratum, Lathyrus laxiflorus, L. vernus, Chelidonium majus, Geum urbanum, Pulmonaria officinalis, Epilobium montanum, Arctium lappa, Trifolium badium, T. pignantii*, as well as the geophytic herbs *Lilium martagon, Fritillaria graeca, Dactylorrhiza sambucina, Cephalanthera rubra*. Early in spring, the renaissance of vegetation begins again underneath the leaf litter layer, with the appearance of early flowering herbaceous species such as *Viola odorata, Primula vulgaris, P. elatior, Cephalanthera longifolia, Gagea arvensis, Erophila verna, Potentilla micrantha*, and is closed by the late flowering autumn species such as the bulbous *Colchicum autumnale, Crocus pulchellus, Cyclamen hederifolium.*

The various floristic combinations of beech forest stands resemble the typical Fagetum moesiacum association, of the Fagion moesiacum alliance, but the extraordinary cases of the plant communities of surroundings, appearing to be rich in illyric elements must not be overlooked.

Beech-fir forests

The mixed beech-fir forest appears in parts of the northeast perimetre of the area. Specifically, this forest constitutes an island-like forest zone up to 1500–1800 m, occurring within the beech forest domain. Dominant constituents are the Central European species *Abies alba* and *Fagus sylvatica*, assisted by the Balkan *Fagus moesiaca*. Its magnificent stands may reach up to 25 m high (Figure 14). Certainly, *Abies alba* stands are relicts of a formerly more extensive forest.

In the floristic combinations of the forest undergrowth various tree species were identified such as *Acer heldreichii, A. platanoides, Sorbus torminalis, Ulmus glabra, Populus tremula, Flaxinus excelsior, Betula pendula, Carpinus betulus*; also the shrubs *Juniperus communis, J. oxyacantha, Rubus hirtus, Sambucus nigra, Chamaecytisus hirsutus* and rarely the shrubs *Crataegus laciniata, C. heldreichii, Daphne mezereum, Rubus scleicheri, Rosa obtusifolia*, and the woody epiphyte *Viscum album* ssp. *abieti.*

The herb layer is represented by *Saxifraga rotundifolia, Veronica chamaedrys, Aremonia agrimonoides, Euphorbia amygdaloides, Mycelis muralis, Cardamine buldiferia, Doronicum austriacum, Myosotis sylvatica, Moenchia trinervia, Melica uniflora, Anemone nemorosa, Poa nemoralis, Veronica urticifolia, Stachys officinalis, Actaea spicata, Symphytum tuberosum* and *Brachypodium sylvaticum.*

The presence of *Abies alba* in combination with the floristic composition of the forest testifies for the presence of the Abieti-Fagetum moesiacum association, which belongs to the Fagion illyricum alliance.

Subalpine vegetation of dwarf shrubs

The subalpine vegetation with its peculiar physiognomy and character extends higher than the upper boundaries of beech in altitudes of 1800 to 2000 m asl. It consists of cold resisting shrubs, chamaephytes and perennial herbs forming a dense and compact layer just 0.30 to 0.50 m high (Figure 15). Its main distribution area is the largest portion of Dikorpho peak (2118 m) at NE and the Falakra (1741 m) elevation in the SE. The most frequent elements, are the dwarfish semi-shrubs *Vaccinium myrtillus, Chamaecitisus polytrichus, Ch. eriocarpus, Juniperus communis* ssp. *nana, Bruckenthalia spiculifera*, the chamaephytes *Genista depressa, G. rumelica, Thymus stojanovii*, the herbs *Trifolium parnassi, T. alpestre, T. velenovskyi, Lotus stenodon, Veronica bellidioides, Scrophularia laciniata,*

Figure 14. Forest dominated by *Abies alba*, and scattered deciduous trees, mostly of the species *Fagus moesiaca, F. sylvatica, Fraxinus excelsior*, and *Acer heldreichii* of the Abieti-Fagetum moesiacum association, in the northeastern perimeter of the National Park.

Galium lucidum and the grasses *Festuca valesiaca, F. laevis, Bellardiochloa violacea, Phleum alpinum, Alopecurus gerardi.* Among the companion species are found the woody *Cotoneaster nebrodensis, Vaccinium uliginosum, Lonicera alpigena* ssp. *formanekiana, Chamaecytisus hirsutus, Ch. heuffelii, Rosa pendulina, Genista lydia* and many herbs, as well as *Potentilla recta, Acinos alpinus, Luzula luzulina,* and *Carex caryophyllea.* According to its composition and physiognomy, this type of vegetation is classified in the *Vaccinium myrtillus-Chamaecitisus polytrichus* association of the Bruckenthalion alliance (Caricetea curvulae class). On the borders and the openings of the vegetation, a variety of cold resisting herbs occur, such as *Filipendula ulmaria, Senecio subalpinus,* and the balkan endemics species *Alyssum corymbosum, Dianthus haematocalyx* and *Dianthus minutifolius.*

On the rocky and stony places of the borders of Dikorpho peak grow the creeping shrubby species *Rubus saxatilis, R. idaeus, Rhamnus saxatilis,* the semi shrub *Daphne mezereum,* the chasmophytes *Sedum ochroleucum, S. dasyphyllum, Saxifraga sempervirens, S. chrysosplenium, Silene lerchelfendiana, Anthemis pindicola, the rock-loving Asplenium septendrionale, A. trichomanes, Dianthus perviformis, Centaurea deustiformis, Sempervivum reginae-amaliae, Arabis caucasica,* and a numerous members of herbs, such as *Polygonum alpinum, Lilium jankae, Betonica scandica, Euphorbia apios, Rumex scutatus, Achillea grandifolia, Geranium macrorrhizum,* and *Doronicum columnae.* The floristic analysis of this peculiar vegetation of rocky places leads to the identification of the rock-loving associations of the Asplenietea rupestria class.

On the rocky slopes of Triklarion mountain (1600–1700 m), a rudimentary subalpine vegetation of a loose form occurs, consisting of the shrubs *Juniperus communis* ssp. *nana, Cotoneaster nebrodensis, Rosa pendulina,* and various herbs, such as *Carex caryophyllea, Minuartia verna* ssp. *idaea, Achillea coarctata, Helianthemum canum, Alyssum alyssoides, Hypericum rumeliacum, Salvia argentea, Sideritis syriaca, Seseli annuum, Stachys cassia, St. tymphaea, Thymus cherlerioides* and many others.

Vegetation of alpine meadows

This meadow vegetation of alpine character is characterised by the absence of woody elements and the dominance of perennial cold resisting plants. It covers, exclusively, the higher parts of Dikorpho peak and extend eastwards to the extensive mountain meadows of Varnous mt. which are grazed by livestock.

The main feature of this type of vegetation is the presence of densely growing grasses to about 0.30 m high (Figure 15), and the local appearance of a cushion-

Figure 15. A view of the timberline and sub-alpine vegetation at the eastern marginal zone of the National Park. At right the last trees of *Fagus sylvatica*, lower are dwarf shrubby creeping clusters of vegetation.

like herb vegetation. Numerous species participate in the composition of these meadows, forming a variety of combinations, but grammineous plants give the physiognomy to the vegetation. The representative species, are the grasses *Festuca valesiaca, F. paniculata, F. circummediterranea, Narduus stricta, Avenula pubescens, Koeleria splendens, Stipa pulcherima, Poa cenisia, Trisetum flavescens*, and *Anthoxanthum odoratum.* There are also the rush-like species *Luzula spicata, L. campestris, Carex atrata, the pteridophytes Asplenium septedrionale, Cystopteris fragilis*, and many other herbs such as *Arabis alpina, Alchemilla alpina, Aster alpinus, Achillea atrata, Hypericum barbatum, Asyneuma limonifolium, Silene ventricosa, Potentilla aurea, Veronica kindlii, Anthemis pindicola, Dianthus myrtinervius, Geranium subcaulescens, Knautia magnifica, Viola gracilis, Jasione orbiculata, Crocus veluchensis, Gentiana punctata, Epilobium collinum, Saxifraga exarata, Campanula sibthorpiana, C. spatulata*, and *Plantago holosteum.*

The dominant combinations of species tend to identify the Poeion violaceae alliance, of the Caricetea curvulae class.

At areas with depressions and furrows and on moist soils in the vicinity of springs, tall clusters of aquatic vegetation develop. The most important species here are *Cirsium palustre, C. appendiculata, Geum coccineum, G. rivale, Eriophorum latifolium, Alchemilla acutiloba, Deschampsia caespitosa, Rumex alpinus, Barbarea balcana, Chrysosplenium alternifolium, Juncus thomasii, Glyceria fluitans, Adenostyles alliariae, Silene asterias, Veratrum album, Carex ovalis, Galium palustre, Parnassia palustris, Botrychium lunaria*, and *Caltha laeta.* The composition of certain of these plant communities indicate the presence of the Geion coccinei alliance, of the Betulo-Adenostyletea class.

Conclusion

The overall picture of the vegetation of the National Park does not show serious degradation trends. This is the conclusion of a recent overall survey of the vegetation and is a hopeful prospect for the future of the National Park. Additionally, a more general environmental stability can be seen, despite the fact that the area is inhabited and thus difficult to control. The preservation of the ecosystem in a rather stabile condition is due to the peculiarity of its geo-political position and the mild and reasonable management of its natural resources during the last decade. The local biodiversity succesfully withstands the small scale modern pressures, remaining rich and essentially intact. This is not mentioned to increase our relaxation but our vigilance, care and interest for improvement.

References

Babalonas, D. & G. Pavlides, 1989. Contribution to the vegetation of aquatic plants of lake Mikri Prespa. BIOS Thessaloniki 1: 19–29. (in Greek)

Braun-Blanquet, J., 1964. Pflanzensoziologie. Grundzuge der Vegetationskunde. (1. Aufl. 1928) 3. Aufl. Wien, New York, 865 pp.

Horvat, I., V. Glavac & H. Ellenberg, 1974. Vegetation Sudosteuropas, Stuttgart, 767 pp.

Lavrentiades, G., 1956. On the hydrophytic and helophytic flora of embryophytes in Greek Macedonia. Ph. D. Thesis, Thessaloniki (in Greek), 88 pp.

Lavrentiades, G. & G. Pavlides, 1985. Contribution to the research of hydrophytic and helophytic associations of the lake Mikri Prespa. Proc. 4th Symp. Greek bot. Soc., Thessaloniki (in Greek): 144–155.

Pavlides, G., 1985. Geobotanical Study of the National Park of lakes Prespa (NW Greece). Part A. Ecology, Flora, Phytogeography, Vegetation. Thessaloniki (in Greek), I–XI, 308 pp.

Hydrobiologia **351**: 61–75, 1997.
A. J. Crivelli & G. Catsadorakis (eds), *Lake Prespa, Northwestern Greece.*

Species composition and seasonal cycles of phytoplankton with special reference to the nanoplankton of Lake Mikri Prespa

E. Tryfon & M. Moustaka-Gouni
Aristotle University of Thessaloniki, Institute of Botany, School of Biology, 54006 Thessaloniki, Greece

Key words: phytoplankton, nanoplankton, water quality, species composition, biomass, temporal distribution, Greece

Abstract

The phytoplankton of Lake Mikri Prespa was studied at monthly or biweekly intervals during the period May 1990–September 1992. Its species composition, consisting of a great number of cyanophytes and a very small number of chrysophytes and desmids, may reflect the eutrophic character of the lake. Moreover, the mean annual biomass values (15.0 and 3.2 g m^{-3} in the two years, respectively) and the maximum biomass (38.1, 6.4 and 9.6 g m^{-3}), classify Mikri Prespa as a eutrophic lake. A tendency towards a double-peaked pattern of biomass distribution in time with one peak in autumn, composed mainly of cyanophytes, and another in spring made up of diatoms, was observed. This pattern contrasts with the standard pattern in eutrophic, stratified temperate lakes, which exhibit a third biomass maximum in summer.

Cyanophytes were the most important group in terms of biomass and were dominated by the species *Microcystis aeruginosa, Microcystis wesenbergii, Anabaena lemmermannii* var. *minor* and *Aphanocapsa elachista* var. *conferta.* Diatoms constituted the second most important group, with main representative the species *Cyclotella ocellata.* Cyanophytes, diatoms, chlorophytes and dinophytes revealed annual periodicity whereas the other algal groups did not show any seasonality at all.

The nanoplankton constituted an important part of algal biomass (38.9 and 49.9% in the two years, respectively) and revealed annual periodicity with maximum values in winter and spring, mainly composed of diatoms and cryptophytes. Low temperature, increased rainfall and high DIN concentrations seemed to be the main factors influencing the seasonality. Although the percentage contribution of nanoplankton decreased with the increase in total biomass, justifying the classification of Lake Mikri Prespa among the eutrophic lakes, the nanoplankton biomass did not correlate significantly with total phytoplankton biomass.

Introduction

Phytoplankton consists of microscopic plant organisms adapted to suspension in the water and exposed to passive movement by wind and currents (Reynolds, 1984a). These organisms are unicellular but many of them form colonies and threads ranging in size from <1 μm up to >500 μm.

The temporal and spatial distribution of phytoplankton and the factors influencing them in fresh waters have been thoroughly studied by many researchers (e.g. Nauwerck, 1963; Morris, 1980; Round, 1981; Sommer, 1981; Reynolds, 1984a; Sommer et al., 1986; Pollingher, 1990). The above studies clearly indicate that the repeated seasonality of phytoplankton growth observed in successive years, mainly constitutes a response of phytoplankton to the seasonality of some environmental parameters. According to Reynolds (1984a), the influence of various factors on the seasonal appearance of phytoplankton differs in significance, with physical factors (temperature, water mixing, underwater light climate) being the most important, and chemical (nutrient levels) and biotic factors (grazing and parasitism) being of lesser importance. However, the qualitative and quantitative changes of phytoplankton assemblages do not only form annually repeated cycles but can also exhibit more stable long-term floristic alterations due to morphological, trophic or hydrological changes in the lakes.

Sommer et al. (1986), using data from a significant number of fresh water systems, have proposed a plankton seasonality model that applies to an ideal temperate lake. Eventually, when a large amount of information is available, the periodicity of planktonic microalgae in this kind of lake constitutes a relatively predictable sequence of steps. Unpredictability can be observed in shallow water masses, where rainfall and wind can disturb this predetermined continuity.

Based on the above, it is clear that research on shallow lakes is of particular importance. Lake Mikri Prespa (Greece) is one of the country's larger shallow lakes. Our knowledge on the richness and ecology of its species was until recently restricted. The only information available consisted of small scale qualitative phytoplankton data and information about the main chemical and physical characteristics of the water (Kousouris & Diapoulis, 1983; Mourkides et al., 1978).

The first long term study on the structure and dynamics of phytoplankton assemblages in the lake were the subject of a Ph.D. Thesis (Tryfon, 1994), supported by the Greek Ministry of Agriculture (grant No. 259533 6/2/90). A part of this work has already been published (Tryfon et al., 1994; Moustaka-Gouni et al., 1994). The present paper reviews the physical and chemical properties of the water, the species composition and the seasonal fluctuations of the different taxonomic algal groups. Characteristics of the cell size and the temporal changes of nano- and netplankton biomass are also described. Finally, the influence of water temperature, DIN concentration and rainfall on the different algal size groups is outlined.

Methods

During this study 37 samplings were carried out at monthly intervals from May 1990 to September 1992 or, during the warm periods, at two-weekly intervals. Samples were collected from three stations along the long axis of the lake but since no great differences were recorded, only the results from the representative central station will be described below. A 2 liter, 1 m long, chain driven sampler (Gounis & Moustaka-Gouni, 1990) was used to collect samples at every meter from the surface to the bottom. The biomass values cited are the mean biomass for all the depths counted. Samples were fixed with Lugol's solution and acetic acid and thereafter used for identification and counting of phytoplankton according to Utermoehl's (1958) sedimentation method. An attempt was made to count at least 100 individuals of the most important phytoplankton species and, as a total, at least 400 individuals in every sedimentation chamber. Within confidence intervals of 95% this measurement contains a maximum error of 20% and 10%, respectively (Lund et al., 1958; Willen, 1976a). The determination of the phytoplankton biomass was based on the calculation of the volume of each species using appropriate geometric formulae and assuming a specific gravity of one.

Temperature and oxygen concentrations were measured every 1 m with a YSI Model 33 or a WTW Microprocessor oxymeter. Transparency was estimated with a 20 cm Secchi disk whereas pH measurements were made with a pH M 84 Research pH Meter. Chemical analyses of silicon, phosphate phosphorus, nitrate nitrogen and nitrite nitrogen were made after filtration through Whatman GF/C filters using Strickland and Parson's (1968) methods, whereas for ammonia nitrogen the method of Liddicoat et al. (1976) and for calcium of Allen (1989) were applied.

Background

Rainfall was very variable in the region of Lake Mikri Prespa during the study period (Figure 1). 1991 was the wettest year with 822.9 mm of rain, whereas 1990 and 1992 had lower rainfall values (590.5 mm for 1990 and 288.4 mm for the first nine months of 1992). The highest water temperatures (Figure 2) were recorded in July and August (26.6 °C) and the lowest in December 1991 (1.8 °C). During winter an ice layer of approximately 15 cm in 1991 and 35 cm in 1992, that lasted for one and two months respectively, was formed on the surface of the lake. A continuous decrease in the temperature of the water column with increasing depth was often observed in spring, but real thermal stratification was recorded only in summer 1991. According to our measurements, the thermocline was destroyed and reestablished within a month during that summer. Moreover, its absence during the summer of 1990 may indicate the weakness of the stratification in the lake.

Dissolved oxygen in the surface water fluctuated between 7.0 and 13.6 mg l^{-1} (84.7–120% saturation) and exhibited the lowest values in summer and autumn and the highest in winter (Figure 2). Bottom values ranged from 0.1 to 14.0 mg l^{-1}. A concentration decrease towards the bottom was often observed in

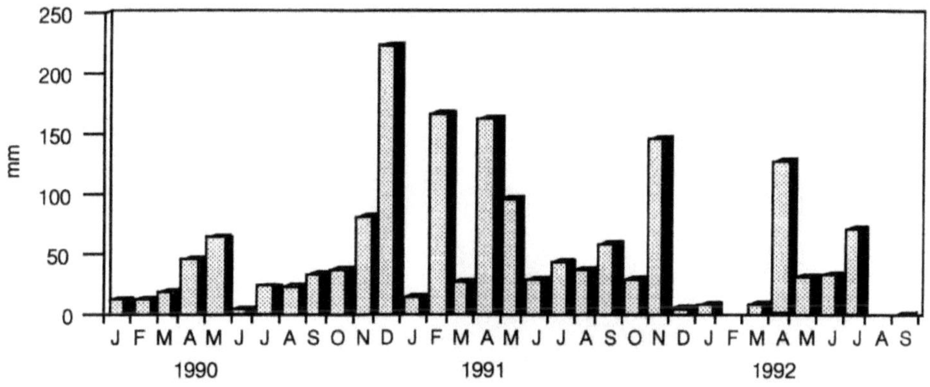

Figure 1. Seasonal variations of monthly precipitation in the catchment area of Lake Mikri Prespa.

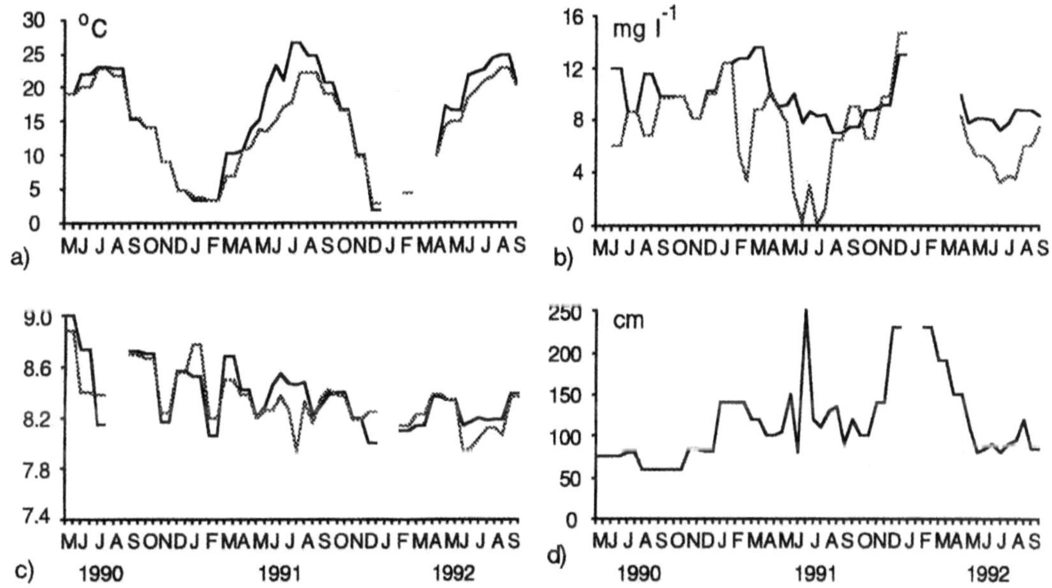

Figure 2. Seasonal variations of (a) temperature, (b) oxygen, (c) pH and (d) water transparency. Solid line indicates surface values, dashed line indicates bottom values.

summer. The pH fluctuated between 7.9–8.7 without any special alterations between surface and bottom values (Figure 2). Water transparency fluctuated between 55 and 285 cm (Figure 2). It was influenced mainly by total phytoplankton biomass and multiplied by a factor of 2.5 it indicated a mean euphotic depth of 2.6 m.

Phosphorus is one of the most common factors limiting phytoplankton growth in temperate lakes (Schindler, 1977) and therefore the study of its annual fluctuations is of particular importance. Orthophosphate ions which form about 5% of the total phosphorus in a lake (Wetzel, 1983) are the most available form for phytoplankton. Soluble reactive phosphorus (SRP) which is measured with the normal laboratory procedures consists mainly of orthophosphate ions and gives an estimate of the available concentrations of the nutrient. SRP concentrations in all samples varied between 0 and 53 $\mu g\,l^{-1}$ and exhibited higher values in September 1990 and March 1991 (Figure 3), which coincided with a storm and the leaching of the catchment area by snow melt, respectively. Water evaporation during the extremely dry period of 1990 must have also contributed to the increased SRP values in that autumn, compared to the following years. The limestone on the bottom of the lake and the high calcium concentration of the water (20–60 mg Ca l^{-1}) must have constituted an important factor for the removal of phosphorus from the overlying water. Isolated indications of phosphorus

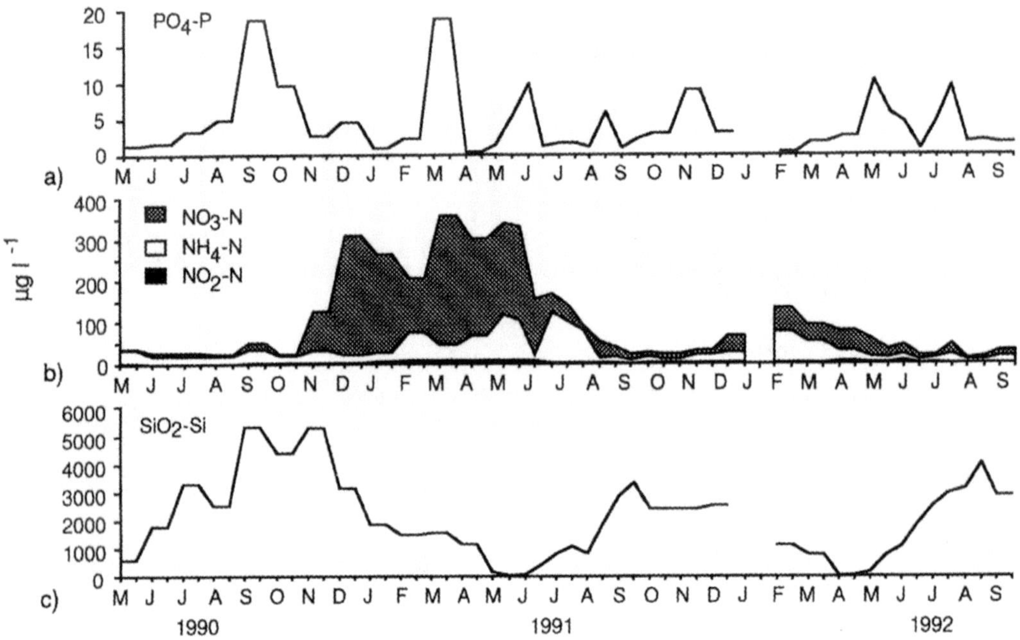

Figure 3. Seasonal variations of mean concentrations in the water column of (a) Soluble Reactive Phosphorus, (b) Dissolved Inorganic Nitrogen and (c) Silicon.

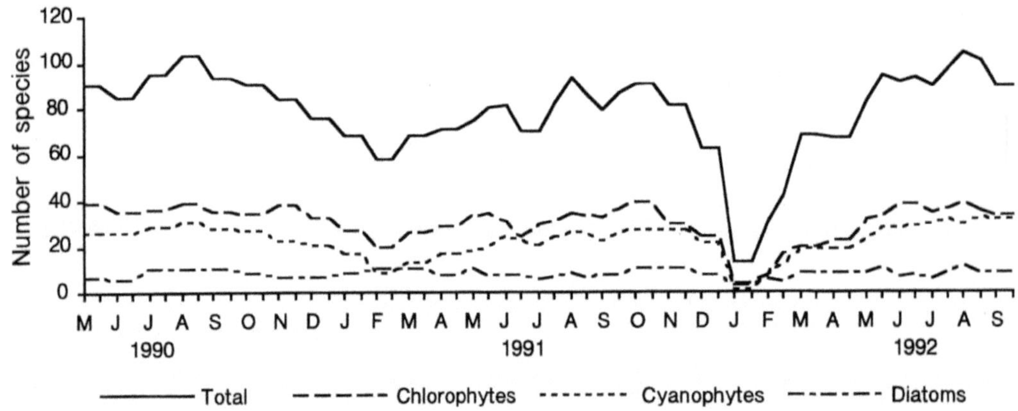

Figure 4. Seasonal variations of total numbers and numbers of chlorophyte, cyanophyte and diatom species.

being released from the sediment were not always coupled with low oxygen concentrations that are believed to be necessary for this liberation, but mainly with storms or wind episodes that could have stirred up the sediment. Moreover, grazing by zooplankton and especially ciliates may have been an important factor for the regeneration of the element. This is supported by the fact that increased phosphorus concentrations in September 1990 and March 1991 coincided with increased abundance of ciliates and of a heterotrophic nanoflagellate (0.44×10^6 ind l^{-1} in September 1990 and 0.16×10^6 ind. l^{-1} in March 1991 for the ciliates and 1.34×10^6 ind. l^{-1} in March 1991 for the nanoflagellate).

Nitrogen exists in many forms in lake water but the most available ones are inorganic nitrate, nitrite and ammonia salts. Nitrate-nitrogen was the most important inorganic nitrogen source in Lake Mikri Prespa (Figure 3). Its concentration in all samples ranged between 0–810 μg NO_3-N l^{-1} and exhibited higher values in winter and spring, especially during 1991. Rainfall, which was extremely high during this peri-

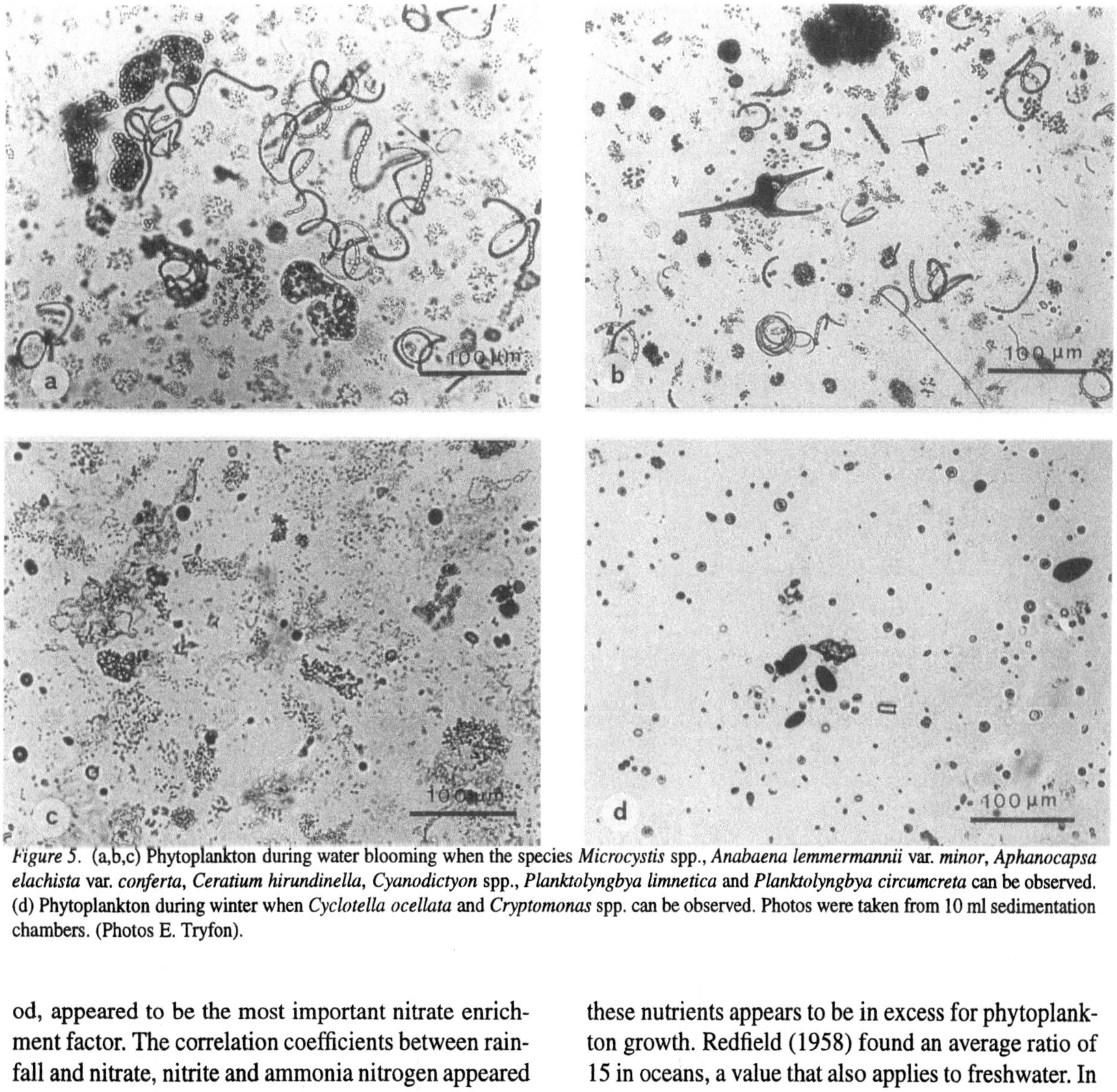

Figure 5. (a,b,c) Phytoplankton during water blooming when the species *Microcystis* spp., *Anabaena lemmermannii* var. *minor*, *Aphanocapsa elachista* var. *conferta*, *Ceratium hirundinella*, *Cyanodictyon* spp., *Planktolyngbya limnetica* and *Planktolyngbya circumcreta* can be observed. (d) Phytoplankton during winter when *Cyclotella ocellata* and *Cryptomonas* spp. can be observed. Photos were taken from 10 ml sedimentation chambers. (Photos E. Tryfon).

od, appeared to be the most important nitrate enrichment factor. The correlation coefficients between rainfall and nitrate, nitrite and ammonia nitrogen appeared to be higher for the first two forms and lower for the third ($r = 0.38$, 0.44 and 0.22, $n = 93$, $p < 0.05$). Nitrite-nitrogen followed the seasonal changes of nitrate but its concentrations remained below 12 μg NO_2-N l^{-1}. Ammonia-nitrogen ranged from 3 to 360 μg NH_4-N l^{-1} and had higher values in winter and spring. Increased concentrations measured above the sediment in summer 1991 indicated that low oxygen concentrations may have contributed to the release of ammonia from the sediment. Total DIN concentrations therefore ranged between 3 and 831 μg l^{-1}, with highest values in winter and spring, particularly in 1991.

The nitrogen:phosphorus atomic ratio in the water is an important variable since it can denote which of these nutrients appears to be in excess for phytoplankton growth. Redfield (1958) found an average ratio of 15 in oceans, a value that also applies to freshwater. In Lake Mikri Prespa, the ratio ranged between 4–1664 with a mean value of 55, thus showing that phosphorus rather than nitrogen could have been the main nutrient limiting phytoplankton growth.

Silicon concentrations in all samples varied between 19–5600 μg SiO_2-Si l^{-1} and showed the highest values in autumn and the lowest in spring (Figure 3). The importance of the element lies in its significance for the construction of the cell wall of diatoms.

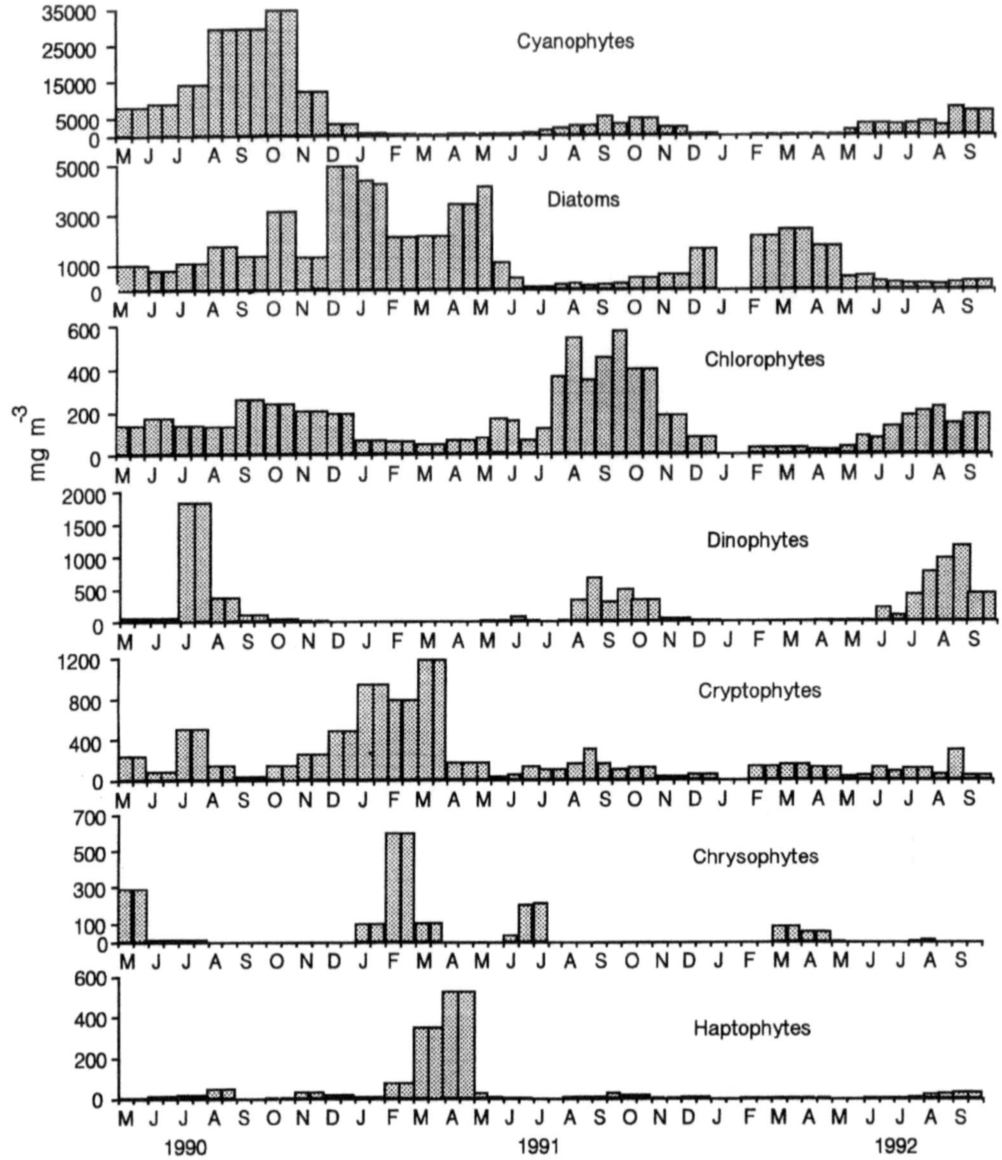

Figure 6. Seasonal variations of the biomass of all algal groups.

Phytoplankton species composition

A total of 138 taxa have been identified in the lake, most of which belong to chlorophytes and cyanophytes (Table 1). The detailed list of the species is given elsewhere (Tryfon et al., 1994). The phytoplankton composition was characterized by the presence of many cosmopolitan species (e.g. *Microcystis aeruginosa, Microcystis wesenbergii, Aulacoseira granulata, Rhodomonas minuta* var. *nannoplanktica* and *Closterium aciculare*) but also of forms of special taxonomic interest (e.g. many chroococcalean cyanophyte species, the most common diatom *Cyclotella ocellata*, the chlorophyte *Radiococcus* cf. *wildemannii* and the chrysophyte *Ochromonas* sp.). The great number of small chroococcal cyanophytes and the large contribution of chlorococcalean species among chlorophytes were also an important characteristic of the lake's phytoplankton.

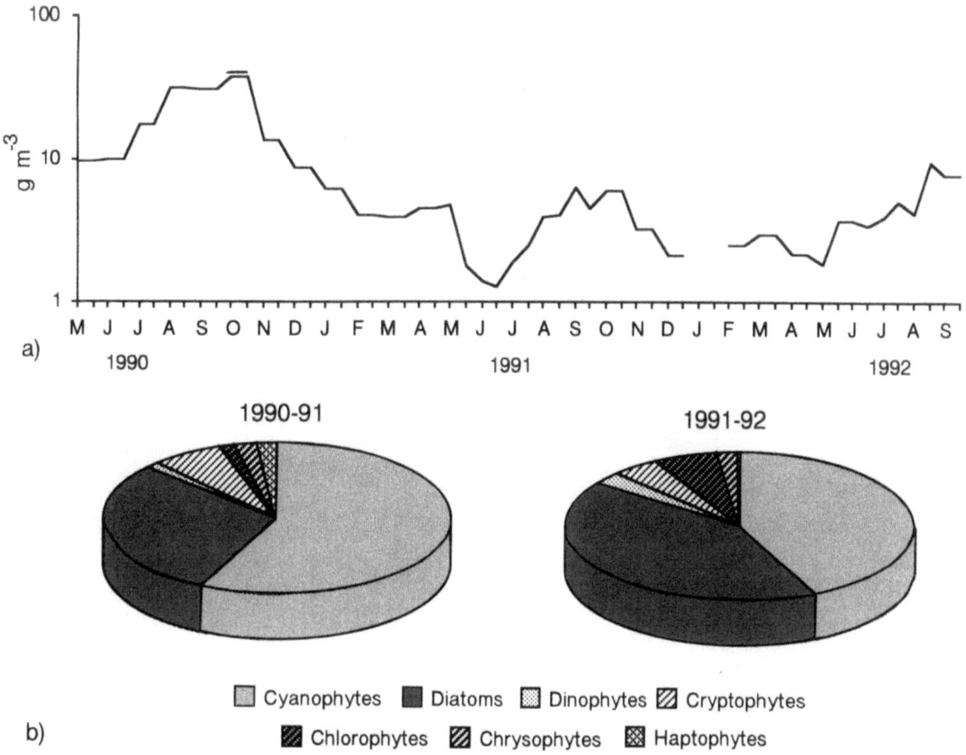

Figure 7. (a) Seasonal variations of total phytoplankton biomass and (b) annual percentage contribution of each algal group to the total phytoplankton biomass.

Table 1. Distribution of phytoplankton species among the algal groups of Lake Mikri Prespa.

Group	Number of species	Percentage (%)
Cyanophyceae	33	23.9
Chlorophyceae	52	37.6
Bacillariophyceae	13	9.4
Chlysophyceae	12	8.6
Dinophyceae	12	8.6
Cryptophyceae	7	5.0
Euglenophyceae	5	3.6
Xanthophyceae	3	2.1
Haptophyceae	1	0.7
Total	138	100
Nanoplankton	49	35.5
Netplankton	89	64.4

The richness of the phytoplankton species and the number of taxa of the various algal classes have often been used as an indicator of the trophic state of a lake. In eutrophic waters, a large number of cyanophyte species is usually present, whereas in oligotrophic lakes desmids and chrysophytes dominate (Round, 1981; Eloranta, 1986; Eloranta & Palomaki, 1986; Willen, 1976b). The low mumber of desmids and chrysophytes and the high number of cyanophyte species reflect the eutrophic character of Lake Mikri Prespa. A similar species richness among the algal classes of phytoplankton has been reported from other eutrophic lakes of northern Greece (Moustaka-Gouni, 1988; Moustaka-Gouni & Nikolaidis, 1990).

The temporal distribution of the total number of species showed annual variations with maximum values in summer and autumn and minimum values in winter. Similar patterns were shown by chlorophytes and cyanophytes, whereas the other algal groups did not show any seasonality (Figure 4). It seemed that the spring increase in temperature had a positive effect on the number of species, the correlation only being non-significant with diatoms and negative with cryptophytes ($r = -0.40$, $n = 34$, $p < 0.05$). On the contrary, rainfall seemed to have a positive effect on the number of cryptophytes ($r = 0.43$, $n = 34$, $p < 0.05$).

The classification of phytoplanktonic organisms in two subgroups according to their size provides an alter-

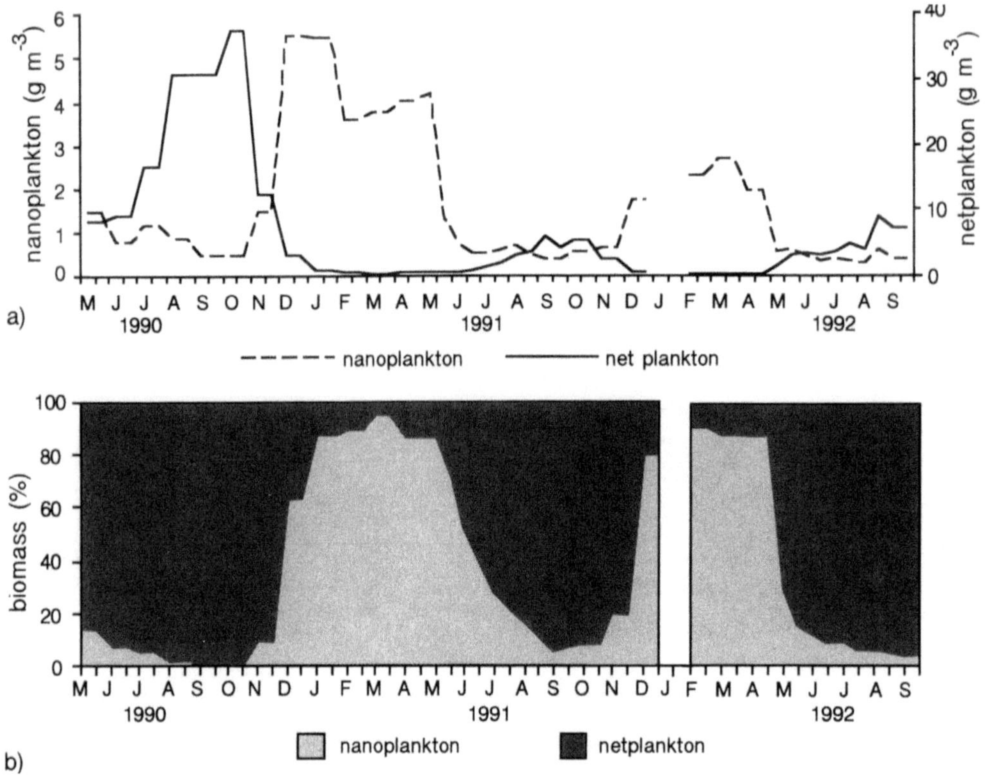

Figure 8. Seasonal variations of (a) netplankton and nanoplankton biomass and (b) their percentage contribution to the total phytoplankton biomass.

native approach to traditional studies, based on taxonomic groups. Whereas cell size plays an important role for the persistence of algae in the euphotic zone, their response to environmental factors and the dynamics of food webs in water systems, taxonomical grouping is not necessarily related with the above functions.

The dimensions of phytoplankton cells in Lake Mikri Prespa varied between 0.6–120.0 μm. Species having cell sizes with their longest dimension smaller than 20 μm can be easily grazed by zooplankton and are grouped under the term nanoplankton. Cells having their longest dimensions from 20 to 200 μm can be classified as netplankton (Morris, 1980). From the 138 taxa identified in the lake, 49 belonged to the nanoplankton and 89 to netplankton (Table 1). Most of the nanoplanktonic species belonged to chlorophytes and cryptophytes, whereas the majority of cyanophyte cells formed colonies and threads far beyond the size limit for nanoplankton. The dimensions and volumes of some of the most important algal species are given in Table 2. *Cyclotella* cf. *distinguenda*, *Cryptomonas* cf. *reflexa* and *Radiococcus* cf. *wildemannii* were, however, in spite of their somewhat longer dimensions (up to 24.0 μm) recorded among the nanoplanktonic species.

Nanoplanktonic species seemed to be affected by temperature but less than the netplankton species ($r=0.50$ and 0.79, $n=34$, $p<0.05$ for nanoplankton and netplankton, respectively). The positive effect of temperature on netplankton growth was expected since larger cells gain a metabolic advantage at high temperatures (Reynolds, 1984a). The positive correlation of nanoplanktonic species with temperature in Lake Mikri Prespa can be mainly attributed to the increase of chlorophyte species composed mainly of small chlorococcalean algae in summer.

Temporal distribution of planktonic algae

Cyanophytes

Cyanophytes appeared to be the most important algal group since they accounted for about 80% of the total biomass during the first year of the study and for 50%

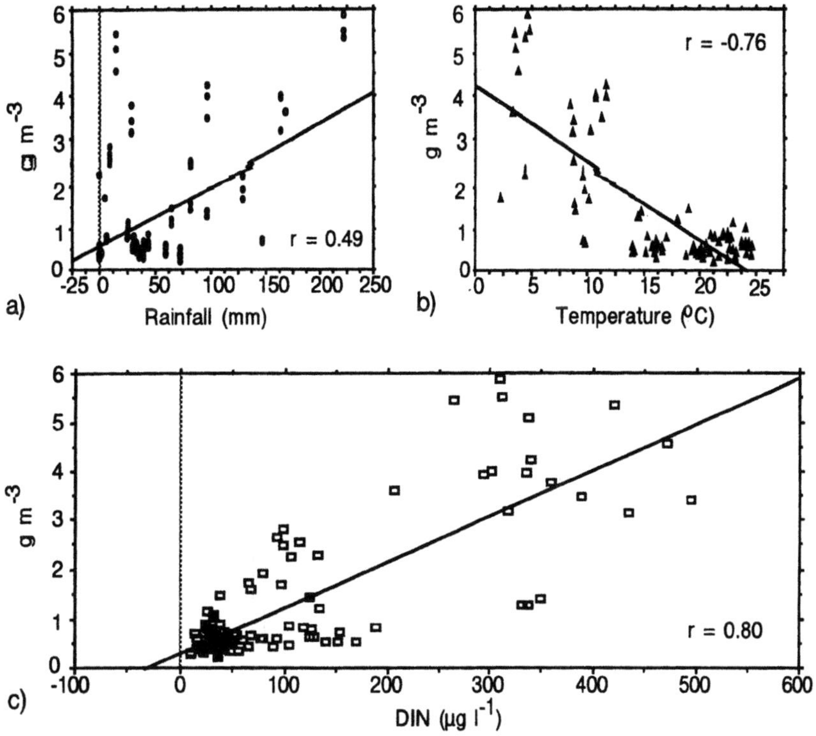

Figure 9. Linear regression of nanoplankton biomass on (a) precipitation, (b) water temperature and (c) dissolved inorganic nitrogen.

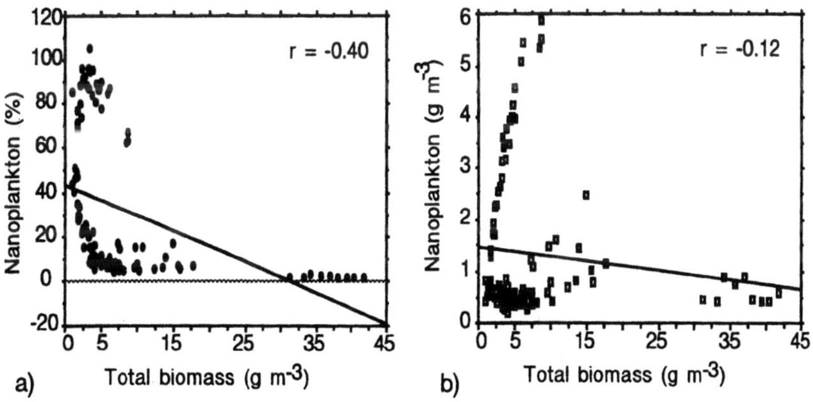

Figure 10. Linear regression of (a) percentage nanoplankton contribution to the total biomass and (b) nanoplankton biomass against total phytoplankton biomass.

during the second. The most important cyanophyte species were *Microcystis aeruginosa* and *Microcystis wesenbergii* which contributed a great deal to the pea soup appearance of the water during the warm period. The species *Anabaena lemmermannii* var. *minor*, *Aphanocapsa elachista* var. *conferta* and *Microcystis viridis* increased substantially, but to a lesser extent (Figure 5a, b, c). Higher biomass values were always observed during summer and autumn (Figure 6). However, maximum values (34 470 mg m^{-3}) were recorded in October 1990, whereas during the subsequent years of the study the biomass values were lower (5227 and 7660 mg m^{-3} in the two years, respectively). In terms of abundance, the most important species were *Aphanocapsa elachista* with 50.5×10^5 cells ml^{-1} in October 1990 and *Cyanodictyon imperfectum* with 6.7×10^5 cells ml^{-1} in May 1990.

An increase of cyanophyte biomass was observed at temperatures greater than 16 °C and a decrease at temperatures below 14 °C. Moreover, during the period of high growth rates of cyanophytes, nitrogen fluctuated at low concentrations (<100 μg l^{-1}), whereas the

Table 2. Mean size values of some phytoplankton species. (*) = nanoplanktonic species.

	Diameter μm	Length μm	Width μm	Height μm	Cell volume μm^3	Colony diameter μm	Thread length μm	Thread volume μm
A. elachista	1.64				2.33	12–80		
A. cf incerta	1.25				1.1	30–85		
A. clathrata		1.75	0.9		0.82	20–350		
A. minutissima		1.5	0.9		0.64	15–65		
C. imperfectum	0.95				0.45	20–120		
C. planktonicum		1.28	0.71		0.36	20–70		
M. aeruginosa	4.84				59.3	85–250		
M. viridis	5.53				88.39	40–150		
M. wesenbergii	4.97				68.81	35–300		
R. geminata		2.4	1.87		48.26	30–60		
S. lacustris		2.8	2.3		7.76	15–35		
A. lemmermannii		5.83	3.64		27.83	>100		
A. mucosa	5.79				101.48	>100		
* *S. cf. spinosus*		6.45	2.42		20.3	<20		
* *R. cf. wildemanii*	3.57				24	10–40		
C. reticulatum	4.72				55.33	10–50		
P. duplex				11.65	50811(1)	60–90		
P. cf. circumcreta	1.57						32.64(2)	609.17
A. cf. solitaria	6.42						275.56	8906.9
A. gracile	3.36						214.16	1902.24
P. limnetica	2.05						386.79	1337.64
A. granulata	2.21						136.22	7184.91
A. gr. angustissima	3.29						137.27	1184.63
* *O. cf. parva*		8.74	5.56		150.12			
* *C. acutum variabile*	43.59	3.51		260.98				
* *M. cf. nanum*		7.35	3.46		46.52			
* *Ochromonas* sp.3		7.9	5.08		107.24			
* *C. cf. sphaeridia*	9.35				428.44			
* *C. cf. reflexa*		22.79	10.96		1466.86			
* *C. marssonii*		15.5	7.8		370.32			
* *R. minuta nannoplanktica*	8.49	4.03		55.4				
* *C. parva*		3.56	2.17		14.79			
* *C. ocellata*	8.26			3.8	147.8			
* *C. cf. distinguenda*	23.69			5.93	2617.71			
C. hirundinella		170	50.5	25	43928.5			
C. apiculata		42.74	35.59		28672.23			
Peridinium spp.	43.19				42209.45			

(1) = volume of colony, (2) = diameter of loop.

highest biomass values were observed simultaneously with the highest phosphate phosphorus concentrations. According to Tilman et al. (1986), many cyanophyte species are good competitors for nitrogen but poor competitors for phosphorus, their better competitive ability being revealed at temperatures above 20 °C. The above statement seems to be confirmed by observations on Lake Mikri Prespa. The highest biomass during 1990 coincided with a period of very dry weather, a situation also observed in another northern Greek lake (Moustaka-Gouni & Tsekos, 1989).

Cyanophytes ability to grow under low concentrations of nitrogen can be clearly demonstrated by the high biomass of the nitrogen fixing *Anabaena lemmermannii* var. *minor*, *Anabaena* cf. *solitaria* and *Aphanizomenon gracile* in summer and autumn 1990 (when the nitrogen:phosphorus ratio was lower than 13). The absence of any increase in 1991 coincided with heavy rainfall whereas their reappearance in 1992 at higher values of nitrogen:phosphorus ratio (5–40) coincided with almost no precipitation. According to Padisak et al. (1990), storms have a negative effect on growth of thread-forming cyanophytes, a fact also observed by other researchers (Moustaka-Gouni, 1993). In Lake Mikri Prespa, the rainfall during August–November 1991 (274 mm) was higher than that during the corresponding period of 1990 (178 mm), whereas during August and September 1992 rainfall was insignificant (1 mm).

The high biomass of cyanophytes in the lake can also be attributed to their ability to increase under low light conditions (Tilman et al., 1986). According to Oskam (1978), when the ratio of the euphotic to mixing depth is less than 0.4, light becomes a limiting factor for growth. During summer and autumn 1990 this ratio was very low in Mikri Prespa (0.25–0.43) and since the circulation of the water column was constant, the algae had to spend long periods of their life time in the dark. On the contrary, in 1991 and 1992 when this ratio was higher (0.38–1.08 and 0.29–2.25, respectively) cyanophytes did not develop as much as in 1990.

Moreover, the reduced grazing pressure from zooplankton, since most of cyanophytes formed colonies and threads too big to be easily grazed, and the relatively high pH values of the water may have functioned as supplementary selecting factors (Shapiro, 1973, 1984).

Diatoms

Diatoms, the second important algal group in terms of biomass, had their highest biomass values in winter and spring (4971 mg m^{-3} in December 1990, 2406 mg m^{-3} in March 1992, Figure 6). The most important representative of the group was the small centric *Cyclotella ocellata* which remained dominant from November till June (Figure 5d). The percentage contribution of diatoms fluctuated from 14 to 30% in the two years, respectively. The increase of the group at high nitrogen concentrations and low temperatures agrees with the findings of Tilman et al. (1986), who stated that diatoms are good competitors for phosphorus but poor competitors for nitrogen, whereas their better competitive ability is gained at low temperatures. The relationship between diatom biomass and DIN was positive ($r=0.64$, $n=93$, $p< 0.05$) while a significant negative correlation was observed between diatom biomass and temperature ($r=-0.73$, $n=93$, $p<0.05$). Fluctuations in silicon concentrations significantly influenced the changes in diatom biomass. High concentrations of the nutrient (>2500 μg Si l^{-1}) seemed to be important for the induction of growth whereas a fall in levels (<33 μg Si l^{-1}) was followed by a decline of diatoms a month later. The lower biomass values in winter and spring 1991–1992 compared to the previous year might be attributed to the lower silicon and nitrogen concentrations and to the extended period of ice cover on the surface of the lake, which would prevent intense mixing of the water column.

Rainfall which generally influenced positively the growth of *Cyclotella ocellata*, by helping the cells to remain in suspension, seemed however to have a negative effect on the growth of *Aulacoseira granulata* and *Aulacoseira granulata* var. *angustissima*. These algae grew significantly only in 1990 during the prolonged dry period. Limitation of growth of other phytoplankton species due to high water inflow has been previously observed in other lakes (e.g. Berman & Pollingher, 1974).

Other plankton groups

Chlorophytes always developed from the end of summer till autumn, the highest biomass values (659 mg m^{-3}) being recorded in September 1991 (Figure 6). Their annual percentage contribution remained low during the study (1.0–6.5%). Among the group's representatives only *Coelastrum reticulatum* and *Radiococcus* cf. *wildemannii* dominated for a short time in the phytoplankton community. *Coelastrum reticulatum* was only sporadically recorded during 1990 when *Aphanocapsa elachista*, *Microcystis viridis*, *Aulacoseira granulata* and *Aulacoseira granulata* var. *angustissima* developed substantially, whereas its increase in the following years coincided with lower biomass values of the latter four species. This situation might indicate a case of allelopathy between these algae. According to Tilman et al. (1986), chlorophytes increase under high temperatures and dominate in lakes when nitrogen, silicon or light are limiting for growth. In Lake Mikri Prespa, the group's biomass revealed a slight positive correlation with temperature and a negative one with DIN ($r = 0.398$ and -0.354, respectively, $n = 93$, $p < 0.05$), whereas light penetration expressed as euphotic to mixing depth was relatively low during the time of chlorophyte development.

Dinophytes also developed during summer and autumn and exhibited higher biomass values in 1990 (1831 mg m^{-3}, Figure 6). Their annual percentage contribution remained low (0.36 and 1.77% in the two years, respectively). The group's seasonal pattern of distribution was influenced mainly by the biomass fluctuations of *Ceratium hirundinella* which appeared to be the main representative. On the contrary to the available data that correlate the increase of this species with the development of thermal stratification (e.g. Reynolds, 1982; Sommer et al., 1986), the development of the thermocline in June and July 1991 in Lake Mikri Prespa, seemed to have a negative effect on *Ceratium hirundinella* and on other dinophytes (*Peridinium cinctum* and *Peridinium gatunense*). An increase in *Ceratium hirundinella* under mixing conditions was also reported by Klemer & Barko (1991) and Nicholls et al. (1980). Dinophytes revealed a slight negative correlation with DIN ($r = -0.35$, $n = 93$, $p < 0.05$) which might indicate their good competitive ability in absorbing the nutrient. Generally, dinophytes are negatively correlated with SRP concentrations (Pollingher, 1981; Pant et al., 1985) rather than with DIN. It is possible that in Lake Mikri Prespa phosphate-phosphorus was in excess for their growth.

Cryptophytes did not show any clear seasonal pattern and exhibited higher biomass values in summer 1990 (500 mg m^{-3}) and spring 1991 (1177 mg m^{-3}, Figure 6). *Cryptomonas* cf. *reflexa* appeared to be most important in terms of biomass, whereas *Rhodomonas minuta* var. *nannoplanktica* dominated in terms of abundance. Cryptophytes are known to develop substantially under low temperature and light conditions (Brown & Richardson, 1968, from Morgan & Kalff, 1979). Indeed, the group's increase under corresponding physical conditions in Lake Mikri Prespa (December 1990–March 1991) verify the above findings. On the contrary, the low biomass values during spring 1992, when the ice layer was thicker and lasted for two months, might indicate that low light intensity for long periods limits many of these species (except *Rhodomonas minuta* var. *nannoplanktica*). The increased rainfall during winter and spring 1990–1991 might have supported the vernal increase of cryptophytes, since storms are believed to enhance the growth of the small r-selected species (Padisak et al., 1990).

Generally, cryptophytes are believed to increase during periods of disintegration of other algal groups possibly due to their ability to assimilate dissolved organic substances (Steward & Wetzel, 1985). In Lake Mikri Prespa a positive effect of inorganic nutrients ($r = 0.33$ and 0.61 for SRP and DIN, respectively, $n = 93$, $p < 0.05$) was observed. The drop in biomass in August 1990 and April 1991 coincided with high numbers of Ciliates and Rotifers (especially *Polyarthra* and *Keratella*) which are believed to apply high grazing pressure on small cryptophytes (Nauwerck, 1963; Klaveness, 1989).

Chrysophytes did not reveal any annual pattern of abundance, a situation that might indicate their opportunistic strategy of exploiting their environment. Highest biomass values were reported in February 1991 (602 mg m^{-3}). Increased growth in winter seems to be relatively common in temperate lakes (e.g. Steward & Wetzel, 1985). The peaks of *Chromulina* cf. *sphaeridia* and *Ochromonas* sp. observed in February and June 1991, respectively, coincided with a reduction in the biomass of other groups possibly due to increased grazing. This fact might indicate that the two species have a very high growth rate, sufficient to overcome the grazing pressure on them.

Haptophytes were represented only by the small-celled *Chrysochromulina parva*. This species had high biomass values in March 1991 (526 mg m^{-3}, Figure 6) but was constantly present in the plankton. It seems that the combination of some environmental factors such as high nitrogen concentrations, low temperature and high rainfall had a positive effect on its growth.

Temporal distribution of total biomass

Seventy-six phytoplankton species that were abundant enough to be counted with the usual methodology composed the total phytoplankton biomass in Lake Mikri Prespa. Fifteen were recorded as being dominant (comprising more than 10% of the total biomass, Sommer, 1981) whereas for the remaining, lower values of fresh weight were determined. The temporal distribution of total biomass revealed a trend towards a double-peaked distribution pattern with one peak in autumn, composed mainly of cyanophytes and another in spring, comprised mainly of diatoms and cryptophytes (Figure 7). Maximum values were recorded in October 1990 (38.06 g m^{-3}), whereas during the rest of the study period the peaks of biomass remained lower (6.39 g m^{-3} in 1991 and 9.57 g m^{-3} in 1992).

According to Vollenweider (1968), eutrophic lakes reveal peaks of biomass with values higher than 10 g m^{-3}, whereas according to measurements by Rott in the vicinity of Tyrolo (Rott, 1984), the mean annual biomass in this type of lakes ranges between 0.5–2 g m^{-3}. Reynolds (1984a, b) and Rott (1984) have also emphasized the increased abundance of cyanophytes and especially of *Microcystis* in late summer and of diatoms in spring in eutrophic water systems. According to the above observations, the mean annual biomass values (15.0 and 3.2 g m^{-3} in 1990–91 and 1991–92, respectively), the height of the maxima and the interchange between diatoms and cyanophytes classify Lake Mikri Prespa among the eutrophic lakes. The trend, however, towards two annual maxima contradicts the model for eutrophic, stratified, temperate lakes which usually exhibit a third biomass peak in summer (Sommer et al., 1986). This disagreement seems to be attributed to the shallowness of the lake and to the absence of stable thermal stratification.

After the second fortnight of May 1991, the total biomass dropped to the lowest observed values (1.3 g m^{-3}). This decrease, which lasted for one and a half months, was followed by an increase in water transparency (except on 5th of June due to heavy rainfall, Figure 1) and is known as the 'clear water phase'. This phase is attributed to high grazing pressure by zooplankton (Sommer et al., 1986). In 1992 this phase was not detected in Lake Mikri Prespa.

Nano- and netplankton

Nano- and netplankton contributed almost equally to the total phytoplankton biomass of Lake Mikri Prespa with nanoplankton comprising 38.9% and 49.9% of the total biomass in 1990–91 and 1991–92, respectively. Netplankton biomass formed peaks in autumn whereas nanoplankton peaked in spring (Figure 8). The maxima of netplankton in autumn consisted mainly of cyanophytes and dinophytes, whereas these of nanoplankton were made up of diatoms and cryptophytes. The percentage contribution of nanoplankton to the total phytoplankton biomass exhibited also seasonal periodicity and revealed higher values in winter and spring (>85% from January to April, Figure 8).

Nanoplankton increased in winter and spring when low temperature, high DIN concentrations and heavy rainfall prevailed. It can be seen from Figure 9 that high levels of DIN and low temperature had a strong positive effect on nanoplankton growth, whereas the correlation between nanoplankton biomass and rainfall was also found to be positive but to a lesser extent.

Temperature is known to increase the metabolism of phytoplankton by acting in favor of large-celled phytoplankters (Reynolds, 1984a). Moreover, small-celled organisms having high growth rates (defined as r-strategists) have increased demands for nutrients compared to the larger, slower growing ones (defined as K-strategists). Indeed, in many lakes nanoplanktonic species increase mainly in spring when low temperatures usually prevail and the nutrient concentrations are high (e.g. Kristiansen, 1971; Gelin, 1975; Gaedke, 1992; Moustaka-Gouni, 1993). The influence of storms on the growth of small-celled algae has been thoroughly studied by Padisak et al. (1988). These writers stated that the positive effect of rain on the nanoplankton in shallow lakes could be attributed to the same reasons as that of spring circulation.

Based on the above, the high nanoplankton biomass during winter and spring in Lake Mikri Prespa must have been influenced significantly by low temperature, high DIN concentrations and increased rainfall (Figure 9). The higher nanoplankton biomass in 1991 compared to that of 1992, could be attributed to the higher levels of DIN and rainfall of that winter.

Although nanoplankton biomass seems to be positively influenced by nutrients, a high biomass of the group has been often observed in oligotrophic lakes (Morris, 1980; Echevarria et al., 1990), or after a decrease in nutrient levels (Cronberg, 1982), or at limiting concentrations of SRP (Pollingher, 1981). In spite of the conflicting points of view, the quantity of nanoplankton has been often used as an indicator of a lake's trophic state. Pavoni (1963) stated that a very high percentage contribution of nanoplankton (100%) was observed in oligotrophic and hypertrophic waters in lakes of Switzerland, whereas lower values (0–25%) were observed in eutrophic ones. Kristiansen (1971) also recorded a mean annual contribution of nanoplankton of 25% in eutrophic waters of Denmark. According to Watson and Kalff (1981) nanoplankton biomass increases linearly with the increasing trophy of a lake but its percentage contribution decreases.

The linear regression of nanoplankton biomass and its percentage contribution to total phytoplankton biomass in Lake Mikri Prespa is presented in Figure 10. This figure shows that the nanoplankton's percentage contribution decreased with the increase in total biomass but that its absolute values were not significantly correlated with the total biomass.

Acknowledgments

This research was supported by grant No. 259533 6/2/90 from the Greek Ministry of Agriculture. The authors would also like to thank Mrs A. Kiratzidou-Dimopoulou for her valuable comments on the English language.

References

Allen, S. E., 1989. Chemical analyses of ecological materials. Black. Sci. Publ., 367 pp.

Berman, T. & U. Pollingher, 1974. Annual and seasonal variations of phytoplankton, chlorophyll, and photosynthesis in Lake Kinneret. Limnol. Oceanogr. 19: 31–54.

Cronberg, G., 1982. Phytoplankton changes in Lake Trummen induced by restoration. Long-term whole-lake studies and food-web experiments. Folia Limnol. Scand. Lund, 119 pp.

Echevarria, F., P. Carrillo, F. Jimenez, P. Sanchez-Castillo, L. Cruz-Pizarro & J. Rodriguez, 1990. The size-abundance distribution and taxonomic composition of plankton in an oligotrophic, high mountain lake (La Caldera, Sierra Nevada, Spain). J. Plankton Res. 12: 415–422.

Eloranta, P., 1986. The phytoplankton of some subarctic subalpine lakes in Finnish Lapland. Mem. Soc. Fauna Flora Fenn. 62: 41–57.

Eloranta, P. & A. Palomaki, 1986. Phytoplankton in Lake Konnevesi with special reference to eutrophication of the lake by fish farming. Aquat. Fenn. 16: 37–45.

Gaedke, U., 1992. The size distribution of plankton biomass in a large lake and its seasonal variability. Limnol. Oceanogr. 37: 1202–1220.

Gelin, G., 1975. Nutrients, biomass and primary productivity of nanoplankton in eutrophic Lake Vombsjon, Sweden. Oikos 26: 121–139.

Gounis, N. & M. Moustaka-Gouni, 1990. Simple sampler for the research of vertical distribution of phytoplankton. 3rd Panhellenic Symp. Oceanogr. Fish. Athens: 453–460 (in Greek).

Klaveness, D., 1989. Biology and ecology of the Cryptophyceae: Status and challenges. Biol. Oceanogr. 6: 257–270.

Klemer, A. & J. Barko, 1991. Effects of mixing and silica enrichment on phytoplankton seasonal succession. Hydrobiologia 210: 171–181.

Kousouris, T. & A. Diapoulis, 1983. Utilization and protection of land freshwaters I Lake Mikri Prespa. Institute of Oceanographic and Fisheries Research. Special Publ. 6: 1–89 (in Greek).

Kristiansen, J., 1971. Phytoplankton of two Danish lakes, with special reference to seasonal cycles of the nanoplankton. Mitt. int. Ver. Limnol. 19: 253–265.

Liddicoat, M. I., S. Tibbits & M. I. Butler, 1976. The determination of ammonia in natural waters. Wat. Res. 10: 567–568.

Lund, J. W., G. Kipling & E. D. Le Cren, 1958. The inverted microscope method of estimating algal numbers and the statistical basis of estimations by counting. Hydrobiologia 11: 143–170.

Morgan, K. C. & J. Kalff, 1979. Effect of light and temperature interactions on growth of *Cryptomonas erosa* (Cryptophyceae). J. Phycol. 15: 127–134.

Morris, L., 1980. The physiological ecology of phytoplankton. University of California Press, 491 pp.

Mourkides, G., G. E. Tsikritsis, S. E. Tsiouris & U. Menkisoglou, 1978. The Lakes of northern Greece I. Trophic status. Sci. Ann. School Agricult. For. 21: 93–131 (in Greek).

Moustaka, M., 1988. Seasonal variations, annual periodicity and spatial distribution of phytoplankton in Lake Volvi. Doctoral Dissertation, University of Thessaloniki, 230 pp (in Greek).

Moustaka-Gouni, M., 1988. The structure and dynamics of the phytoplankton assemblages in Lake Volvi, Greece I. Phytoplankton composition and abundance during the period March 1984–March 1985. Arch. Hydrobiol. 112: 251–264.

Moustaka-Gouni, M., 1993. Phytoplankton sucession and diversity in a warm monomictic, relatively shallow lake: Lake Volvi, Macedonia, Greece. Hydrobiologia 249: 33–42.

Moustaka-Gouni, M. & I. Tsekos, 1989. The structure and dynamics of the phytoplankton assemblages in Lake Volvi, Greece II. Phytoplankton biomass and environmental factors. Arch. Hydrobiol. 115: 575–588.

Moustaka-Gouni, M. & G. Nikolaidis, 1990. Phytoplankton of a warm monomictic lake – Lake Vegoritis, Greece. Arch. Hydrobiol. 119: 299–313.

Moustaka-Gouni, M., G. Nikolaidis & E. Tryfon, 1994. Chlorophytes and their ecology in three Macedonian lakes, Greece. Biologia 4: 1–10.

Nauwerck, A., 1963. Die Beziehungen zwishen Zooplankton und Phytoplankton in See Erken. Symbolae Botanikae Upsaliensis 17: 1–163.

Nicholls, K. H., W. Kennedy & C. Hammett, 1980. A fish-kill in Heart Lake, Ontario, associated with the collapse of a massive population of *Ceratium hirundinella* (Dinophyceae). Freshwat. Biol. 10: 553–561.

Oskam, G., 1978. Light and zooplankton as algae regulating factors in eutrophic Biesbosch reservoirs. Verh. int. Ver. Limnol. 20: 1612–1618.

Padisak, J., L. Toth & M. Rajczy, 1988. The role of storms in the summer sucession of the phytoplankton community in a shallow lake (Lake Balaton, Hungary). J. Plankton Res. 10: 249–265.

Padisak, J., L. Toth & M. Rajczy, 1990. Stirr-up effects of wind on a more-or-less stratified shallow lake phytoplankton community, Lake Balaton, Hungary. Hydrobiologia 191: 249–254.

Pant, M. C., S. Jaiswal & A. P. Sharma, 1985. A compositional and structural analysis of phytoplankton in Lake Khurpatal. (U.P.), India. Int. Revue ges. Hydrobiol. 70: 269–280.

Pavoni, M., 1963. Die Bedeutung des Nannoplanktons im Vergleich zum Netzplankton. Schweitz. Z. Hydrol. 25: 219–341.

Pollingher, U., 1981. The structure and dynamics of the phytoplankton assemblages in Lake Kinneret, Israel. J. Plankton Res. 3: 93–105.

Pollingher, U., 1990. Effects of latitude on phytoplankton composition and abundance in large lakes. In Tilzer, M. M. & C. Serruya (eds), Large Lakes, Ecological Structure and Function. Springer Verlag, Berlin: 368–402.

Redfield, A. C., 1958. The biological control of chemical factors in the environment. Am. Sci. 46: 205–210.

Reynolds, C. S., 1982. Phytoplankton periodicity: its motivation, mechanisms and manipulation. Freshwat. Biol. Ass., Annual Report 60–75.

Reynolds, C. S., 1984a. The ecology of freshwater phytoplankton. Cambridge Univ. Press., Cambridge, 383 pp.

Reynolds, C. S., 1984b. Phytoplankton periodicity: the interactions of form, function and environmental variability. Freshwat. Biol. 14: 111–142.

Rott, E., 1984. Phytoplankton as biological parameter for the trophic characterisation of lakes. Verh. int. Ver. Limnol. 22: 1078–1085.

Round, F. E., 1981. The Ecology of Algae. Cambridge University Press, 653 pp.

Schindler, D. W., 1977. Evolution of phosphorus limitation in lakes. Science 195: 260–262.

Shapiro, J., 1973. Blue-green algae: Why they become dominant. Science 179: 382–384.

Shapiro, J., 1984. Blue-green dominance in lakes: The role and management significance of pH and CO_2. Int. Revue ges. Hydrobiol. 69: 765–780.

Sommer, U., 1981. The role of r- and K-selection in the sucession of phytoplankton in Lake Constance. Acta Oecologica 2: 327–342.

Sommer, U., Z. Gliwicz, W. Lampert & A. Duncan, 1986. The PEG-model of seasonal succession of planktonic events in fresh waters. Arch. Hydrobiol. 106: 433–471.

Stewart, A. J. & R. G. Wetzel, 1985. Cryptophytes and other microflagellates as couplers in planktonic community dynamics. Arch. Hydrobiol. 106: 1–19.

Strickland, J. D. H. & T. R. Parsons, 1968. A practical handbook of seawater analysis. Bull. Fish. Res. Bd Can 167: 1–311.

Tilman, D., R. Kiesling, R. Sterner, S. S. Kilham & F. A. Johnson, 1986. Green, bluegreen and diatom algae: Taxonomic differences in competitive ability for phosphorus, silicon and nitrogen. Arch. Hydrobiol. 106: 473–485.

Tryfon, E., 1994. Structure and dynamics of phytoplancton assemblages in Lake Mikri Prespa. Doctoral Dissertation, University of Thessaloniki, (in Greek), 270 pp.

Tryfon, E., M. Moustaka-Gouni, G. Nikolaidis & I. Tsekos, 1994. I. Phytoplankton and physical-chemical features of the shallow Lake Mikri Prespa, Macedonia, Greece. Arch. Hydrobiol. 131: 477–494.

Utermoehl, H., 1958. Zur Vervollkommung der quantitativen Phytoplankton – Methodik. Mitt. int. Ver. Theor. Angew. Limnol. 9: 1–38.

Vollenweider, R. A., 1968. Scientific Fundamentals of the Eutrophication of Lakes and Flowing Waters with Particular Reference to Nitrogen and Phosphorus as Factors in Eutrophication. Paris, OECD. 159 pp.

Watson, S. W. & J. Kalff, 1981. Relationships between nanoplankton and lake trophic status. Can. J. Fish. aquat. Sci. 38: 960–967.

Wetzel, R. G., 1983. Limnology. 2nd edn., Sanders College Publishing, 753 pp.

Willen, E., 1976a. Simplified method of phytoplankton counting. Br. Phycol. J. 11: 265–278.

Willen, E., 1976b. Phytoplankton in Lake Hjalmaren. Acta Universitalis Uppsaliensis 387: 3–18.

Hydrobiologia **351**: 77–94, 1997.
A. J. Crivelli & G. Catsadorakis (eds), Lake Prespa, Northwestern Greece.

The zooplankton of Lake Mikri Prespa

E. Michaloudi, M. Zarfdjian & P. S. Economidis
Aristotle University of Thessaloniki, Department of Zoology, P.O. Box 134, 54006 Thessaloniki, Greece

Key words: zooplankton, species composition, endemism, seasonal variability, Greece

Abstract

A study from June 1990 to October 1992 revealed that the pelagic zooplankton community of Lake Mikri Prespa consists of 45 invertebrate species (28 Rotifera, 11 Cladocera, 6 Copepoda and 1 mollusc). It includes *A. steindachneri,* an endemic calanoid of the Western Balkans. Half of the species are first records for the lake. The total abundance and biomass ranged from 61 to 905 ind l^{-1} and 58 to 646 μg l^{-1}, respectively. Seasonal fluctuations agreed with the PEG-model. Copepods prevailed throughout the year. The species composition and seasonal variability were affected by temperature-dependent food conditions, competition and predation. These factors induced a seasonal succession which involved a replacement of larger species or individuals by smaller ones during summer. The cladoceran population, as a whole, was characterised by small size classes of the species present, probably due to fish predation. An examination of the horizontal distribution showed that, although seasonal variability was the same at all stations sampled, the central station showed higher abundance. Inshore-offshore diel migrations of juvenile fish is considered to account for this uneven distribution. Shannon's diversity index ranged from 0.98 to 2.95 and evenness from 0.3 to 0.9. Both were influenced by temperature.

Introduction

Our knowledge of the zooplankton of Lake Mikri Prespa is very limited. The first data date back to the early 1970s and are of a qualitative nature (Stathatos et al., 1972; Karvounaris, 1979; Koussouris & Diapoulis, 1983; Koutsoubidis, 1989). In contrast, no data are available concerning the quantitative aspects of the zooplankton community.

The first thorough study was carried out during 1990–92 as a part of a limnological project funded by the Greek Ministry of Agriculture. The purpose of the survey was to study the composition and distribution of the zooplankton community and their relationship to abiotic and biotic factors of Mikri Prespa lake (Michaloudi, 1997). The present paper describes the abundance and distribution of the dominant pelagic zooplankton species and their seasonal variability.

Methods

Integrated samples (0–7 m) were collected monthly during ice free seasons, at 3 stations (Figure 1), using a 2-l Niskin sampler. At least 30 l of the water column were filtered each time (through a net of 50 μm mesh size) and preserved in 4% formalin final dilution. In addition, temperature, oxygen, pH and transparency were also measured (see Tryfon & Moustaka-Gouni, 1997). For each sample (total volume of 100 ml), five counts of 1 ml subsamples were made on a Sedwick-Rafter cell. The individual dry weights for each species, sex, size class or developmental stage were estimated after weighing 30–100 individuals, dried at 60 °C for 48 h (Michaloudi, unpublished). All statistical analyses were performed on log-transformed data.

Diversity was calculated using the Shannon-Wiener diversity index H, and evenness using Pielou's J (Maguran, 1988): $H = -\Sigma p_i \times \mathrm{Log}_2 p_i$ where H is the diversity index, and p_i the relative abundance of each species in a sample. $J = H/H_{\max}$ where $H_{\max} = \mathrm{Log}_2 S$

Table 1. List of planktonic invertebrates of Lake Mikri Prespa noted during the present and previous studies. (1) Stathatos et al., 1972, (2) Karvounaris, 1979, (3) Koussouris, and Diapoulis, 1983, (4) Koutsoubidis, 1989.

	Rotifera	(1)	(2)	(3)	(4)	Present study	Indicators of eutrophy
	Anuraeopsis fissa Gosse			+			
*	*Ascomorpha ecaudis* Perty			+		+	
*	*A. saltans* Bartsch					+	
	Asplanchna priodonta Gosse					+	
	Brachionus angularis Gosse					+	†
	B. diversicornis Daday			+		+	
*	*B. forficula* Wierz.					+	
	B. havanensis Rouss.				+		
*	*B. urceolaris* Müller					+	
*	*Collotheca* sp.					+	
*	*Conochilus hippocrepis* (Schrank)					+	
	Conochilus sp.			+			
	Filinia longiseta (Ehrb.)			+		+	†
#	*Gastropus hyptopus* (Ehrb.)					+	
*	*G. stylifer* Imhof					+	
	Hexarthra mira (Huds.)			+			
	Kellicottia longispina (Kellicott)					+	
	Keratella cochlearis (Gosse)	+				+	
	K. c. v. *tecta* Gosse					+	†
*	*K. c.* v. *hispida* (Lauter.)			+		+	
	K. quadrata (Müller)			+		+	†
*	*Lecane bulla* (Gosse)					+	
*	*Lecane* sp.					+	
*	*Lepadella ovalis* (O.F.M.)					+	
*	*Notholca squamula* (Müller)					+	
	Polyarthra minor Voigt			+			
	P. vulgaris Carl.			+		+	
	Polyarthra ssp.					+	
	Pompholyx sulcata Hudson					+	†
	Synchaeta pectinata Ehrb.					+	
	Trichocerca birostris (Mink.)			+			
	T. capucina (W. & Zach.)			+		+	
*	*T. cylindrica cylindrica* (Imhof)			+		+	†
*	*T. pusilla* (Jenn.)					+	†
*	*T. similis* (Wierz.)					+	
	T. stylata (Gosse)			+			
#	*Trichotria* sp.					+	
	Mollusca						
	Dreissena polymorpha Pal.			+		+	

is the maximum possible diversity for any given number of species.

Zooplankton community

Species composition

Overall, 45 planktonic invertebrate species were identified (28 Rotifera, 11 Cladocera, six Copepoda and

Table 1. Continued.

	Cladocera	(1)	(2)	(3)	(4)	present study	indicators of eutrophy
	Acroperus harpae (Baird)		+	+			
	Alona affinis (Leydig)		+				
	A. rectangula Sars		+	+		+	
#	*Biapertura intermedia* (Sars)					+	
	Bosmina coregoni Baird				+		
	B. longirostris (O.F.M.)		+	+		+	†
	Ceriodaphnia pulchella Sars		+			+	
	C. quadrangula (O.F.M.)	+	+				
	Chydorus latus Sars		+				
	C. ovalis Kurz		+	+			
	C. sphaericus (O.F.M.)					+	†
	Daphnia cucullata Sars		+	+	+	+	†
*	*D. galeata* Sars					+	
	D. hyalina Leydig		+	+	+		
	D. pulex Leydig	+					
	Diaphanosoma brachyurum (Liev.)		+	+	+		
	Diaphanosoma cf. *mongolianum*					+	
	Eurycercus lamellatus (O.F.M.)	+					
	Graptoleberis testudinaria (Fisch.)		+				
*	*Illyocryptus agilis* Kurz					+	
	Leptodora kindtii (Focke)		+	+		+	
*	*Leydigia leydigi* (Schoedler)					+	
	Pleuroxus denticulatus Birge		+				
	P. laevis Sars		+				
	P. truncatus (Jurine)		+				
	Scapholeberis mucronata (O.F.M.)		+				
	Sida crystallina (O.F.M.)		+	+			
	Simocephalus vetulus Schoedler		+				
	Copepoda						
♦	*Arctodiaptomus steindachneri* Rich.		+	+		+	
	Cyclops strenuus Fisch.	+			+		
	Cyclops vicinus Ulj.		+	+		+	
	Diaptomus graciloides (Lillj.)	+					
	Eucyclops macruroides (Lillj.)		+	+		+	
	E. serrulatus (Fisch.)					+	
	Eudiaptomus gracillis (Sars)				+		
	Macrocyclops albidus (Jurine)		+	+		+	
	Mesocyclops leuckarti (Claus)		+	+		+	
	Thermocyclops crassus (Fisch.)		+				

* Rare in Greek lakes # new for the Greek fauna ♦ endemic in Western Balkan
† Species considered to be indicators of eutrophic conditions according to Gannon & Sternberger (1978), Geller & Müller (1981) and Pejler (1983)

one planktonic Mollusc larva: Table 1), 50% of which were recorded for the first time in Lake Mikri Prespa. The differences in the number of species recorded in previous studies and in our study (Table 1) could be either attributed to different sampling methods or to taxonomical problems. The present study concerns the pelagic zone. Hence, littoral species such as *Acroperus harpae*, *Alona affinis*, *Chydorus latus*, *Eurycer-*

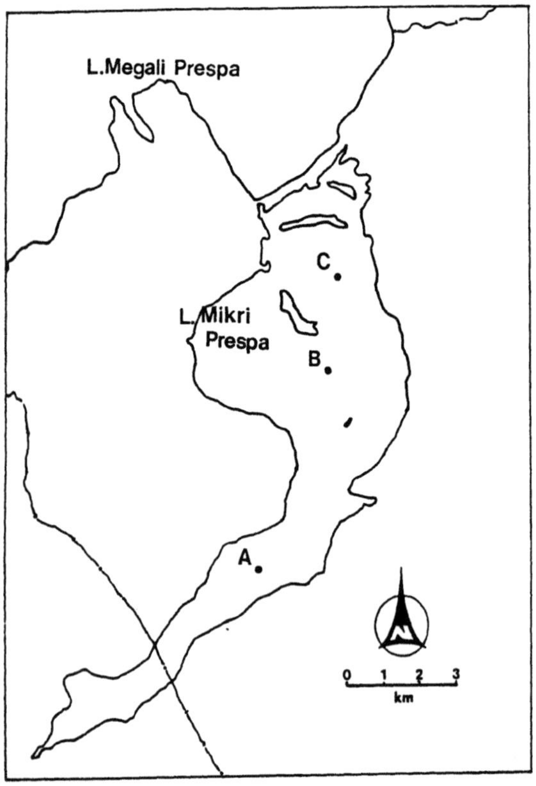

Figure 1. Map of Lake Mikri Prespa showing the location of the three sampling stations.

Table 2. Comparison of the number of euplanktonic species recorded in different lakes of the Balkan Peninsula.

Lake	Depth (m)	Trophic condition	Number of species
Ohrid[1]	286	oligotrophic	15
Megali Prespa[2]	52	>>	25
Vegoritis[3]	45	mesotrophic	28
Volvi[4]	23	eutrophic	32
Doirani[5]	9.5	>>	40
Skadar[6]	8.3	>>	45
Mikri Prespa	7	>>	33

[1] Stankovic, 1960, [2] Serafimova-Hadzisce, 1954 1958, [3] Zarfdjian unpublished, [4] Zarfdjian et al. 1990, [5] Popovska-Stankovic, 1954, [6] Gannon and Stemberger, 1981.

Table 3. Percentage participation (%) of calanoid copepods to the crustacean community and the ratio of calanoid copepods to cyclopoid copepods plus cladocerans (data refer to adults).

	%	Ratio
J 1990	9.58	0.11
J	7.10	0.08
A	3.79	0.04
S	0.17	0.0017
O	0.00	0.00
N	0.13	0.0013
D	0.45	0.0045
J 1991	7.72	0.08
F	2.34	0.02
M	0.56	0.01
A	2.83	0.03
M	13.33	0.15
J	25.25	0.34
J	47.22	0.89
A	8.33	0.09
S	13.17	0.15
0	8.63	0.09
N	29.70	0.42
D	82.19	4.62
J 1992		
F	88.04	7.36
M	86.62	6.48
A	48.64	0.95
M	9.87	0.11
J	18.15	0.22
J	7.82	0.08
A	6.91	0.07
S	11.74	0.13
O	12.13	0.14

cus lamellatus, *Graptoleberis testudinaria*, were not found. Taxonomical problems exist with *Trichocerca birostris* and *T. stylata* [possibly synonymous to *T. similis* and *T. pusilla*, respectively (Ruttner-Kolisko, 1974)], while the *Diaphanosoma* sp. in Mikri Prespa is *D.* cf. *mongolianum* (N. Korovchinsky, pers. comm.). In addition, the identification of some species could be erroneous when one considers the geographical distribution and ecological demands, i.e. *Brachionus havanensis* is only found in South America at temperatures ranging from 28 to 31 °C (Ruttner-Kolisko, 1974); *Chydorus ovalis* is encountered in acid waters of oligotrophic and mesotrophic lakes (Amoros, 1984); while *Daphnia hyalina* is a species of glacial lakes and is not found in Greece (Hrbacek et al., 1978). On the other hand, the absence of some common pelagic species from our samples (i.e. *Bosmina coregoni, Ceriodaphnia quadrangula, Daphnia pulex, Cyclops strenuus, Diaptomus graciloides, Eudiaptomus gracilis* and *Thermocyclops crassus*) cannot be explained.

Considering only the species identified in the present study, the rotifers *Gastropus hyptopus, Trichotria* sp. and the cladoceran *Biapertura intermedia* are recorded for the first time in the Greek fauna. *G. hyptopus,* a typical winter species (occurring at 6–10 °C), was found in February and March. *Trichotria* sp., a littoral rotifer occasionally found in open waters (Ruttner-Kolisko, 1974), is rarely found in the Balkans while it occurs quite frequently in North and Central Europe. In Mikri Prespa it was found in August 1992. *B. intermedia* is a benthic-littoral species and was encountered accidentally in a sample including bottom mud in September 1991.

Nineteen rotifers and five cladocerans were recorded for the first time in the lake (Table 1). Overall, the copepod community consists of 10 rather common species, six of which have been identified in the present study. *Arctodiaptomus steindachneri* is a typical endemic species of the Western Balkans and was initially described by Richard (1897) in Lake Ioannina (Greece). In Mikri Prespa it was first recorded by Karvounaris (1979) and according to our study it was present throughout the year. It has also been found in: (a) the lakes of Kerkyra island (var. *kerkyrensis*), absent during summer (Stephanides, 1948); (b) Lake Ohrid, throughout the year (Serafimova-Hadzisce, 1975); (c) Megali Prespa lake (Serafimova-Hadzisce, 1958); (d) Yliki & Paralimni Lakes, occasionally (Zarfdjian & Michaloudi, unpublished).

In Mikri Prespa ten species can be classified as indicators of eutrophic conditions (Table 1). Serafimova-Hadzisce (1958) maintained that the number of eulimnetic species increases with the trophy of the lake. From Table 2, where the number of eulimnetic species in various lakes of the Balkan Peninsula are compared, it becomes evident that: (a) this number does increase with trophy and decrease with depth and (b) in character Mikri Prespa is eutrophic.

Seasonal variability

Total zooplankton

Total abundance ranged from 61 to 905 ind l^{-1}, with a mean value of 386.9 ± 239.2, exhibiting a major peak in spring and a minor one in late summer (Figure 2a). Total biomass ranged from 58 to 646 $\mu g\ l^{-1}$, with a mean value of 233.9 ± 153.7 and displayed similar seasonal fluctuations (Figure 2b). The spring maximum is mainly attributed to rotifers, which accounted for 36–80% of the total abundance, whereas the summer peak was dominated by copepods (22–69%) and cladocerans (20–48%) (Figure 3a). In terms of biomass (Figure 3b) the situation is completely different since rotifers, because of their small body weight, make a very limited contribution to the total biomass in all seasons.

The seasonal variability in zooplankton abundance and biomass is in accordance with the PEG-model proposed for eutrophic lakes by Sommer et al. (1986). The rapid increase in rotifer numbers (Figure 2a) may be attributed to their intrinsic high fecundity, supported by favourable food and environmental conditions (Edmondson, 1965; Allan, 1976; Dumont, 1977; Lynch, 1979). At the same time, copepod abundance increased (Figure 2a), probably because of a combination of high fecundity of the overwintering population and the hatching of resting eggs (Hutchinson, 1967; Wetzel, 1983; Brandl et al., 1989). Cladocerans, on the other hand, built up to a maximum mainly during summer (Figure 2a), although there was a smaller spring peak of *Daphnia cucullata*, which followed the rotifers and copepods in time (Figure 4). Since the spring maximum consisted of filter feeders, including the main filter feeders of the population (*Daphnia cucullata, Arctodiaptomus steindachneri*) there was a consequent reduction in nanoplankton biomass (Tryfon, 1994). Following this there was a decline in herbivore numbers (Figure 4) which is thought to be reinforced by fish predation. Predation pressure leads to a shift towards smaller-sized crustaceans and rotifers, which can also be seen in Mikri Prespa (Figure 4). In autumn with the reduction of predation pressure and the increase of nanoplankton another peak with large-bodied zooplankton species is expected but this was not profound in our case. The autumn peak was mainly a continuation of the summer species extending their presence.

The annual composition of rotifers (Figure 4), which were clearly separated with each season, was mainly controlled by temperature. During summer cladocerans exhibited a specific succession pattern (Figure 4), well described by DeMott (1989). The summer population started with Daphniidae: the larger species (*Daphnia cucullata*) first, followed by the smaller one (*Ceriodaphnia pulchella*). In turn, *Diaphanosoma* started building up its population with its peak coinciding with that of temperature. The decline of Daphniidae is known as the midsummer decline and has been described by several researchers (Larsson et al., 1985; Threlkeld, 1985; Sommer et al., 1986). In addition, the overall Cladocera species com-

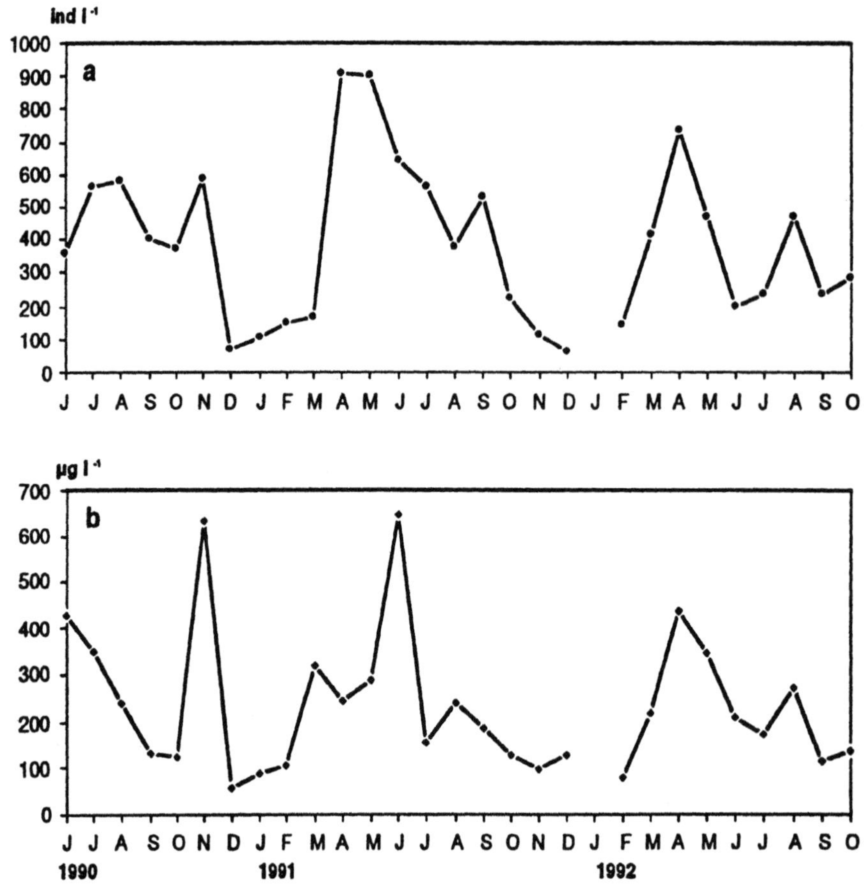

Figure 2. Seasonal variations of the total (a) abundance (ind l^{-1}) and (b) biomass ($\mu g\ l^{-1}$).

position seems to be controlled by the individual ability of each species to consume bacteria. According to Geller & Müller (1981) all the above species are highly efficient bacteriofeeders and their presence is expected during summer. However, in spring and autumn bacteriofeeders of low efficiency exist. One such species is *Bosmina longirostris*, whose appearance in Mikri Prespa (Figure 4) agrees with the above pattern.

In the crustacean community the adult calanoid copepods were relatively less abundant than adult cyclopoid copepods and cladocerans (Figure 5). The calanoids mostly comprised <10% of the crustacean population (Table 3). According to Gliwicz (1969) and Gannon and Stemberger (1978), the percentage of calanoids in the crustacean community tends to decrease with increasing trophy. Thus the ratio calanoida/(cyclopoida+cladocera) is a useful indicator of trophic conditions. This ratio in Mikri Prespa ranged from 0.0013 to 7.36 (Table 3). With the exception of winter 1992, when the highest values were recorded, the remaining values were <1 (Table 3), indicating eutrophic conditions.

Species occurrence

The zooplankton species can be classified into three main groups according to their time of appearance (Table 4): perennial, species present during warm months and species present during cold months. The remaining species occurred occasionally at very low densities.

Sex ratio

The sex ratio (% males) was examined for the dominant crustacean species only (Table 5). For *A. steindachneri* and *M. leuckarti*, the sex ratio did not differ significantly ($p>0.05$) from the theoretical 1:1. This is to be expected and implies that there are no major stress factors influencing the populations (Pourriot et al., 1982). For *D. cucullata* males were absent throughout the study period with the exception of November 1991, when their numbers were very low and the sex ratio

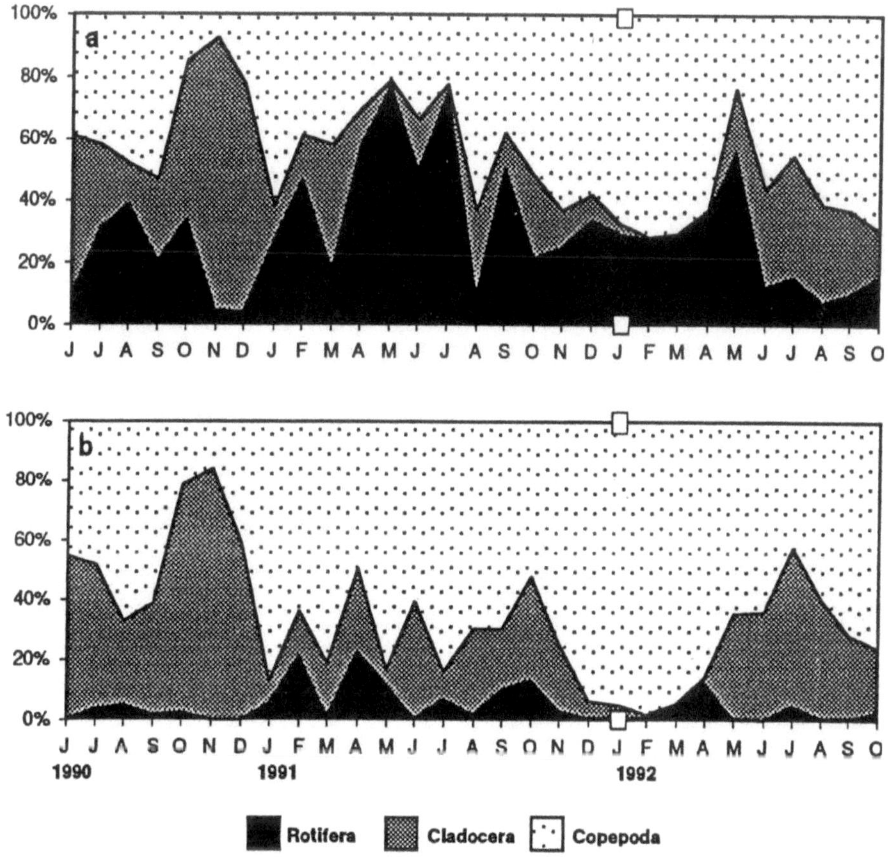

Figure 3. Seasonal variations of the percentage contribution of the main zooplanktonic groups to the total (a) abundance (ind l^{-1}) and (b) biomass ($\mu g\ l^{-1}$). Squares indicate missing values.

differed significantly ($p<0.05$) from 1:1. On the other hand, for *Diaphanosoma* males occurred in October and the sex ratio did not differ significantly ($p>0.05$) from 1:1.

Factors affecting seasonal variability

Seasonal variability must be approached in a multi-factor rather than in a single-factor framework (Bengtsson, 1987). It is affected by abiotic factors, food conditions, prey selection, competition and predation.

Abiotic factors

The correlation matrix between the 15 most important species and temperature and oxygen content is shown in Table 6, from which it is evident that most of the species exhibit a rather strong relationship with temperature but not with oxygen. Almost all crustaceans are positively correlated with temperature. In contrast, *B. longirostris* was negatively correlated with temperature, a fact most probably indicating a preference for lower temperatures. Apart from temperature, other factors (e.g. food), which co-fluctuate with temperature, also affect seasonal variability. However, temperature is an important limiting factor for the presence of species (Berzins & Pejler, 1989) also affecting several of their biological mechanisms (Edmondson, 1965; Hutchinson, 1967; Gophen, 1976; Wetzel, 1983). Moreover, abiotic factors can induce unpredictability in plankton events in a lake which deviates (e.g. shallow lakes) from the ideal temperate type of the PEG-model (Sommer et al., 1986).

Food and competition

In the spring population, the first herbivores to take advantage of the available nanoplankton biomass and to quickly build up high numbers were *K. cochlearis* and *Polyarthra.* They were followed by the calanoid

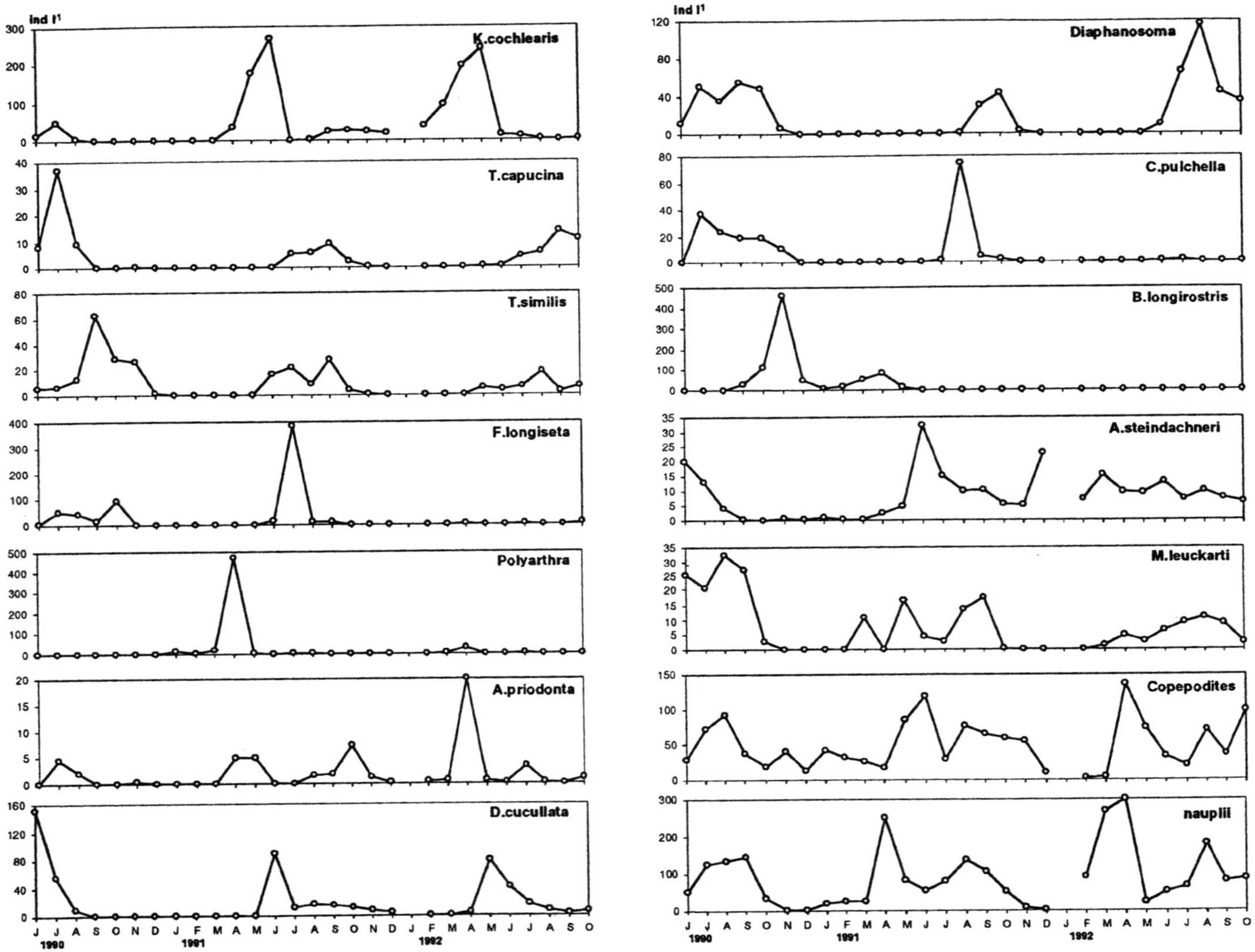

Figure 4. Seasonal abundance (ind l^{-1}) variations of the predominate species in Lake Mikri Prespa.

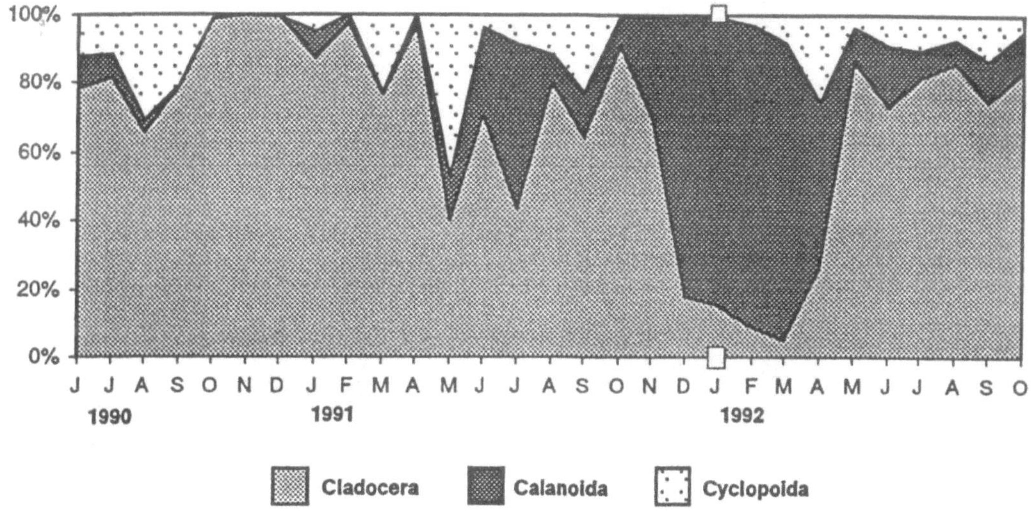

Figure 5. Percentage composition of the planktonic crustacean groups. Squares indicate missing values.

A. steindachneri and copepod nauplii and later by *D. cucullata* (Figure 4), the presence of which coincided with the phytoplankton decrease (Tryfon, 1994). Following this decrease in nanoplankton biomass other herbivores started to increase in numbers (*F. longiseta, C. pulchella, Diaphanosoma*). These probably feed on bacteria, considering the eutrophic character of the lake. According to Gliwicz (1969), bacteria can become even more important than phytoplankton in eutrophic lakes as a source of food for zooplankton. This is also supported by the PEG-model (see also Geller & Müller, 1981).

During the first study year, *T. similis* (Figure 4) took advantage of the increase in Cryptophyceae biomass (Tryfon, 1994) which, according to Pourriot (1970), is its main food item. *B. longirostris* (Figure 4) exploited the available Chlorophyceae, Cryptophyceae and *Microcystis* colonies, its main food (Gliwicz, 1977; DeMott, 1982, 1985; DeMott & Kerfoot, 1982; Jarvis et al., 1987), and reached high numbers. In the second study year none of the above mentioned food sources attained an important biomass, probably affecting the abundance of *B. longirostris*.

The decline in abundance of *K. cochlearis* coincided with the appearance of *D. cucullata* (Figure 4) which has the same food preferences, thus reducing probably the available amount of food, although predation may also be involved (see below). In addition, Gilbert & Stemberger (1985) and Gilbert (1988) noted that, apart from food competition, *K. cochlearis* populations can also be suppressed by mechanical interference competition by *Daphnia cucullata* and *Diaphanosoma* which appeared in succession (Figure 4), also demonstrate competitive interactions (Matveev, 1987; Kerfoot & DeMott, 1985). This kind of interaction can also be noted for *D. cucullata* and *C. pulchella*. When the latter started to increase, *D. cucullata* declined (Figure 4), as has also been observed by Lynch (1978). The above competitions contribute to the occurrence of the Cladocera summer succession but temperature and food composition should also be taken into account. *Daphnia* decreased when temperatures ranged between 20 °C and 25 °C. According to Lynch (1978), the feeding efficiency of *Daphnia* drops dramatically in this temperature range. Moreover, in midsummer the algal community of Mikri Prespa is dominated by Cyanophyceae (Tryfon, 1994), which being inedible, affect the growth, reproduction and survival of *Daphnia* (Porter & Orcutt, 1980; De Bernardi et al., 1981). In general, the effects of Cyanophyceae on the zooplankton community was most profound during the first study year when their high numbers resulted in the temporary disappearance from the population of otherwise perennial species such as *K. cochlearis, D. cucullata* (Figure 4).

Competitive interactions also exist between *Daphnia* and calanoid copepods, favouring the former in eutrophic conditions. Calanoids are better adapted to low food concentrations than *Daphnia* (McNaught, 1975; Muck & Lampert, 1984). However, Richman & Dodson (1983) pointed out that they also have a competitive advantage over *Daphnia* under

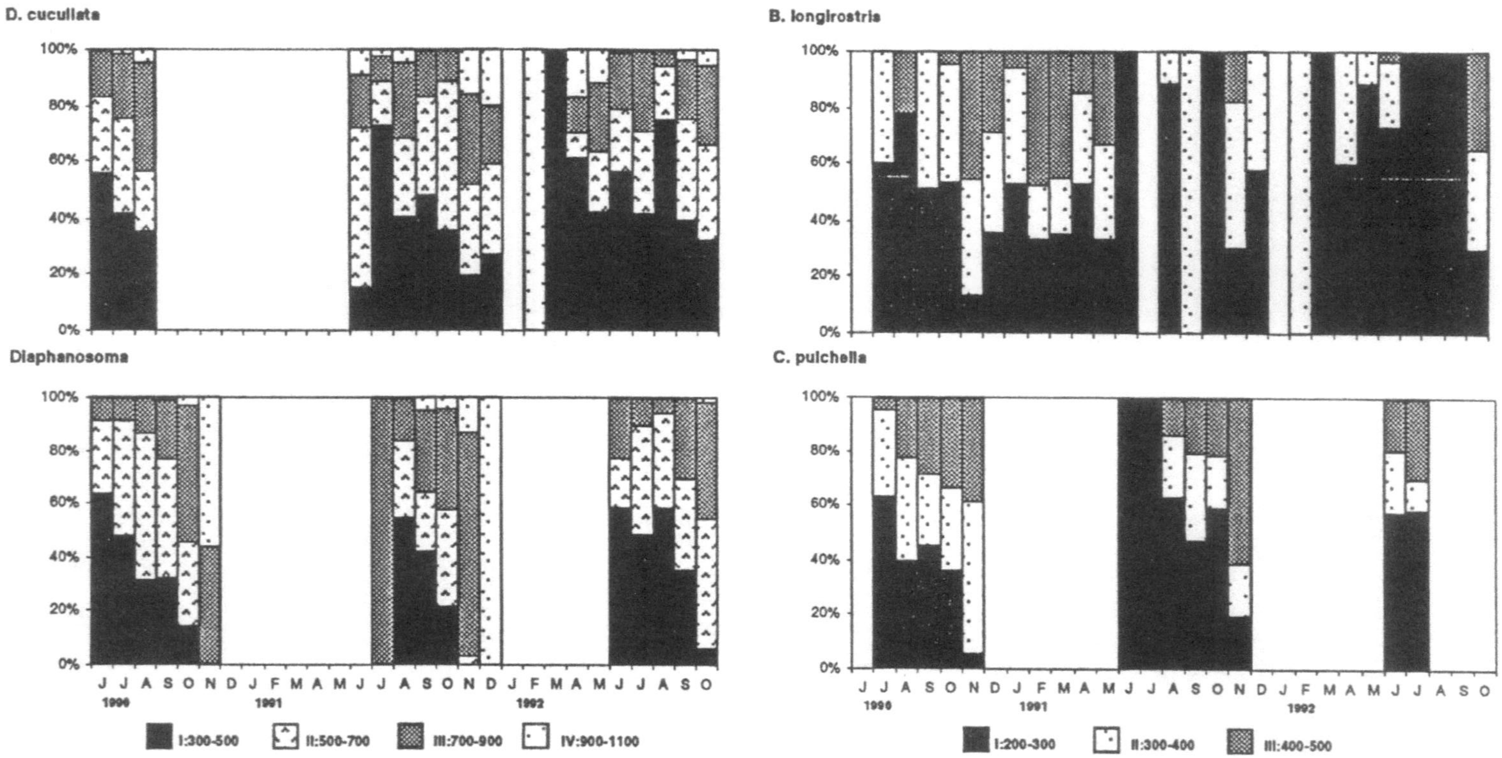

Figure 6. Seasonal changes in the percentage composition of the size classes (in μm) of the predominate cladoceran species.

Table 4. Occurence time of the zooplanktonic species of Mikri Prespa identified in the present study and the min, max and mean values of density (ind l^{-1}) and biomass ($\mu g\ l^{-1}$) for species with a regular appearance. Species for which values are not noted were found only occasionally.

													Density			Biomass		
	J	F	M	A	M	J	J	A	S	O	N	D	min	max	mean	min	max	mean
Ascomorpha ecaudis							+											
A. saltans				+	+													
Asplanchna priodonta		+	+	+	+	+	+	+	+	+	+	+	0.12	19.9	1.99	0.27	45.16	5.53
Brachionus angularis						+	+	+	+	+	+	+	0.04	122.59	4.52	0.001	3.19	0.12
B. diversicornis							+	+	+	+			0.1	13.35	1.49	0.02	2.51	0.28
B. forficula								+		+								
B. urceolaris			+															
Collotheca sp.					+		+	+	+	+								
Conochilus hippocrepis		+	+		+	+	+				+	+	0.02	497.62	17.05	0.001	21.39	0.73
Filinia longiseta			+	+	+	+	+	+	+	+		+	0.13	384.73	20.63	0.003	9.23	0.49
Gastropus hyptopus		+																
G. stylifer					+	+	+											
Kellicottia longispina						+	+											
Keratella cochlearis		+	+	+	+	+	+	+	+	+	+	+	1.18	277.91	53.12	0.003	1.39	0.25
K. quadrata			+	+		+	+											
Lecane bulla							+											
Lepadella ovalis			+															
Notholca squamula			+															
Polyarthra	+	+	+	+	+	+	+	+	+	+			0.22	471.43	18.51	0.017	42.43	1.67
Pompholyx sulcata						+	+	+	+	+	+							
Synchaeta pectinata	+	+	+	+	+							+	0.07	64.29	3.95	0.025	22.05	1.35
Trichocerca capucina					+	+	+	+	+	+	+		0.36	37.04	4.05	0.012	1.22	0.13
T. cylindrica									+	+	+	+	0.09	173.54	6.19	0.006	10.76	0.38
T. pusilla							+	+	+	+								
T. similis					+	+	+	+	+	+	+	+	0.14	61.78	8.95	0.002	1.05	0.15
Dreissena polymorpha				+	+	+	+	+	+	+	+	+	0.02	38.18	9.36	0.01	15.54	3.81
Alona rectangula											+	+						
Bosmina longirostris	+	+	+	+	+	+	+	+	+	+	+	+	0.46	461.42	26.42	0.04	506.79	24.24
Ceriodaphnia pulchella						+	+	+	+	+	+		0.15	74.92	6.19	0.08	34.02	2.89
Chydorus sphaericus					+													
Daphnia cucullata		+	+	+	+	+	+	+	+	+	+	+	0.02	151.1	19.61	0.07	246.05	31.61
D. galeata		+		+							+							
Diaphanosoma cf. *mongolianum*						+	+	+	+	+	+	+	0.04	115.48	21.34	0.1	80.0	14.89
Leptodora kindtii					+	+	+	+	+	+	+		0.12	0.93	0.11	0.49	12.35	1.32
Leydigia leydigi					+													
Arctodiaptomus steindachneri	+	+	+	+	+	+	+	+	+	+	+	+	0.31	14.02	4.01	0.77	162.27	40.75
Cyclops vicinus		+	+								+	+	0.02	5.17	0.18	0.53	130.88	4.39
Eucyclops macruroides	+	+																
E. serrulatus	+	+			+	+						+						
Macrocyclops albidus					+							+						
Mesocyclops leuckarti			+	+	+	+	+	+	+	+			0.08	32.21	7.76	0.08	53.91	12.11

extreme eutrophic conditions when large proportions of Cyanophyceae appear in the algal community. Our findings show that although there is a general prevalence of cladocerans over calanoids (Figure 5), *A. steindachneri* and *D. cucullata* do coexist. In general, the possibility of coexistence, increases when two competing species have the same ability to compete and the same food requirements (Ghilarov, 1984).

Table 5. Sex ratio (% males) of the dominant crustaceans and the χ^2 test values. For cladocera, values are given only for the months males were present. * indicates a deviation from the 1:1 sex ratio at the 95% significance level (Elliot, 1983).

	A. steindachneri		*M. leuckarti*		*Diaphanosoma*		*D. cucullata*	
	% males	χ^2	% males	χ^2	% males	χ^2	% males	χ^2
J 1990	50.00	0.00	43.48	0.43				
J	46.51	0.06	65.02	1.91				
A	39.57	0.17	74.13	7.50*				
S	0.00	0.22	70.49	4.55 *				
O	–	-	70.17	0.44	20.13	17.45 *		
N	56.83	0.01	–	–				
D	0.00	0.22	–	–				
J 1991	66.67	0.10	–	–				
F	0.00	0.48	–	–				
M	100.00	0.39	63.03	0.72				
A	100.00	2.36	–	–				
M	100.00	4.76 *	28.57	3.06				
J	56.38	0.52	38.75	0.22				
J	40.00	0.60	56.84	0.05				
A	57.44	0.21	45.31	0.11				
S	39.86	0.42	53.24	0.07				
O	73.34	1.17	0.00	0.25	23.62	11.91 *		
N	54.90	0.05	0.00	0.09	44.81	0.03	2.63	4.51 *
D	45.42	0.19	–	–				
J 1992								
F	47.33	0.02	100.00	0.08				
M	49.40	0.002	33.45	0.16				
A	41.99	0.23	64.17	0.39				
M	55.07	0.09	53.59	0.01				
J	46.89	0.05	65.04	0.57				
J	76.25	1.92	38.96	0.45				
A	61.33	0.51	46.55	0:05				
S	68.57	1.05	51.82	0.01				
O	50.00	0.00	47.32	0.006	41.28	0.53		

However, this cannot be confirmed for Mikri Prespa since there is no available information on the feeding habits of *A. steindachneri*. Nevertheless, coexistence of Calanoida and *Daphnia* has also been recorded in other lakes of the world (Herzig, 1979; Gannon et al., 1984; Zankai & Ponyi, 1986).

Invertebrate predation

The most important invertebrate predators in Mikri Prespa were: *A. priodonta*, *L. kindtii* and *M. leuckarti*. Their main prey are: (a) for *A. priodonta*: *Keratella* and *Polyarthra*; (b) for *L. kindtii*: *Conochilus, Daphnia, Diaphanosoma* and *Bosmina* and (c) for *M. leuckarti*: *Diaphanosoma* and *Ceriodaphnia* (Comita, 1972; Ejsmont-Karabin, 1974; Karabin, 1974; Dumont, 1977; Gophen, 1977; Pourriot et al., 1982, Herzig & Auer, 1990).

From the seasonal fluctuations of the predators and prey (Figure 4, Table 4) one could speculate that the most affected prey were *Polyarthra, Conochilus, Daphnia* and *Diaphanosoma*.

Fish predation

B. longirostris, in contrast to the remaining cladocerans attained maximum density in winter (Figure 4). The same is also true of Lake Volvi (Greece) and was attributed to fish predation avoidance (Zarfdjian, 1989). *C. vicinus* (Table 4) also appeared in winter and

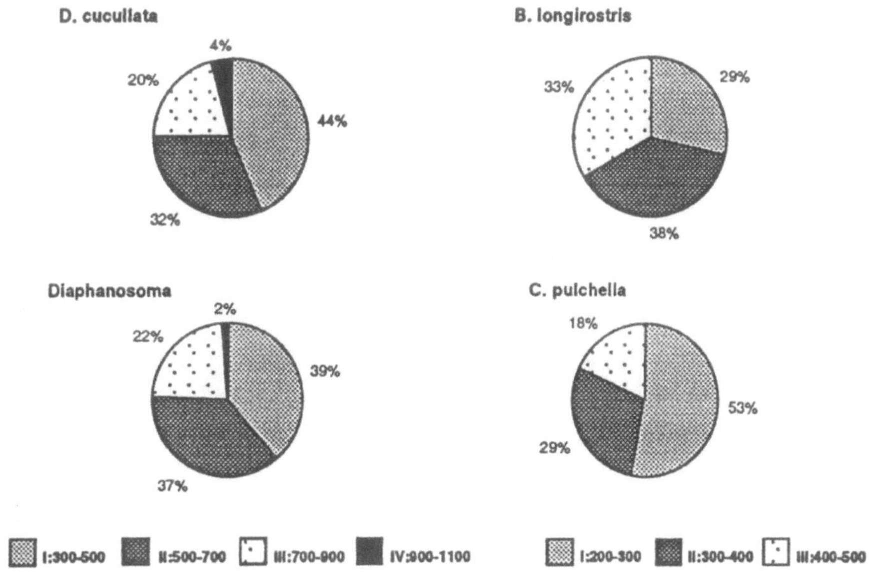

Figure 7. Percentage contribution of the size classes (in μm) to the mean total abundance (ind l^{-1}) of each predominate cladoceran species.

this winter appearance may also be attributed to fish predation avoidance (see also Maier, 1989).

Most freshwater fish feed on zooplankton at some stage of their life. In Mikri Prespa, the reproduction of the most important fish (*Alburnoides bipunctatus prespensis, Rutilus ohridanus prespensis, Chalcaburnus belvica, Leuciscus cephalus, Cyprinus carpio, Carassius auratus gibelio, Cobitis meridionalis*) takes place in late spring and early summer. At the time of hatching there is a decline in the numbers of herbivores such as *K. cochlearis, Polyarthra* and *D. cucullata* (Figure 4). Such a decline may also be the result of predation by newly hatched fish. In addition to these species, nauplii, which may almost exclusively become food for fry (Keast, 1980; Hartmann, 1983), and early copepodite stages are also affected. This is particularly evident when comparing the number of the developmental stages (300 ind l^{-1} nauplii, 150 ind l^{-1} copepodites) with that of adults (32 ind l^{-1}). It is at this time that the appearance of smaller cladoceran species (e.g. *C. pulchella*) occurs (Figure 4). This is considered to be the indirect effect that planktivores have on the cladoceran population, which alters the competition forces within the community (De Bernardi et al., 1987).

On the other hand, fish predation contributes significantly to the population structure through size selective predation (Brooks & Dodson, 1965; Zaret, 1980; Lampert, 1988). In Mikri Prespa the population of cladocerans, during summer, consisted of the small size classes of the predominant species while during winter the percentage of larger individuals increased. The populations of the two small species *B. longirostris* and *C. pulchella* consisted of individuals <300 μm in summer and of larger individuals (300–500 μm) in winter (Figure 6). The small size classes made up 29% and 53% of their mean total abundance, respectively (Figure 7). *Diaphanosoma* and *D. cucullata* showed a similar pattern, with more individuals <700 μm during summer and more relatively larger ones (700–1100 μm) in winter (Figure 6), although, smaller individuals dominated throughout the year. The small size classes constituted 76% and 75%, respectively of their mean total abundance (Figure 7). The same was also true of Lake Volvi (Zarfdjian, 1989).

Horizontal distribution

The seasonal fluctuations in the total abundance for rotifers, cladocerans and copepods did not differ spatially (Figure 8).

One-way analysis of variance indicated significant ($p<0.05$) differences between the mean density of rotifers and copepods with station but not for cladocerans (Table 7). Since no spatial differences were found for water physicochemical factors and phytoplankton biomass (Tryfon, 1994), the explanation may possibly lie in predation. Juvenile fish undertake an inshore-

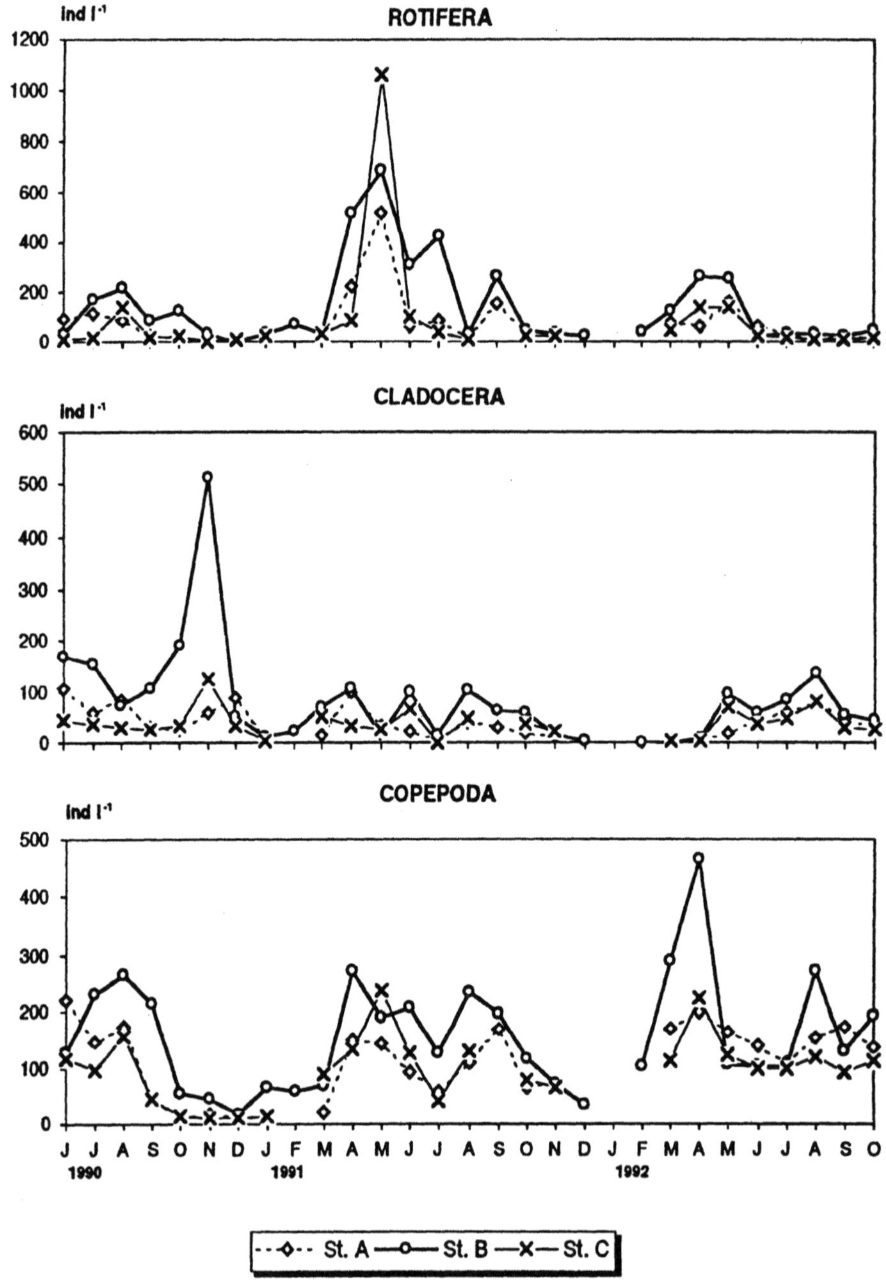

Figure 8. Monthly changes in the total abundance (ind l^{-1}) of rotifers, cladocerans and copepods at the three sampling stations.

offshore migration as a result of the possible threat caused by the presence of higher predators (such as *Anguilla anguilla* and birds) in a lake (Gliwicz & Jachner, 1992). Hence, planktivorous juvenile fish from the littoral zone may migrate up to a certain distance (600–800 m) offshore. Since station B is located some 2 km from the shore whereas station C 1 km we can assume that station B is not as greatly affected by juvenile fish predation as station C. Although station A is located nearer to the shore than station B (Figure 1), its density did not differ significantly (p>0.05) from that of station B, probably because of its lack of vegetation resulting in few or no fish spawning in the surrounding area.

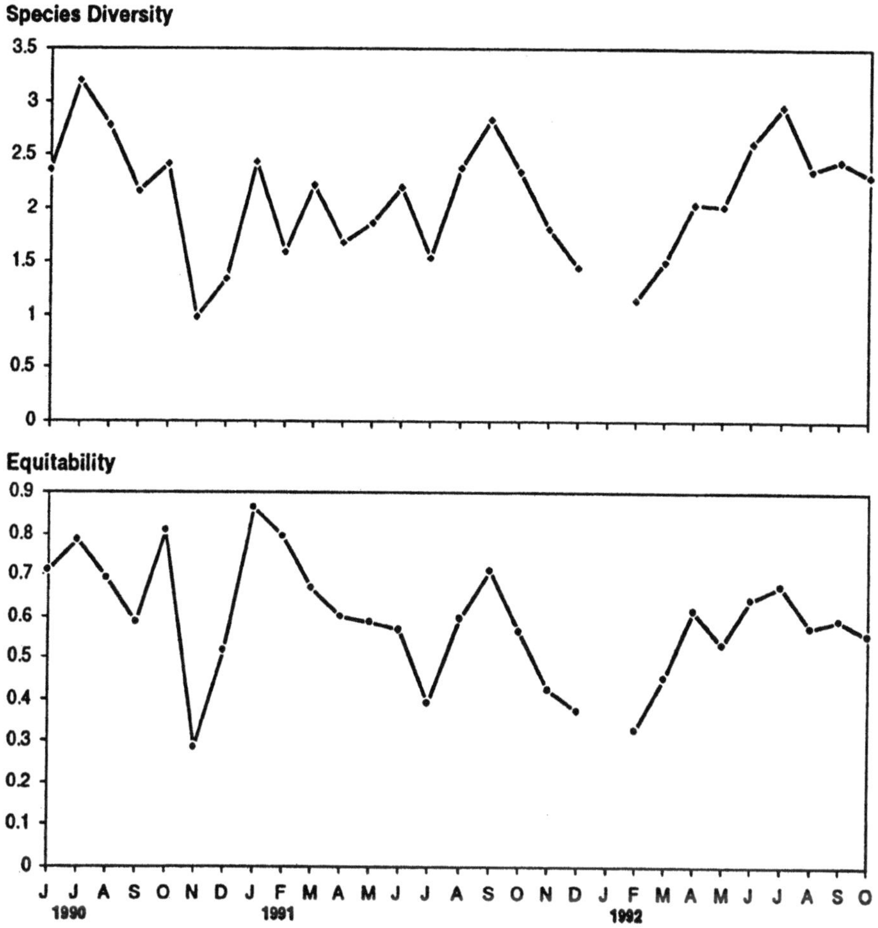

Figure 9. Monthly changes in the species diversity index H (upper plot) and species evenness index J (lower plot).

Table 6. Correlation (r) of species density with temperature and oxygen content.

	Temperature	Oxygen
K. cochlearis	−0.161	−0.133
T. capucina	0.624***	−0.146
T. similis	0.514***	−0.172*
F. longiseta	0.428***	−0.22*
Polyarthra	0.055	−0.148
A. priodonta	−0.023	0.042
D. polymorpha	0.628***	−0.036
D. cucullata	0.52***	−0.127
Diaphanosoma	0.487***	0.038
C. pulchella	0.26**	0.078
B. longirostris	−0.388***	0.085
A. steindachneri	0.282**	−0.223*
M. leuckarti	0.613***	−0.051
copepodites	0.227**	−0.112
nauplii	0.643***	0.073

* $p<0.05$; ** $p<0.01$; *** $p<0.001$

Table 7. Mean annual abundance (ind l^{-1}) for the main zooplanktonic groups at 3 sampling stations. Mean values underlined by a common line do not differ significantly. Fisher's least significant difference test, applied on log-transformed data (p=0.05).

	Mean density		
	St.C	St.A	St.B
Rotifera	83.885	75.949	140.562
Cladocera	36.525	40.424	82.409
Copepoda	97.026	110.332	155.644

Zooplankton diversity

Diversity indicates the degree of complexity in a community. It is a function of the number of species (richness) and equitability (evenness) with which the individuals are distributed among the species.

Diversity in Mikri Prespa ranged from 0.98 to 2.95 and was generally lower during autumn and winter and higher during the warm months (Figure 9). The low winter values are due to lower temperatures which in conjunction with lower food concentrations do not favour the appearance of many species and their reproduction. Hofmann (1975) and Omori & Ikeda (1984) also found a positive relationship between diversity and temperature. Furthermore, oxygen content also seems to affect diversity inasmuch as H was lower in 1991 when oxygen concentration was also lower when compared to 1990 and 1992. A positive relationship between diversity and oxygen has also been noticed by Hakkari (1978).

Evenness ranged from 0.3 to 0.9, attaining the lowest value in November 1990 and the highest one in January 1991 (Figure 9). The lower values were noted in winter, when environmental conditions favour one or two species in particularly. Following Odum's (1975) classification, Mikri Prespa may be classified as a relatively stable ecosystem not subjected to severe stress.

Acknowledgments

The authors thank the Greek Ministry of Agriculture for financial support and Prof. I. Tsekos for the realisation of the project. Thanks are due to Dr M. Vranonvsky, from the Slovak Akademy of Science, for his help with the identification of the copepod species; and to Dr N. M. Korovchinsky, from the USSR Academy of Science, for his help with the identification of *Diaphanosoma* species. The authors are also indebted to Prof. P. Larsson, from the University of Bergen, Dr K. I. Stergiou and Dr H. J. Dumont for their very helpful remarks on the manuscript.

References

Allan, J. D., 1976. Life history patterns in zooplankton. Am. Nat. 110: 165–180.

Amoros, C., 1984. Crustacés Cladocères. Bull. mens. Soc. Linn. Lyon 3/4: 63 pp.

Bengtsson, J., 1987. Competitive dominance among Cladocera: Are single-factor explanations enough? Hydrobiologia 145: 245–257.

Berzins, B. & B. Pejler, 1989. Rotifer occurrence in relation to temperature. Hydrobiologia 175: 223–231.

Brandl, Z., B. Desortova, J. Hrbacek, V. Vyhnalek, J. Seda & M. Straskraba, 1989. Seasonal changes in zooplankton and phytoplankton and their mutual relations in some Czechoslovak reservoirs. Arch. Hydrobiol. 33: 597–604.

Brooks, J. & I. Dodson, 1965. Predation, body size and composition of plankton. Science 150: 28–35.

Comita, G. W., 1972. The seasonal zooplankton cycles, production and transformations of energy in Severson Lake, Minnesota. Arch. Hydrobiol. 70: 14–66.

De Bernardi, R., G. Guissani & M. Manca, 1987. Cladocera: Predators and prey. Hydrobiologia 145: 225–243.

De Bernardi, R., G. Guissani & E. L. Pedretti, 1981. The significance of blue-green algae as food for filterfeeding zooplankton: experimental studies on *Daphnia* spp. fed by *Microcystis aeruginosa*. Verh. int. Ver. Limnol. 21: 477–483.

DeMott, W. R., 1982. Feeding selectivities and relative ingestion rates of *Daphnia* and *Bosmina*. Limnol. Oceanogr. 27: 518–527.

DeMott, W. R., 1985. Relations between filter mesh-size, feeding mode, and capture efficiency for cladocerans feeding on ultrafine particles. Arch. Hydrobiol. 21: 125–134.

DeMott, W. R., 1989. The role of competition in zooplankton succession. In Sommer, U. (ed.), Plankton Ecology. Succession in Plankton Communities, 195–252.

DeMott, W. R. & W. C. Kerfoot, 1982. Competition among Cladocerans: nature of the interaction between *Bosmina* and *Daphnia*. Ecology 63: 1949–1966.

Dumont, H. J., 1977. Biotic factors in the population dynamics of rotifers. Arch. Hydrobiol. 8: 98–122.

Edmondson, W. T., 1965. Reproductive rate of planktonic rotifers is related to food and temperature in nature. Ecol. Monogr. 35: 61–111.

Ejsmont-Karabin, J., 1974. Studies on the feeding of planktonic polyphage *Asplanchna priodonta* Gosse (Rotatoria). Ekol. Pol. 22: 311–317.

Elliot, J. M., 1983. Some methods for the statistical analysis of samples of benthic invertebrates. Freshwater Biological Association. Sci. Publ. 25, 159 pp.

Gannon, J. E. & R. S. Stemberger, 1978. Zooplankton (especially crustaceans and rotifers) as indicators of water quality. Trans. am. Micros. Soc. 97: 16–35.

Gannon, J. E. & R. S. Stemberger, 1981. Impact of influent rivers on the distribution of zooplankton in Lake Skadar. In Karaman, G. S. & A. M. Beeton (eds), The Biota and Limnology of Lake Skadar: 199–216.

Gannon, J. E., R. K. Cap, K. M. Stewart & W. A. Stewart, 1984. Zooplankton community composition in Richter Lake, Wisconsin. Verh. int. Ver. Limnol. 22: 525–532.

Geller, W. & H. Müller, 1981. The filtration apparatus of Cladocera: filter mesh-sizes and their implications on food selectivity. Oecologia 49: 316–321.

Ghilarov, A. M., 1984. The paradox of the plankton reconsidered; or, why do species coexist? Oikos 43: 46–52.

Gilbert, J. J., 1988. Suppression of rotifer populations by *Daphnia*: A review of the evidence, the mechanisms, and the effects on zooplankton community structure. Limnol. Oceanogr. 33: 1286–1303.

Gilbert, J. J. & R. S. Stemberger, 1985. Control of *Keratella* populations by interference competition from *Daphnia*. Limnol. Oceanogr. 30: 180–188.

Gliwicz, Z. M., 1969. Studies on the feeding of pelagic zooplankton in lakes with varying trophy. Ekol. Pol. 12: 663–708.

Gliwicz, Z. M., 1977. Food size selection and seasonal succession of filter feeding zooplankton in an eutrophic lake. Ekol. Pol. 25: 179–225.

Gliwicz, Z. M. & A. Jachner, 1992. Diel migrations of juvenile fish: a ghost of predation past or present? Arch. Hydrobiol. 124: 385–410.

Gophen, M., 1976. Temperature effect on lifespan, metabolism, and development time of *Mesocyclops leuckarti* (Claus). Oecologia 25: 271–277.

Gophen, M., 1977. Food and feeding habits of *Mesocyclops leuckarti* (Claus) in Lake Kinneret (Israel). Freshwat. Biol. 7: 513–518.

Hakkari, L., 1978. On the productivity and ecology of zooplankton and its role as food for fish in some lakes in Central Finland. Biol. Res. Rep. Univ. Jyväskylä 4: 3–87.

Hartmann, J., 1983. Two feeding strategies of young fishes. Arch. Hydrobiol. 96: 496–509.

Herzig, A., 1979. The zooplankton of the open lake. In Illies, J. (ed.), Neusiedlersee: The Limnology of a Shallow Lake in Central Europe, Monographiae Biologicae, Vol. 37, Dr W. Junk Publishers, The Hague: 281–335

Herzig, A. & B. Auer, 1990. The feeding behaviour of *Leptodora kindtii* and its impact on the zooplankton community of Neusiedler See (Austria). Hydrobiologia 198: 107–117.

Hofmann, W., 1975. The influence of spring circulation on zooplankton dynamics in the Plußsee. Verh. int. Ver. Limnol. 19: 1241–1250.

Hrbacek, J., V. Korinek & D. G. Frey, 1978. Cladocera. In Illies 'Limnofauna Europaea' (2nd edn.): 189–195.

Hutchinson, G. E., 1967. A Treatise on Limnology. Vol. 2. Introduction to lake biology and the limnoplankton. J. Wiley and Sons, N.Y., 1114 pp.

Jarvis, A. C., R. C. Hart & S. Combrink, 1987. Zooplankton feeding on size fractionated *Microcystis* colonies and *Chlorella* in a hypertrophic lake (Hartbeespoort Dam, South Africa): implications to resource utilization and zooplankton succession. J. Plankton Res. 9: 1231–1249.

Karabin, A., 1974. Studies on the predatory role of the cladoceran *Leptodora kindtii* (Focke) in secondary production of two lakes with different trophy. Ekol. Pol. 22: 295–310.

Karvounaris, D., 1979. Contribution to the knowledge of the planktonic Cladocera and Copepoda in ten natural lakes of Makedonia (Greece). Thesis, Univ. Thessaloniki, Greece, 158 pp. (in Greek, English summary).

Keast, A., 1980. Food and feeding relationships of young fish in the first weeks after the beginning of exogenous feeding in Lake Opinicon, Ontario. Envir. Biol. Fish. 5: 305–314.

Kerfoot, W. C. & W. R. DeMott, 1985. Interactions among cladocerans: food limitation and exploitative competition. Arch. Hydrobiol. 21: 431–451.

Koussouris, Th. & A. Diapoulis, 1983. For the development and protection of freshwater resources in Greece. I. Lake Mikri Prespa. Inst. Oceanogr. Fish. Res., Special Publ. 6, 89 pp. (in Greek, English summary).

Koutsoubidis, E., 1989. Ecological study in lakes and rivers of the prefecture of Florina. Special publication by the prefecture of Florina. 183 pp. (in Greek).

Lampert, W., 1988. The relative importance of food limitation and predation in the seasonal cycle of two *Daphnia* species. Verh. int. Ver. Limnol. 23: 713–718.

Larsson, P., G. Johnsen & A. L. Steigen, 1985. An experimental study of the summer decline in a *Daphnia* population. Verh. int. Ver. Limnol. 22: 3131–3136.

Lynch, M., 1978. Complex interactions between natural coexploiters-*Daphnia* and *Ceriodaphnia*. Ecology 59: 552–564.

Lynch, M., 1979. Predation, competition, and zooplankton community structure: An experimental study. Limnol. Oceanogr. 24: 253–272.

Maguran, A. E., 1988. Ecological Diversity and its Measurement. Groom Helm, London, 179 pp.

Maier, G., 1989. Variable life cycles in the freshwater copepod *Cyclops vicinus* (Uljanin 1985): Support for the predator avoidance hypothesis? Arch. Hydrobiol. 115: 203–219.

Matveev, V. F., 1987. Effect of competition on the demography of planktonic cladocerans-*Daphnia* and *Diaphanosoma*. Oecologia 74: 468–477.

McNaught, D. C., 1975. A hypothesis to explain the succession from calanoids to cladocerans during eutrophication. Verh. int. Ver. Limnol. 19: 724–731.

Michaloudi, E., 1997. Composition, abundance and biomass of the zooplanktonic organisms in Lake Mikri Prespa (Macedonia, Greece). Doctoral dissertation. Aristotle University, Thessaloniki, Greece.

Muck, P. & W. Lampert, 1984. An experimental study on the importance of food conditions for the relative abundance of calanoid copepods and cladocerans. Arch. Hydrobiol. 66: 157–179.

Odum, E., 1975. Ecology. Hold Rinehart & Winston, London, 2nd edn.

Omori, M. & T. Ikeda, 1984. Methods in marine zooplankton ecology. John Wiley (ed.), New York, 332 pp.

Pejler, B., 1983. Zooplanktonic indicators of trophy and their food. Hydrobiologia 101: 111–114.

Popovska-Stankovic, O., 1954. Die Plankton-Production des Dojran-sees, vom Aug. 1951 bis Aug. 1952. Zavod za ribarstvo na N.R.Makedonija, Izdanija (Skopje) 1: 1–20.

Porter, K. G. & J. D. Orcutt, Jr., 1980. Nutritional adequacy, manageability, and toxicity as factors that determine the food quality of green and blue-green algae for *Daphnia*. Am. Soc. Limnol. Oceanogr. Spec. Symp. New England. 3: 268–281.

Pourriot, R., 1970. Quelques *Trichocherca* (Rotifères) et leurs régimes alimentaires. Ann. Hydrobiol. 1: 155–171.

Pourriot, R., J. Capblancq, P. Champ & J. A. Meyer, 1982. Ecologie du plancton des eaux continentales. Masson, Paris, 198 pp.

Richard, J., 1897. Entomostracés, recueillis par M. le Directeur Steindachner dans les lacs de Janina et de Scutari. Ann. naturh. Mus. Wien. XII.

Richman, S. & S. I. Dodson, 1983. The effect of food quality on feeding and respiration by *Daphnia* and *Diaptomus*. Limnol. Oceanogr. 28: 948–956.

Ruttner-Kolisko, A., 1974. Plankton Rotifers. Biology Taxonomy. Binnengewasser 26 suppl., Schweizerbarts'che verlag., Stuttgart, 146 pp.

Serafimova-Hadzisce, J., 1954. Vertikale wanderungen des zooplanktons des Prespasees. Rev. Trav. Stat. Hydrobiol. Ohrid 2: 29–38.

Serafimova-Hadzisce, J., 1958. Particularités du zooplanton du lac de Prespa et aperçu de la composition du zooplankton des grands lacs de la peninsule des Balcans. Rev. Trav. Stat. Hydrobiol. Ohrid 6: 1–18.

Serafimova-Hadzisce, J., 1975. Seasonal variations of the reproductive rate of pelagic copepods of Lake Ohrid. Verh. int. Ver. Limnol. 19: 2976–2982.

Sommer, U., Z. M. Gliwicz, W. Lampert & A. Duncan, 1986. The PEG-model of seasonal succession of planktonic events in fresh waters. Arch. Hydrobiol. 106: 433–471.

Stankovic, S., 1960. The Balkan lake of Ohrid and its living world. In Bodenheimer, F. S. & W. W. Weisbach (eds), Monographiae Biologicae, Vol. 9, Dr W. Junk Publishers, The Hague, 357 pp.

Stathatos, P., J. Barry, M. Christomanou & A. Christomanos, 1972. Beitrag zur planktonkunde des kleinen Prespa sees in Mazedonien (Griechenland). Folia Biochem. biol. graeca 9: 12–26.

Stephanides, Th., 1948. A short note of the freshwater biology of Corfu and certain other regions of Greece. Praktika Hell. Hydr. I. Inst. 2: 1–264.

Threlkeld, S. T., 1985. Resource variation and the initiation of mid-summer declines of cladoceran populations. Arch. Hydrobiol. 21: 333–340.

Tryfon, E., 1994. Structure and dynamics of phytoplankton associations in Lake Mikri Prespa. Doctoral Dissertation, Univ. of Thessaloniki, 270 pp. (in Greek).

Tryfon, E. & M. Moustaka-Gouni, 1997. Species composition and seasonal cycles of phytoplankton with special reference to the nanoplankton of Lake Mikri Prespa. Hydrobiologia 351: 61–75.

Wetzel, R. G., 1983. Limnology. Ed. Sanders College Publishing, New York (2nd ed.), 767 pp.

Zankai, P. N. & J. E. Ponyi, 1986. Composition and feeding of crustacean zooplankton community in a shallow, temperate lake (Lake Balaton, Hungary). Hydrobiologia 135: 131–147.

Zaret, T. M., 1980. Predation and freshwater communities. Yale University Press, New Haven, 187 pp.

Zarfdjian, M. H., 1989. Seasonal variations and spatial distribution of planktonic inverterbrates in lake Volvi (Macedonia, Greece). Doctoral dissertation. Aristotle University, Thessaloniki, Greece, 249 pp. (in Greek).

Zarfdjian, M. H., M. Vranovsky & P. S. Economidis, 1990. Les invertebrés planctoniques du lac Volvi (Macédoine, Grèce). Int. Revue ges. Hydrobiol. 75: 405–412.

Hydrobiologia **351:** 95–105, 1997.
A. J. Crivelli & G. Catsadorakis (eds), Lake Prespa, Northwestern Greece.

The benthic fauna of Lake Mikri Prespa

D. Petridis[1] & A. Sinis[2]
[1]*T.E.I. of Thessaloniki, Department of Food Technology, P.O. Box 14561, 54101 Thessaloniki, Greece*
[2] *Department of Zoology, University of Thessaloniki, P.O. Box 134, 54006 Thessaloniki, Greece*

Key words: benthos, species composition, spatial and temporal distribution, oligochaete, endemism, Greece

Abstract

A brief description of the benthic macrofauna in Lake Mikri Prespa is given. The species composition and abundance were determined. Emphasis was given to the spatial and temporal distribution of the oligochaetes, the most important benthic group in the lake. Four out of five oligochaete species were endemic and the influence of some significant environmental variables: depth, C:N ratio, substrate type and pH, was examined. Indications of a potential for interspecific competition for space between *Psammoryctides* subspecies and other benthic species were detected.

Introduction

Lying in the shadow of the widely investigated, much larger, sister lake Megali Prespa, Mikri Prespa has been the subject of only few studies providing information about the benthos. Neither the name Prespa, nor the short distance between the two lakes should let the reader imagine that the small lake is a natural continuum of the larger one or that they look alike. In fact, the only point in common is the one-way water course from Mikri Prespa to Megali, due to the slightly higher altitude of Mikri Prespa (853 m). Otherwise, the physical dissimilarities are many: the configuration of the shoreline and depth contours (smoother in Mikri Prespa, abnormal in Megali Prespa), maximum depth (Mikri Prespa: shallow-8 m, Megali Prespa: deep, >40 m), trophic level (eutrophy in Mikri Prespa (Petridis & Sinis, 1995), oligotrophy in Megali Prespa (Sapkarev, 1956)).

The basin water comes from various sources, including rainfall, small streams and torrents, springs and subterranean waters. Subterranean channels facilitate the flow of water to Lake Ohrid whose altitude is 160 m lower. Depths greater than 6m occur only in the central part of Mikri Prespa (Figure 1), particularly in the region between the two islands. To the north, massive invasions of reed communities greatly encroach on the shallow water creating floating islands of 'Plau'. Submerged vegetation flourishes extensively only in the southern narrow part at depths of less than 2 m. Industrial effluent, that was formerly discharged into the lake, has now been replaced by diffuse agricultural sources of nutrients from livestock rearing and extensive use of fertilizers on cultivated land.

The purpose of this paper is to study the benthos of Lake Mikri Prespa with an emphasis on the oligochaetes, considering that four taxa out of five are endemic to the Balkans.

Methods

One hundred and twenty seven benthic samples were taken from the whole lake basin from 10 to 13 July, 1990. An Ekman grab (225 cm^2 sampling area) was used from a boat and the collected material was seived with a FBA pond net (500 μm mesh size). Depending on the shape and the quantity of the retained material, the mud residue was assigned to four categories:(1) little detritus with amorphous plant fragments, (3) an enormous amount of material rich in plant fragments, in particular reeds, (2) an intermediate situation between (1) and (2), (4) a mixture of muddy and gravel sediment. The retained material was preserved in 10% formalin for further treatment in the laboratory.

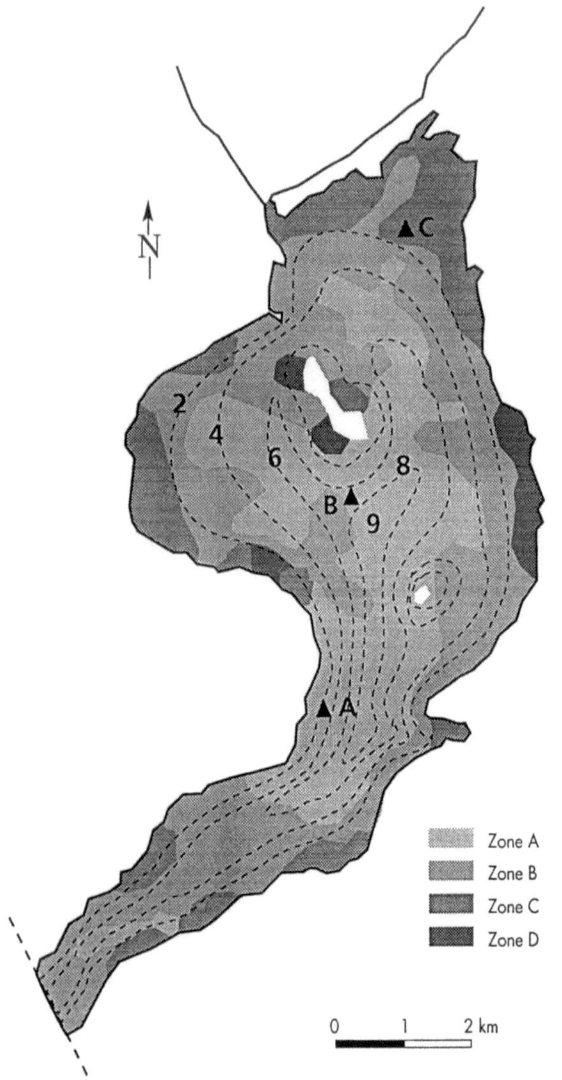

Figure 1. Diagram of Lake Mikri Prespa showing the depth contours, its division in four zones and the sampling stations A, B and C.

Changes in the benthic population densities were monitored monthly, from September 90 to September 91, by sampling at three permanent stations of different depth, A-4 m, B-6 m, C-1.5 m (Figure 1) and taking five replicates per station.

The lake was divided into four zones (Figure 1) according to the TWINSPAN site-grouping analysis (Hill, 1979; Petridis & Sinis, 1995). The grouped sites were subjected to a linear multiple discriminant analysis using six environmental variables: depth, conductivity, pH, substrate category, C:N ratio (carbon:nitrogen, before sieving on a subsample of 5 cm^2) and dissolved oxygen (Table 1). This analysis assigned

Table 1. Linear discriminant function coefficients for environmental variables on TWINSPAN site groups. The number of samples per site group is shown in brackets.

	TWINSPAN group			
	A (n=49)	B (n=49)	C (n=23)	D (n=6)
Constant	−991.20	−976.44	−966.88	−965.25
Depth	9.26	8.23	7.69	10.17
Conductivity	2.15	2.15	2.10	2.10
pH	179.25	177.81	177.93	175.72
Substrate	17.59	17.42	19.03	24.69
C:N	0.29	0.39	0.56	−0.28
Oxygen	−26.46	−26.22	−26.95	−25.74

65.4% of the 127 sites to the correct TWINSPAN site grouping. It was 81.6% successful in correctly predicting between sites in zone A, 78.3% in zone C and 83.3% in zone D, but failed in correctly predicting between sites in zone B (40.8%). Mahalanobi's squared distance (Table 2) showed close values of affinity between zones A and B and very distant values between A and D. Dissolved oxygen and conductivity failed to explain significant differences between site groups, because they gave similar coefficients (Table 1). Depth coefficients increased from zone C to A and substrate coefficients increased inversely from zone A to D; C:N coefficients increased from A to C, but negatively affected zone D. The influence of pH was minor. Thus, the lake is best represented by four zones adequately discriminated by the scaling of three features: depth, substrate type and C:N ratio. Zones A and B occurred in the deep profundal parts of the lake (mean depth plus 95% confidence intervals: 5.3 ± 0.4 and 3.9 ± 0.4, respectively), whereas zone C occupied the shallow littoral part (2.6 m $\pm$ 0.5, substrate category 3 and C:N ration 19.5 ± 1.4). The sparsely distributed zone D (3.9 m $\pm$ 1.0) contains hard substrate, a mixture of little mud and much gravel (substrate value, 4) with low C:N ratio (15 ± 0.8). Mud also covers the deeper parts of the basin, mostly characterized by fine sediment particles with no recognizable plant material (substrate value, 1 and C:N ratio, 17 for zones both A and B).

To investigate potential interactions between species, the index of patchiness, $I_d = \beta / x$, was employed; β denotes mean crowding ($\beta = x + S^2/x - 1$); x denotes mean population density and S^2 variance of the mean. The mean crowding defines the mean number per individual of other individuals in the same sampling unit (Lloyd, 1967). A value equal to 1 indi-

Table 2. Mahalanobi's squared distance between the site groups derived from TWINSPAN analysis.

	TWINSPAN group		
	A	B	C
B	1.48		
C	7.31	2.95	
D	18.63	17.36	14.21

cates random distribution of the population and higher values reflect the degree of animals' aggregation. Low patchiness accompanied by low population numbers indicates predation (Lloyd, 1967) and low dispersion together with high population density indicates competitive effects (Christensen et al., 1977; Petridis, 1990).

Species composition

A total of 10 313 benthic individuals were counted, in all the samples combined, of which 4067 individuals were found in the profundal, 3984 in the zone B, 1972 in C and 290 in the hard substrate. In terms of the mean abundance in each zone (Table 3), it appears that the animals distributed themselves equally between littoral and profundal, but in lower numbers in the zone D. The benthic fauna of the lake consists of five oligochaete and nine chironomid species, together with a few other diptera larvae of which *Chaoborus flavicans* predominates (Table 3). Among the chironomids, the most important species in numbers and occurrence is *Chironomus plumosus* which is dominant in all zones except zone D. The littoral species are *Cladopelma lateralis*, *Tanytarsus* sp. and *Tanypus punctipennis*. The latter also occurs at similar densities in the hard substrate and this may be related to the free-moving way of living (Bryce & Hobart, 1972).

Species such as *Chaoborus flavicans* and *Psychoda* sp. can be used to describe the lake as eutrophic. The former is a common inhabitant of shallow productive lakes and the latter favours very eutrophic or even polluted waters. In the authors' opinion, the lake has recently undergone the transition to a eutrophic lake, during the last decade (Petridis & Sinis, 1995).

Oligochaetes deserve particular mention because all the species, excluding *Potamothrix hammoniensis*, are endemic to the lake. Two peaks are obvious in the seasonal pattern of oligochaete population density (Figure 2), the highest extending from October to December and the lower occurring in June, with minimal values from July to September. Maximal densities approach 5000 ind. m^{-2} and minimal range from 300–1000 m^2. These low population values coincide well with those recorded in the study of the bathymetric distribution of oligochaetes, because the sampling was performed in July 1990 (Figure 3). In this month the highest population densities were observed in the littoral zone and at middepths and the lowest densities in the profundal zone. The vertical (bathymetric) distribution of oligochaetes in July should be viewed with caution, because at other times of the year the population densities that were monitored monthly at the three permanent stations with varying depths, were much higher in the deeper parts of the lake (>4 m), in particular from middle autumn to December (Figure 4). Oligochaetes comprise the most abundant group in the lake, also affecting the variation of the total benthic population in winter. *Chaoborus* population peaks shortly in October, then decreases with time showing also a smaller peak in summer (Figure 5). Chironomids peak in summer with minimal densities in May and September (Figure 6).

In Lake Megali Prespa oligochaetes also displayed two population density maxima, one in the littoral and one in the profundal (3000 ind. m^{-2}) with minimal values in the sublittoral zone (1800 m^2) (Sapkarev, 1963). According to Sapkarev the oligochaetes in Megali Prespa occur at roughly equal densities where the depth is homogeneous, and vary in numbers when the substrate changes having maximum densities in mud rich in organic matter, minimum densities on sandy sediments and disappearing only on solid substrates. They penetrate soft sediments to a depth of 27 cm, the maximum abundance occurring at depths of 6–9 cm below the mud surface. The oligochaetes are, therefore, eurybathic and eurytopic.

Endemic species of the lake

Four environmental variables can best be considered to influence the abundance of the oligochaete species. Apart from depth whose role in species distribution is crucial, the C:N ratio also provides information on the trophic status of the sediment (Hakanson, 1984). The lower the ratio, the more productive the sediment becomes. Structure of bottom could also be important for the species occurrence. Substrate values ranged

Table 3. List of benthic species of lake Mikri Prespa. The species are arranged in order of their numerical abundance (numbers m^{-2} and percentage contribution) and percentage occurrence (number of samples in which each species was present), according to the TWINSPAN division of the lake into the zones A to D. [1] Percentage of numerical abundance.

		Mean number m^{-2}				Percentage of occurrence			
	Species list	A	B	C	D	A	B	C	D
	Oligochaeta								
1	Potamothrix	34	190	525	133	22.4	65.3	100	83.3
	prespaensis	1.5[1]	5.9	15.7	6.0				
2	Potamothrix	1407	2236	1652	200	100	100	100	83.3
	hammoniensis	37.6	61.1	42.8	10.3				
3	Psammoryctides	0	0	89	45.2	0	0	43.5	100
	ochridanus	0.0	0.0	1.8	20.0				
	typica								
4	Psammoryctides	3	0	122	200	4.1	0	39.1	83.3
	ochridanus	0.1	0.0	3.1	9.2				
	variabilis								
5	Spirosperma	2	0	13	7	2	0	8.7	16.7
	tenuis	0.1	0.0	0.3	0.4				
	Chironomidae								
6	Einfeldia	1	9	93	22	2	18.4	69.6	33.3
	dissidens	0.1	0.3	2.7	2.2				
7	Chironomus	650	496	212	0	98	100	87	0
	plumosus	19.6	14.0	5.7	0.0				
8	Chironomus sp.	4	6	8	0	2	4.1	13	0
		0.2	0.3	0.3	0.0				
9	Microchironomus	1	0	13	0	2	0	17.4	0
	tener	0.1	0.0	0.4	0.0				
10	Polypedilum	0	1	4	7	0	2	8.7	16.7
	nubeculosum	0.0	0.1	0.1	0.2				
11	Cladopelma	0	1	29	15	0	2	34.8	33.3
	lateralis	0.0	0.1	0.9	1.7				
12	Tanytarsus sp.	0	3	182	30	0	6.1	65.2	33.3
		0.0	0.1	4.9	1.1				
13	Tanypus	4	69	240	230	2	40.8	95.6	100
	punctipennis	0.1	1.7	6.6	12.3				
14	Procladius	26	16	33	30	37.2	37.2	39.1	50
	(Holotanypus)	0.6	0.4	0.9	1.5				
	Other diptera								
15	Chaoborus	1544	547	508	807	100	95.9	91.3	83.3
	flavicans	39.8	14.6	11.5	34.0				
16	Culex sp.	2	8	21	0	2	4.1	13	0
		0.1	0.1	0.6	0.0				
17	Psychoda sp.	8	2	10	0	16.3	4.1	17.4	0
		0.2	0.1	0.3	0.0				
18	Ceratopogooidae	2	23	29	0	4.1	34.7	39.1	0
		0.1	1.0	0.7	0.0				
	Megalopiera								
19	Sialis sp.	0	1	8	7	0	2	13	16.7
		0.0	0.1	0.2	0.4				
	Amphipoda								
20	Gammarus sp.	0	3	2	0	0	4.1	4.3	0
		0.0	0.1	0.1	0.0				
	Gastropoda								
21	Valvata	1	1	4	7	2	2	8.7	16.7
	piscinalis	0.1	0.1	0.1	0.6				
	Total	3689	3612	3797	2147				

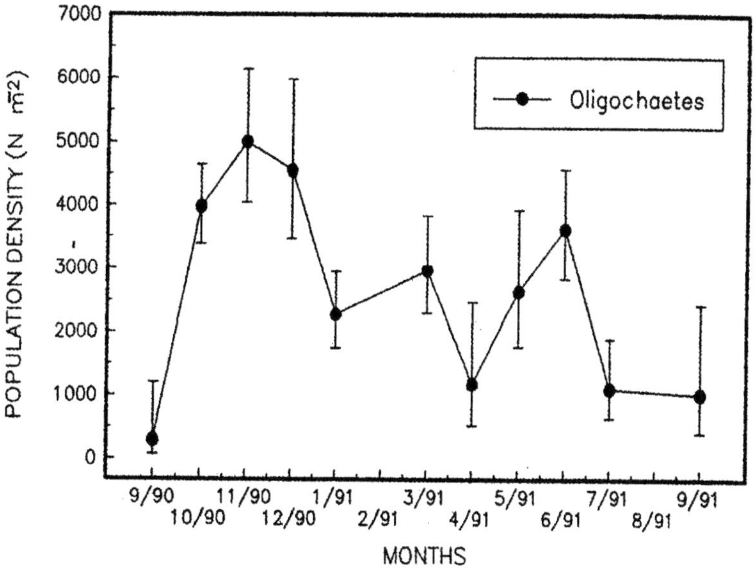

Figure 2. Monthly variation in the oligochaete population. Vertical bars represent the 95% confidence intervals of the mean derived after logarithmic transformation of the data. Fifteen replicates per month were used after combining the samples from the stations A, B and C.

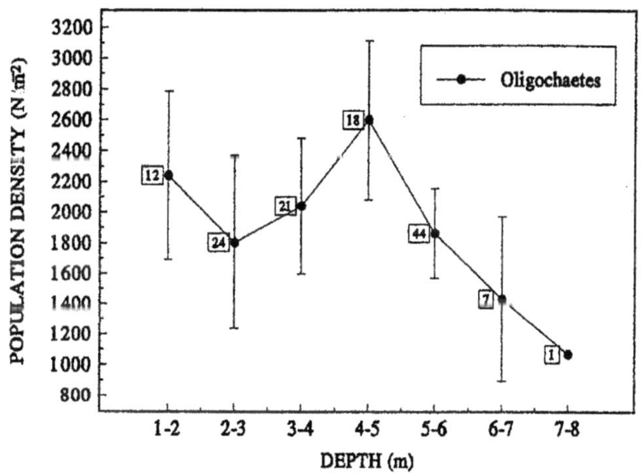

Figure 3. Bathymetric variation in the oligochaete population in July 1990. Vertical bars represent the 95% confidence intervals of the mean. The number of replicates is shown in squares.

from 1 to 4 according to the type of sediment encountered in the lake and pH values ranged from 7.2 to 8.8. Drastic changes in species occurrence due to pH are unlikely to occur because only values down to 4 can lower the species diversity (Lemkukhl, 1979). Examining the effects of the above variables on each species distribution (Figure 7, Table 4), the following results were deduced:

Potamothrix prespaensis: a littoral species, second in abundance after *P. hammoniensis*, occurring down to 5 m. It prefers less productive areas of the lake, with soft sediments rich in decaying reed remnants, and peaks at midrange pH values. This species has been recorded in three lakes, Mikri and Megali Prespa and Lake Skadar in Albania, is no longer considered as an endemic species since it has been discovered in another lake, 400 km away to the south (Lake Lysimachia, Western Greece, (Petridis, 1993)). It is well adapted to the eutrophic conditions of the shallow Lake Lysimachia, occurring in both deep and shallow waters. It

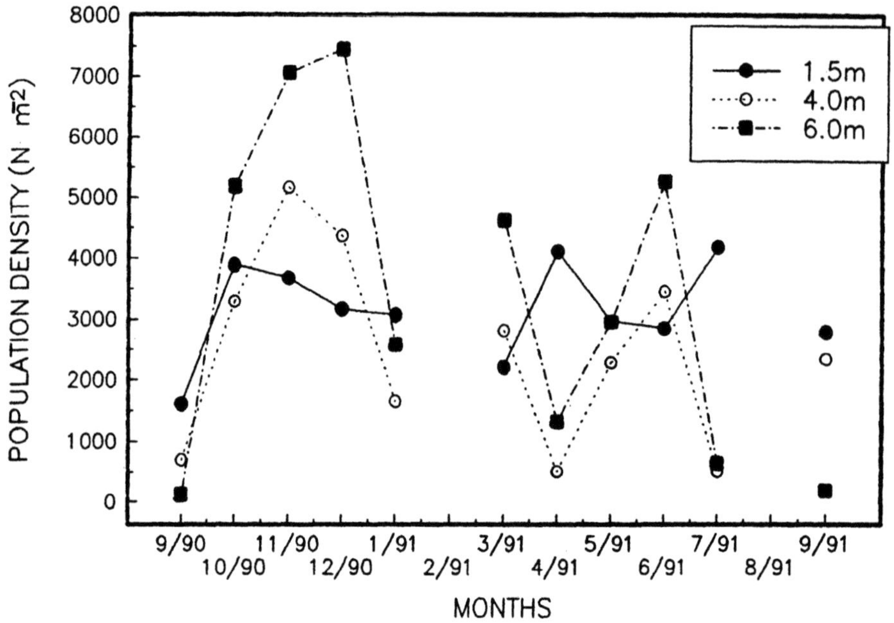

Figure 4. Monthly variation in the oligochaete population at different depths. Each point represents the mean of five replicates per station after logarithmic transformation of the data.

appeared to be sensitive to urban effluents, in particular to high ammonia concentrations in the water and showed a negative correlation with the abundance of the riparian oligochaete species *Limnodrilus hoffmeisteri*, probably due to competition for food or space (Petridis, 1993).

Psammoryctides ohridanus typica: a littoral and sublittoral species, with maximum occurrence in the sublittoral (middepths). It has low population densities, prefers less productive sediments, peaks in abundance at low range of pH values and shows an outstanding preference for gravel substrates.

Psammoryctides ohridanus variabilis: a littoral species not occurring below 5 m, peaking in abundance at midrange pH values. It favours less productive sediments and displays a remarkable preference for gravel sediments. This subspecies has even lower population densities than comparatively to *P. ohridanus typica*, a situation which was also reported in lakes Ohrid and Megali Prespa (Sapkarev, 1963).

The morphological features recorded on *Psammoryctides* were incomplete (absence of genital chaeta) despite the relatively large number of individuals examined (110 for *typica* and 90 for *variabilis*). All the specimens of *P. ohridanus variabilis* were characterized by the great variability in the dorsal chaeta: some were pectinates with short intermediate teeth, others were only bifids. The two *Psammoryctides* subspecies were differentiated mainly by the hair chaeta that were absent in *typica*, but according to Giani (personal communication) the latter form could possibly have been mistaken for the young stage of *variabilis*. Both subspecies also flourish in Lake Ohrid.

Spirosperma tenuis: a rare species, whose low abundance (0.1%) and occurrence (3.1%) makes impossible any conclusion on its habits, has also been found in Lake Ohrid.

P. hammoniensis: a ubiquitous species, showing a preference for soft sediments in deeper waters and for less productive sediments. Depth and pH did not affect its distribution at all (stable performance along the scale of each environmental variable). It is the dominant species in the benthos in both numbers and occurrence, not only of the oligochaetes but also of all the benthic species (Table 3).

All the above oligochaete species were thoroughly studied in Lake Megali Prespa by Sapkarev (1956, 1962, 1963) 30 years ago. In Megali Prespa the species *P. prespaensis* and *P. ohridanus variabilis* were also both littoral, whereas *P. hammoniensis* and *P. ohridanus typica* though abundant in the littoral were even more abundant in the profundal. Both the *Psammoryctides* subspecies had low population abundances in Lake Megali Prespa, reaching bathymetrically, 200–750 ind. m^{-2} for *typica* and 200–650 m^2 for *variabilis*, and horizontally (extending throughout the lake), 300–

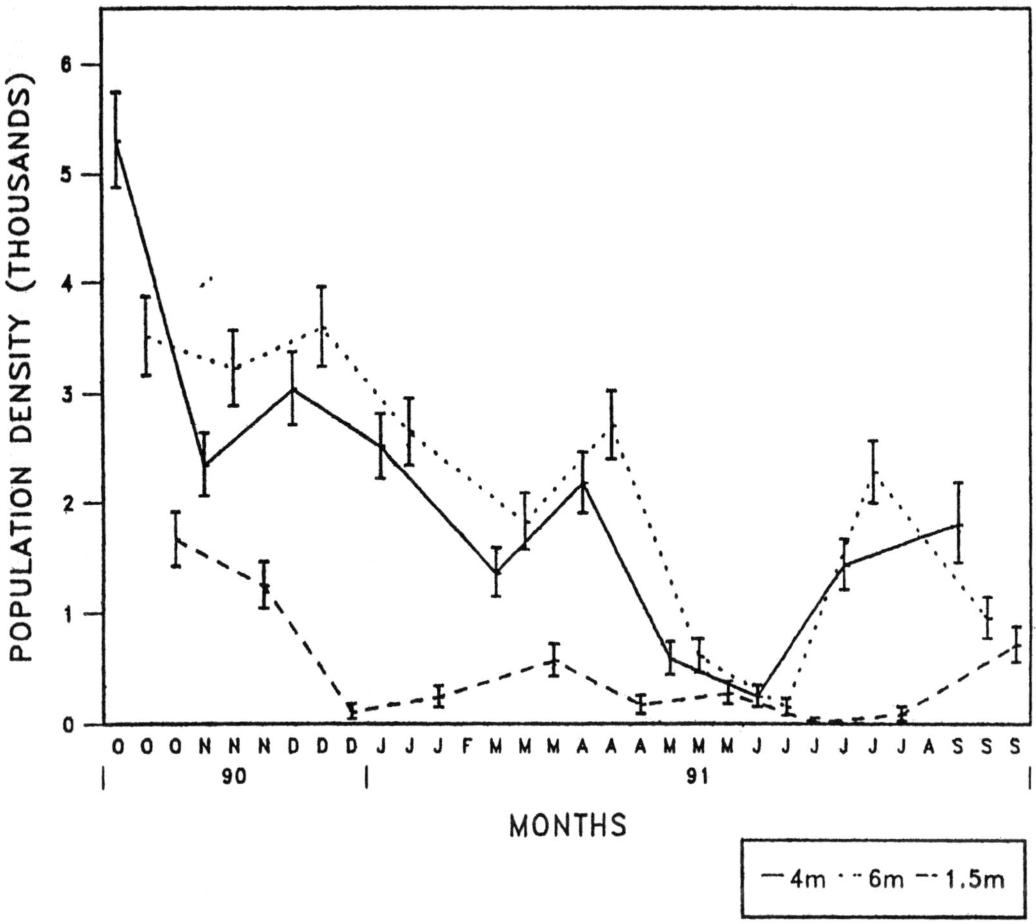

Figure 5. Monthly variation in the *Chaoborus flavicans* population at different depths. Vertical bars represent the 95% confidence intervals of the mean of five replicates per station.

400 ind. m^{-2} and 200–250 m^2 respectively. The distribution of *S. tenuis* showed two equal maxima in the littoral and sublittoral zones. All the above four species had maximum populations in spring and minimum ones in summer and comprised 97% of the total oligochaete fauna. The most dominant species was *P. prespaensis* (37.1% of individuals) closely followed by *P. hammoniensis* (31.8%). The inverse situation holds for lake Mikri Prespa in which *P. hammoniensis* overshadows the numerical abundance of the rest of oligochaete species (86.6% compared to 9.1% for *P. prespaensis*). Both species are also dominant in the eutrophic Lake Lysimachia except that the occurrence of *P. hammoniensis* is unaffected by polluted waters (Petridis, 1993).

Intra and interspecific relationships of endemic species

From the species list in Table 3, it appears that the species distribution is mainly related to two major factors: depth and substrate type, and all the intra-interspecific relationships that may arise due to the potential species interactions between these factors. In particular, the strong preference of *Psammoryctides* subspecies for coarse inorganic sediments is a tendency not observed before in the other neighbouring lakes, and this behaviour may be related to competition for space. This hypothesis is partly supported by the decreased population density of *P. hammoniensis* in zone D (200 ind. m^{-2} compared to more than 1500 m^2 in other substrates) and partly by a similar response of *C. plumosus* which is completely absent from coarse inorganic sediments. On the other hand, the presence of *C. flavicans* in high numbers (807 m^2)

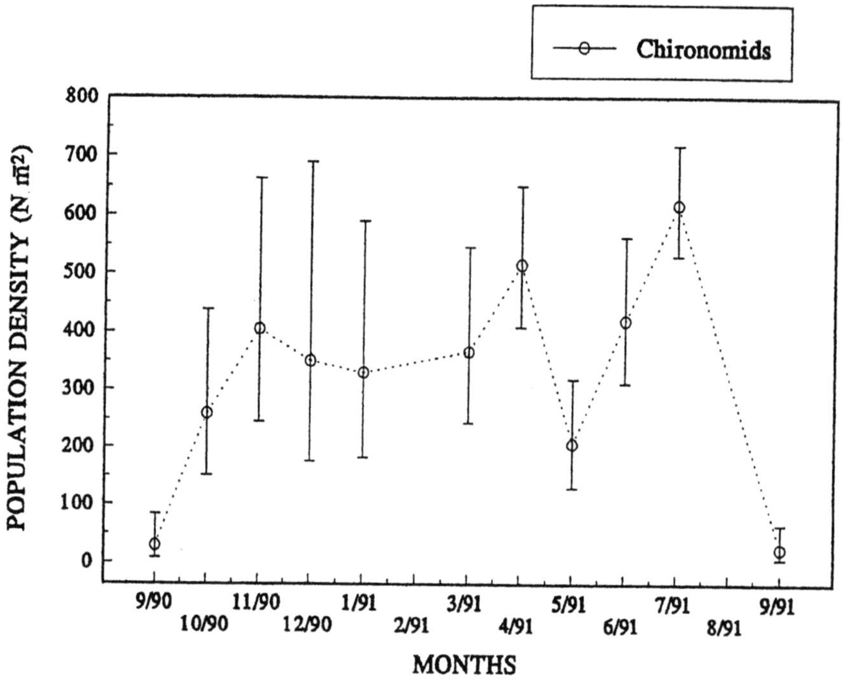

Figure 6. Monthly variation in the chironomid population. Vertical bars represent the 95% confidence intervals of the mean derived after logarithmic transformation of the data. Fifteen replicates per month were used after combining the samples from the stations A, B and C.

on the gravel bottom should not adversely affect the living conditions of *Psammoryctides* since these phantom larvae have an epibenthic mode of living at times of resting (at feeding times they migrate to the water column searching for zooplankton). Only the high density of *T. punctipennis* in the hard substrate (Table 3) could negatively influence the *Psammoryctides* distribution, since being predators of small chironomids and oligochaetes (Bryce & Hobart, 1972), they are able to reduce the density of their prey, a fact that does not agree with *Psammoryctides* abundance, which remains high. The plentiful presence of *T. punctipennis* and also that of *P. Holotanypus)* in the hard substrate has much to do with its feeding mode. It does not penetrate the sediment but moves freely over the surface seeking vulnerable prey. The drift of *Psammoryctides* to the gravel sediment may alleviate the populations from interspecific competition in other substrates but could make them available to *T. punctipennis* as no vertical escape inside the sediment can be sought.

Another questionable point is that whilst both subspecies share the same habitats in Mikri Prespa, littoral and solid bottom, they have completely segregated distributions in Lake Ohrid, the *variabilis* form living in the littoral and the *typica* form in the profundal zone (Sapkarev, 1956). A preference for the same substrates but with the species tending to be eurybathic was evident in Megali Prespa (Sapkarev, 1956). Question can be asked as to why they occur together in Mikri Prespa. Most probably other species are mainly responsible regulating their spatial distribution, in particular *C. plumosus* whose population density shows a strong negative correlation with *Psammoryctides* populations ($r=-0.988$, $p<0.01$, with *typica* in a log-log transformation, and $r=-0.973$, $p<0.05$, with *variabilis*) in terms of their distribution from zone A to D (deep, soft to coarse inorganic sediments). It is likely that although both species effectively penetrate the soft sediment, *C. plumosus* is better adapted to living in the mud, therefore compelling the *Psammoryctides* to move to other habitats in which *C. plumosus* cannot thrive or compete successfully. A similar competitive situation can also be observed for *P. hammoniensis* whose population abundance showed a strong negative correlation with the *Psammoryctides* distribution, in particular with the *typica* form ($r=-0.970$, $p<0.05$, in semilog transformation).

Changes in the spatial distribution of the benthos in the four zones were detected by employing the index of dispersion I_d for the population of each species (Table 5). Comparing the index of patchiness with the mean population density per site group, it can be implied that most of the species display a lower index of dispersion with increasing population numbers, which

Table 4. Mean population densities (number m^{-2}), 95% confidencc intervals (C.I.) of the mean and number of replicates (N) of the oligochaete species distribution in four environmental variable categories, after logarithmic transformation of the data.

		P. prespaensis		*P. hammoniensis*		*Ps. ochridanus* *typica*		Variabilis	
	N	Mean	95% C.I.	Mean	95% C.I.	Mean	95% C.I.	Mean	95% C.I
Depth (m)									
0.9–1.5	10	400	259–618	1666	1281–2172	14	3–51	5	0–41
1.5–2.1	5	274	115–649	934	386–2264	2	*	0	*
2.1–2.7	16	168	53–529	870	491–1541	8	2–24	3	0–15
2.7–3.3	13	52	12–218	1345	840–2157	3	1–7	2	0–8
3.3–3.	13	133	45–387	1166	596–2285	8	1–29	2	0–13
3.9–4.5	13	135	33–546	1640	970–2772	4	1–13	1	0–6
4.5–5.1	19	6	1–19	1402	582–3380	2	2–4	0	*
5.1–5.7	23	1	0–1	1584	1270–1971	2	*	1	0–1
5.7–6.3	9	0	*	1207	633–2295	0	*	0	*
6.3–7.6	6	0	*	1006	549–1844	0	*	0	*
C:N ratio									
7–10	2	24	*	1212	799–6137	2	*	0	*
10–13	5	229	115–454	531	132–2127	3	0–17	2	0–13
13–16	18	28	6–118	1827	360–2448	2	*	0	*
16–19	73	9	4–18	1166	886–1534	4	2–6	1	0–2
9–22	15	35	5–199	1527	1046–2227	5	2–12	1	0–2
22–25	10	106	17–639	1655	1155–2522	5	1–13	1	0–7
25–28	4	239	65–874	1551	435–5545	7	0–217	11	0–354
Substrate									
1	72	4	2–7	636	1425–1874	2	*	0	*
2	26	181	80–409	1509	1221–1870	3	2–6	1	0–2
3	25	165	69–387	1020	666–1566	11	4–25	6	1–20
4	4	46	0–863	37	0–883	232	19–715	56	0–374
pH									
7.2–7.4	1	0	*	3555	*	0	*	0	*
7.4–7.6	6	142	10–1844	790	294–2117	6	0–61	1	0–17
7.6–7.8	11	35	4–266	1232	629–2420	11	1–62	3	0–24
7.8–8.0	20	32	7–136	1682	1329–2132	7	2–17	3	0–12
8.0–8.2	39	15	5–42	1317	1015–1713	3	2–4	1	0–2
8.2–8.4	39	18	7–45	1116	697–1785	3	2–4	1	0–1
8.4–8.6	11	10	1–49	1725	1421–2093	3	1–7	1	0–3

Table 5. Index of patchiness (I_d) and mean number of individuals ($\bar{x}$) of selected benthic species arranged according to the division of the lake into zones A to D.

	Zone							
	A		B		C		D	
Species	I_d	$\bar{x}$	I_d	$\bar{x}$	I_d	$\bar{x}$	I_d	$\bar{x}$
P. prespaensis	7.07	0.8	2.13	4.3	1.11	11.8	1.24	3.0
P. hammoniensis	1.28	31.6	1.16	50.3	1.24	37.2	1.54	4.5
P. ohridanus typica	–	0.0	–	0.0	9.59	2.0	2.11	10.2
P. ohridanus variabilis	11.67	0.1	–	0.0	4.49	2.7	1.15	4.5
C. plumosus	1.21	14.6	1.25	11.2	1.59	4.8	–	0.0
T. punctipetuus	4.14	0.1	4.83	1.5	1.36	5.4	1.74	5.2
C. flavicans	1.43	34.7	1.54	12.3	2.74	11.4	1.78	18.2

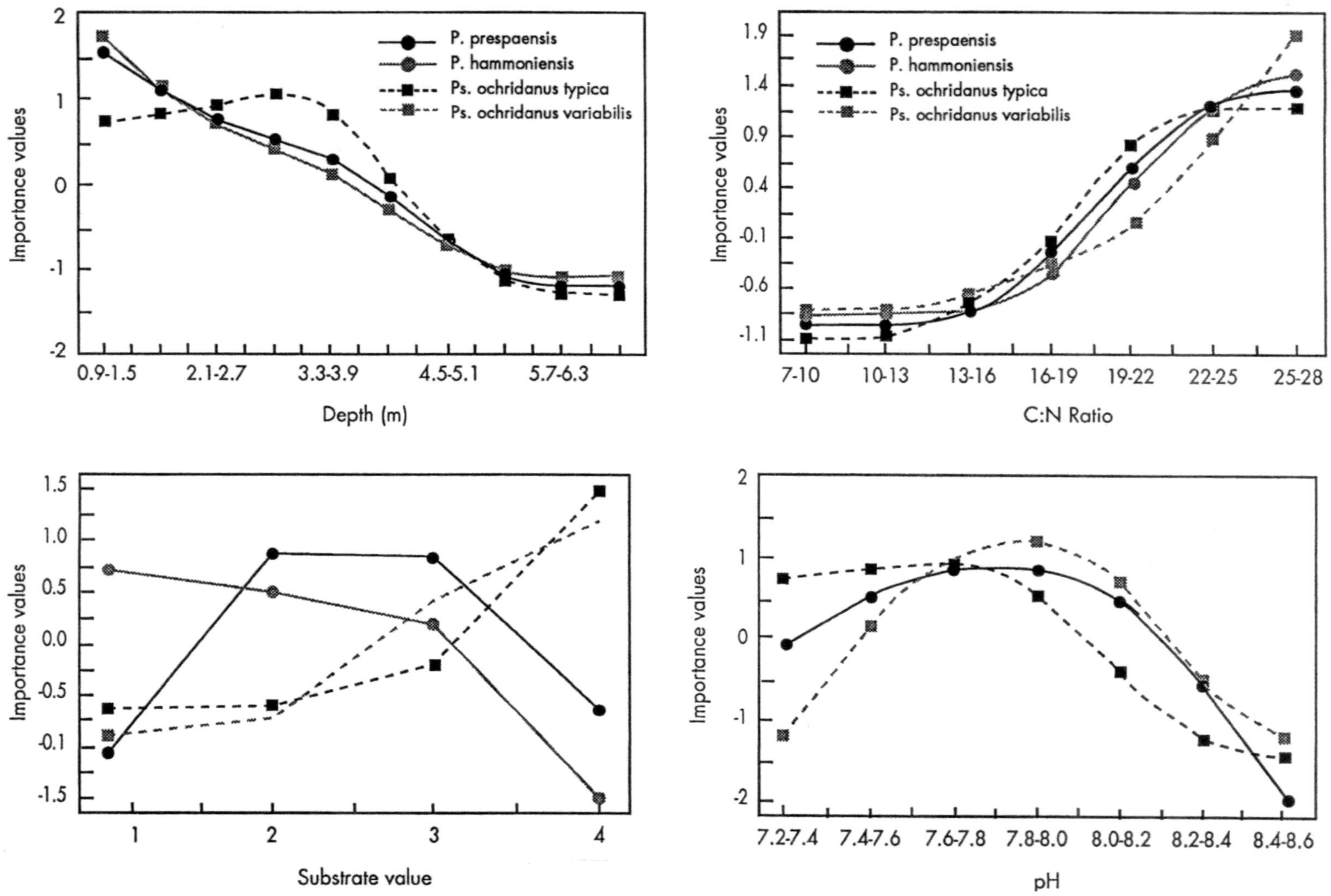

Figure 7. The influence of four environmental variables on oligochaete species distribution. The importance values were derived from (variable minus mean)/standard deviation, to eliminate the scaling effects of the widely varying numerical abundance of the species, and they were smoothed to reduce noise in the data.

is an indication of competition for food or space. If predation occurs, then this situation is reflected by a low population density together with a low index of patchiness (Lloyd, 1967). Thus, there is some evidence that intraspecific competition may exist between the individuals of *P. hammoniensis* and *C. plumosus* and that this event prevails over the potential interspecific competition for space with *Psammoryctides*, thereby leading to the displacement of *Psammoryctides* from the deeper parts of the lake and perhaps from many localities in the littoral. *T. punctipennis* and *C. flavicans* do not seem to influence the *Psammoryctides* distribution on gravel beds, for different reasons of course as previously mentioned, despite their high population densities since they are not accompanied by low indices of dispersion. In contrast to the above, the high index of patchiness for *Psammoryctides* suggests that individuals thriving in the littoral are more tolerant, although occurring at lower densities than in gravel beds, in which the index declines dramatically as it is impelled to do so by the high population level.

Another strategy for escaping from conditions of overcrowding is vertical migration down into the soft sediment. According to Sapkarev (1961) all oligochaetes in Lake Megali Prespa penetrate the sediment down to 27 cm. At times when high population densities develop, oligochaetes migrate deeply into the sediment, whereas at times of population scarcity the vertical distribution is limited to the upper layers of the mud. Different depth optima for such a vertical migration were obtained by Sapkarev for each species; thus, *P. hammoniensis* prefers depths of 6–15 cm peaking at 9–12 cm, *P. prespaensis* prefers depths of 0–9 cm peaking at 3–6 cm, *P. ohridanus typica* peaks at 9–12 cm and specifically at 15–18 cm and *S. tenuis* inhabits the uppermost layer of the sediment. In fact, *P. ohridanus typica* shows maximum penetration at the deeper parts of the sediment, perhaps as a result of having the least ability to adapt and compete effectively. Its absence from the profundal region of Mikri Prespa may also be connected with the lack of oxygen in the deeper parts of the sediment, which occasionally occurs at depths greater than 5 m, particularly in times of calm weather combined with no wave activity.

Conclusion

The above picture of the benthos of Lake Mikri Prespa presents a first approach to this topic with particular emphasis on the role of the oligochaete species for two main reasons: (a) their uniqueness, and (b) the atypical performance they exhibit in some responses such as spatial and vertical distribution not encountered in other nearby lakes.

Two statistical techniques, the correlation coefficient and the index of patchiness, were used to highlight the potential competitive effects exerted on the endemic species *Psammoryctides* by the two dominant species of the lake, *Potamothrix hammoniensis* and *Chironomus plumosus*.

References

Bryce, D. & A. Hobart, 1972. The biology and identification of the larvae of the Chironomidae (Diptera). Ent. Gaz. 23: 175–217.

Christensen, J. B., A. P. Gutienez, W. R. Cothran & S. G. Summers, 1977. The within field spatial pattern of the larval Egyptian alfalfal weevil, *Hypera brunneipennis*. An application of parameters estimates in simulation. Can. Ent. 109: 1599–1604.

Hakanson, A. L., 1984. On the relation between lake trophic level and lake sediments. Wat. Res. 18: 303–314.

Hill, M. O., 1979. TWINSPAN: a FORTRAN program for arranging multivariate data in an ordered two-way table by classification of the individuals and attributes. Section of Ecology and Systematics, Cornell University, Ithaca, New York.

Lehmkuhl, D. M., 1979. Environmental disturbance and life histories: principles and examples. J. Fish. Res. Bd Can. 36: 329–334.

Lloyd, M., 1967. Mean crowding. J. anim. Ecol. 36: 1–30.

Petridis, D., 1990. Influence of grass carp and tench on the ecology of *Asselus aquaticus*. Arch. Hydrobiol. 118: 105–124.

Petridis, D., 1993. Macroinvertebrate distribution along an organic pollution gradient in Lake Lysimachia, Western Greece. Arch. Hydrobiol. 128: 367–384.

Petridis, D. & A. Sinis, 1994. Benthos of Lake Mikri Prespa (North Greece). Hydrobiologia 304: 185–196.

Sapkarev, J., 1956. Contribution a la connaissance de la faune des oligochètes des trois grands lacs Yougaslaves, Prespa, Dojran et Skadar. Arch. biol. nauka VIII, 1–2: 135–144 (In Yugoslavian, with French summary).

Sapkarev, J., 1961. Die vertikale Verteilung des Benthos in den Sedimenten der Profundalregion des Prespa-Sees. Ann. Fac. Sci. Univ. Skopje (1959), 8: 139–151. (In Russian with German summary).

Sapkarev, J., 1962. Ecologie de *Ilyodrilus prespaensis* Hrabe (Oligochaeta) du Lac de Prespa (Macedoine, Yougoslavie. Fac. Sci. Univ. Skopje, Biologie: 29–37. (In Yugoslavian, with French summary.)

Sapkarev, J., 1963. Quantitative zusammensetzung der Oligochaetenfauna des Prespa-Sees. Ann. Fac. Sci. Univ. Skopje 14 (1962): 31–77. (In Yugoslavian, with German summary).

Hydrobiologia **351**: 107–125, 1997.
A. J. Crivelli & G. Catsadorakis (eds), *Lake Prespa, Northwestern Greece.*

Fish and fisheries of the Prespa lakes

A. J. Crivelli[1], G. Catsadorakis[2], M. Malakou[2] & E. Rosecchi[1]
[1]*Station biologique de la Tour du Valat, Le Sambuc, 13200 Arles, France*
[2] *Society for the Protection of Prespa, 530 77 Agios Germanos, Greece*

Key words: fish communities, trends, traditional fisheries, management, endemism, Greece

Abstract

Twenty-three taxa of fish have been identified from the Prespa lakes. Eleven of these have been introduced or translocated, and 7 of the remaining 12 are endemic to the Prespa lakes. This high proportion of endemism requires confirmation because the systematic position of several of the species and subspecies remains uncertain. The absence of a predatory fish in Mikri Prespa, the occurrence of trout in Megali Prespa, the presence of eels in both lakes and the occurrence of many rheophilous fish species typifies the fish communities of these two lakes. Monitoring of the fish population in Mikri Prespa, from 1984 to 1994, showed apparent stability or a slight increase in catches during this period. Not all the introduced species have been able to establish populations and their impact on the indigenous populations is yet to be defined. A spectacular re-establishment of the carp population, following the crash of the stock of this species in the 1970s–1980s, was recorded in the spring of 1991. The past and present situation of the fisheries in lakes Mikri and Megali Prespa is described, and their productivity and yields are analysed. Particular attention is given to a traditional fishing method unique to the Prespa region: "pelaizia" fishing. The factors involved in the temporal trends in the fisheries are identified and analysed. Recommendations for the management of the fish populations and the fisheries are proposed, so that the sustainability of this resource and the survival of the endemic fish species can be ensured.

Introduction

Greece is the north-Mediterranean country that harbours the most diversified freshwater fish fauna with the highest proportions of endemism (Crivelli, 1996). It is also one of the few Mediterranean countries to have still retained an important inland fishery (Crivelli, 1990,1992). Among the Greek lakes, the Prespa lakes are the most unique and interesting, since unlike all the others, they lie within the Adriatic basin and not Aegean basin and they are probably among the most ancient lakes in Europe, dating from the Tertiary era.

The Prespa lakes lie within three different countries, have been relatively little studied, and it is only within the last twenty years that scientists and conservationists have undertaken studies at Prespa, especially on the bird fauna of this region (see Catsadorakis this volume). Studies have been conducted on the fish in the Yugoslavian part of Megali Prespa. These have dealt with the fisheries (Stankovic, 1929; Apostolski, 1972), the biology of some species (Tocko, 1958; Petrovski & Sidorovski, 1971; Jovanovic, 1972; Popovska-Stankovic, 1972) and the parasites (Hristovski, 1976). Others studies were done on the Greek part of the lake, especially on fisheries (Athanassoupolos 1922, 1923) and on the systematics (Karakousis & Triantaphyllidis, 1990; Karakousis et al., 1991). Several studies have been undertaken on the systematics, biology, ecology and parasites in Mikri Prespa (Crivelli, 1984; Dupont & Lambert, 1986; Crivelli & Dupont, 1987; Dupont & Crivelli, 1988; Bianco, 1988; Berrebi et al., 1989; Crivelli, 1990, 1992; Rosecchi et al., 1993; Economidis, 1993; Crivelli, 1995; Catsadorakis et al., 1996a; Crivelli et al., 1996).

The aim of the present work is to review current knowledge of the fish populations, their structure and trends, and of the fisheries in these lakes, on the basis of published works and particularly using our own data from monitoring of the fish populations of Mikri Prespa over the last ten years.

Study area and methods

Lake Mikri Prespa (4841 ha) lies mostly in northwestern Greece, but a small section stretches into Albania (500 ha). It is linked by a narrow channel to Lake Megali Prespa which lies between Greece (3764 ha), FYROM (Former Yugoslavian Republic of Macedonia; 18 820 ha) and Albania (6780 ha). Both lakes were formed during the Tertiary period (Cvijic, 1911; Stankovic, 1931). Details of their hydrology (Hollis & Stevenson, 1997) and their limnology can be found in Petridis & Sinis (1997), Tryfon & Moustaka-Kouni (1997) and Michaloudi et al. (1997).

A sampling programme on the fish of Lake Mikri Prespa, using experimental gill-nets, was carried out, monthly in spring (April–June) in 1984 and 1985 and then from 1990 to 1994. Two sets of 50 m monofilament gill-nets with 10 mesh sizes from 10 to 60 mm (stretched mesh) were set at three sampling stations. Two stations were close to the shore: one (Station 1) was situated in relatively deep (2–5 m) water over a muddy substrate with a high density of macrophytes and bordering a reedbed; the other (Station 2) was in a shallower area (0.5–5 m) with a gravelly substrate and very scattered vegetation. The third station was pelagic in the deepest part of the lake (>5 m), where nets were set from the surface down to a depth of 2 m. Nets were set for two to four consecutive days in 1984 and 1985, and then for two consecutive days from 1990 to 1994 and were visited daily. Unfortunately no sampling could take place from 1986 to 1989. The selectivity of the nets was investigated (Boy & Crivelli, 1988) and it was shown that three taxa were incorrectly sampled using this method. These were two small species *Cobitis* and *Paraphoxinus*, of which only the largest individuals were taken in the 10 mm mesh nets, and the eel. To obtain better information on these species, traps and fyke nets were set, but not at regular intervals. More than 40 000 fish were caught in more than 200 hauls. Fish mere measured (fork length in mm), weighed (g) and sexed. The length at sexual maturity, duration of the spawning period and diets were also studied (see Crivelli, 1984; Crivelli & Dupont, 1987; Dupont & Crivelli, 1988; Crivelli, 1990; Rosecchi et al., 1993).

Secchi disk measurements (20 cm diameter disc, ten replicates each time) were made at each sampling station at monthly intervals. There was a significant decrease ($r = -0.58$, $F_{1,17} = 8.54$, $p < 0.009$) in Secchi disk transparency (Stations 1 and 2) between April 1977 and 1994 (Figure 1). Possible eutrophication of Lake Mikri Prespa, which remains a debatable issue, may explain this trend in turbidity (see Stevenson & Flower, 1991; Hollis & Stevenson, 1997; Moss et al., 1994; Anderson, 1995; Crivelli, 1995).

Catches were standardised per unit effort (CPUE): CPUE = total biomass captured per hour per m^2 of net, and were then log transformed (log $n + 1$ CPUE). Non-parametric statistics were mainly used. Catch data (CPUE) can be considered to be a relative estimate of population size and are a reliable index for monitoring changes in the abundance of fish populations (Cowx, 1991).

Fishery statistics were obtained from the fishery office at the Nomarchia of Florina. After 1990, the collection of fishery statistics stopped for unknown reasons.

The fish community

In the 1920s, Karaman (1924) recorded eleven taxa, in the first inventory of the fish species in the Prespa lakes. In 1984 twelve taxa were recorded (Crivelli, 1984), then Bianco (1988) added a thirteenth species. In 1984–1986, Crivelli (1990) recorded 14 taxa and Economidis (1993) has recently counted 15 species in the Prespa lakes. Between 1990–1996, we recorded 23 taxa and two hybrids (Table 1). Eleven of these 23 taxa are exotic species, five (*Carassius auratus gibelio*, *Ctenopharyngodon idella*, *Hypophthalmichthys molitrix*, *Parabramis pekinensis*, *Pseudorasbora parva*) are of Asiatic origin introduced into the lakes, either by the Albanians at the end of the 1970s, or during stocking with carp carried out in 1983, 1984 and 1986 in the Greek part of the lake. A sixth species, *Tinca tinca* coming from Lake Kastoria, was secretly introduced in 1983, together with pike (*Esox lucius*) which has apparently not survived.

Two (*Salmo letnica*, *Silurus glanis*) are of European origin and three of North American origin (*Oncorhynchus mykiss*, *Lepomis gibbosus*, *Gambusia affinis*). *Salmo letnica* has been introduced from Ohrid lake (Anonymous, 1985, page 67). *Silurus glanis* and two species of North American origin (*Lepomis gibbosus* and *Oncorhynchus mykiss*) have probably been introduced recently into Lake Megali Prespa by the inhabitants of FYROM. However, Athanassopou-

Table 1. The fish community of the Prespa lakes. * Introduced species with date of introduction

Taxa	Greek Red Data Book Status (Economidis 1992)	Mikri Prespa	Megali Prespa
Salmonidae			
* Salmo letnica (Karaman, 1924)	??		+
Salmo trutta peristericus Karaman 1937	Endangered		+
* Oncorhynchus mykiss (Walbaum, 1792)	??		+
Cyprinidae			
Alburnoîdes bipunctatus prespensis (Karaman, 1924)	–	+	+
Barbus prespensis Karaman, 1924	Endangered	+	+
* Carassius auratus gibelio (Bloch, 1783)	1970s	+	+
Chalcalbumus belvica (Karaman, 1924)	–	+	+
Chondrostoma prespensis Karaman, 1924	–	+	+
Cyprinus carpio Linnaeus 1758	–	+	+
* Ctenopharyngodon idella (Valenciennes,1844)	ca 1988	+	+
* Hypophthalmichthys molitrix (Valenciennes, 1844)	ca 1987	+	+
Leuciscus cephalus (Linnaeus, 1758)	–	+	+
* Parabramis pekinensis (Basilwesky, 1855)	1970s	+	+
Paraphoxinus epiroticus prespensis Karaman, 1924	–	+	+
* Pseudorasbora parva (Temminck & Schlegel, 1846)	1970s	+	+
Rhodeus amarus (Bloch, 1785)	–		+
Rutilus ohridanus prespensis (Karaman, 1924)	–	+	+
* Tinca tinca (Linnaeus, 1758)	1983	+	+
Cobitidae			
Cobitis meridionalis Karaman, 1924	–	+	+
Siluridae			
* Silurus glanis Linnaeus, 1758	ca 1989?		+
Anguillidae			
Anguilla anguilla (Linnaeus, 1758)	–	+	+
Centrarchidae			
* Lepomis gibbosus (Linnaeus, 1758)	??	+	+
Poeciliidae			
* Gambusia affinis Baird et Girard, 1853•	ca1995–1996?	+	
Hybrids			
Chalcalburnus belvica × Alburnoides bipunctatus••		+	
Rutilus ohridanus × Chalcalburnus belvica• • •		+	+

• E. de Vries and F. Willems (pers. communication)
•• Crivelli (unpublished data)
• • • Crivelli & Dupont (1987)

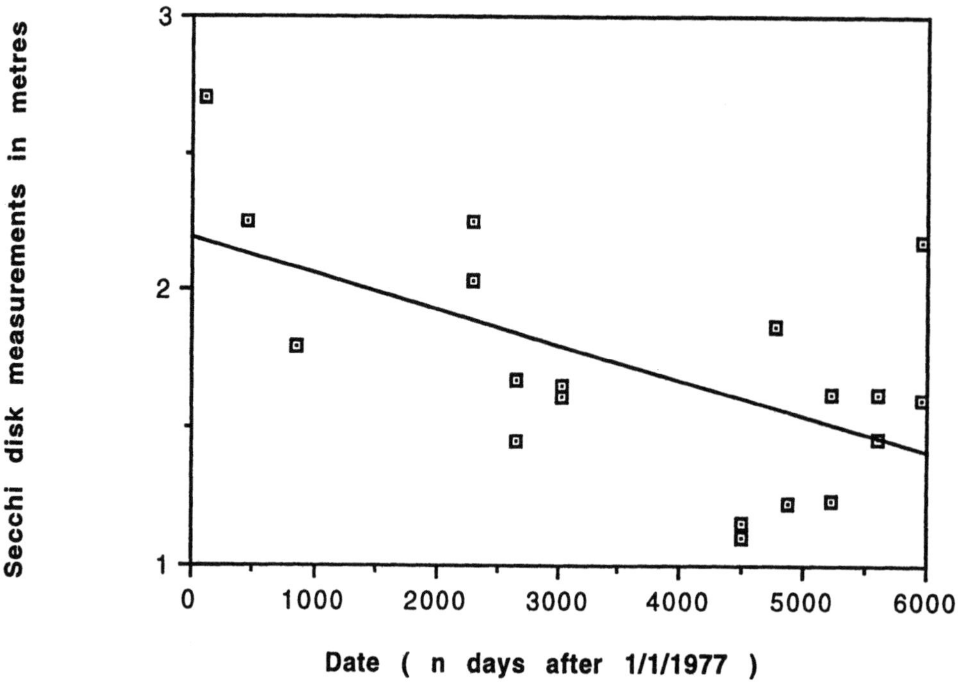

Figure 1. April Secchi disk measurements at Stations 1 and 2 at Lake Mikri Prespa from 1977 to 1994.

los (1922) described how the wels (*Silurus glanis*) of Megali Prespa and those of Veggoritis were renowned and could weigh up to 100 kg, but in a later article (Athanassopoulos, 1923) he no longer mentioned the species. Stankovic (1929) and Apostolski (1972) made no mention of wels in Megali Prespa. Since the wels is a Danube species (Banarescu, 1964) and is absent from Lakes Ohrid (Stankovic, 1960) and Skadar (Ivanovic, 1973), which form part of the same Adriatic basin as the Prespa lakes, we think that this species never occurred at Prespa formerly.

The third species of North American origin, *Gambusia affinis*, was found in Lake Mikri Prespa in spring 1996 (E. de Vries & F. Willems, pers. comm.).

Of the remaining 12 taxa, four species (*Barbus prespensis, Chondrostoma prespensis, Chalcalburnus belvica* and *Cobitis meridionalis*) and three subspecies (*Alburnoides bipunctatus prespensis, Paraphoxinus epiroticus prespensis* and *Rutilus ohridanus prespensis*) are endemic to Lakes Prespa or the Balkans. To these must be added two hybrids: one very common (*Rutilus ohridanus prespensis* × *Chalcalburnus belvica*, Crivelli & Dupont 1987; Dupont & Crivelli, 1988, Berrebi et al., 1989) and the other which was only found once, in May 1985 in Mikri Prespa (*Chalcalburnus belvica* × *Alburnoides bipunctatus prespensis*, Crivelli, unpubl. data).

Scardinius erythropthalmus has been mentioned as occurring in the Prespa lakes (Koussouris & Diapoulis, 1983; Athanassopoulos, 1922, 1923). But it is doubtful that the rudd has ever existed in the Prespa lakes.

At first sight, the proportion of endemism in the fish populations of the Prespa lakes seems remarkable. It should be mentioned however, that, in our opinion, the taxonomic position of a number of taxa occurring in the Prespa lakes remains doubtful. In future, many new taxonomic studies need to be undertaken using various methods (morphometry, parasites, enzyme polymorphism, DNA, etc.) with samples taken from throughout the Balkans, in order to confirm or refute the systematics (Crivelli, 1996). The doubtful taxa are: *Chondrostoma prespensis* (Elvira, 1987), *Chalcalburnus belvica* (Economidis, 1986), *Cobitis meridionalis* (Economidis, 1991), *Rutilus ohridanus prespensis, Alburnoides bipunctatus prespensis* (Daget et al., 1977), *Paraphoxinus epiroticus prespensis* and *Salmo trutta peristericus* (Karakousis & Triantaphyllidis, 1990; Karakousis et al., 1991). At present, only the barbel *Barbus prespensis* would appear to be undoubtedly endemic to the Prespa lakes (Dupont & Lambert, 1986; Economidis, 1989; Catsadorakis et al., 1996; Crivelli et al., 1996). As far as *Paraphoxinus epiroticus prespensis* and *Rutilus ohridanus prespensis* are concerned, whatever their true taxonomic status, they are probably endemic to the Balkans (Crivelli, 1996).

In addition to the endemism of certain of the taxa, the fish populations of the Prespa lakes are typified by: (a) the absence of a predator in Mikri Prespa, (b) the presence of trout in Megali Prespa, (c) the presence of the eel *Anguilla anguilla*, despite the fact that apparently these lakes have no direct outlet to the sea, and (d) the presence of a number of taxa that are considered to be typically rheophilous.

The absence of a predator is a feature frequently encountered in fish populations in the Balkans and in Turkey (Crivelli, 1995). In Lake Mikri Prespa, Crivelli (1984) and our more recent studies have shown that, to some extent, *Leuciscus cephalus* plays the role of predator in the fish population. The percentage occurrence of fish in the diet of chub was 45.5% for those between 83 and 199 mm fork length ($N = 23$) and 38.2% in those between 200 and 403 mm in length ($N = 33$). The rest of the diet was composed of invertebrates and plants. In contrast to other studies of the diet of this species (Hellawell, 1971; Mann, 1976) which showed that only the oldest individuals ate fish, chub in Prespa eat a large proportion of fish from an early age.

In contrast to Mikri Prespa, Megali Prespa does have a native predator: *Salmo trutta peristericus.* This is the only population of native trout occurring in a lake in Greece (Economidis, 1991). However, there are doubts as to whether the trout presently found in Megali Prespa do indeed belong to this subspecies and not to *Salmo letnica* introduced from Ohrid lake to Megali Prespa lake several times at unknown dates or to hybrids *Salmo letnica* × *Salmo trutta peristericus* (Anonymous, 1985). It has to be noted that the work of Karakousis and Triantaphyllidis (1990) and Karakousis et al. (1991) was carried out on trout caught in the upstream part of the Agios Germanos stream, an inflow of Lake Megali Prespa, isolated from the lake by weirs since the late 1960s (Catsadorakis et al., 1996a). Consequently, it is feasible, if the introduction of *Salmo letnica* took place after the 1960s, that the trout found in the Agios Germanos stream are still the native trout of the Megali Prespa.

The presence of eels in the Prespa lakes may seem somewhat surprising. The eels caught at Prespa have nevertheless arrived naturally after a long journey: from the Adriatic they enter the River Drin in Albania and ascend as far as Lake Ohrid. From Ohrid, the eels then use subterranean connections to reach Megali Prespa. Nowadays it seems doubtful if eels can still ascend the River Drin because of the many hydroelectric dams that have been built on this river in the last decade. Since the 1960s however, a second route enabling eels to colonise the Prespa lakes has existed; the Albanians built a canal between Mikri Prespa and the River Devoll (Hollis & Stevenson, 1997) enabling eels to ascend the Devoll and thus reach Mikri Prespa. It would appear that the great majority of eels that manage to enter the Prespa lakes are unable to find their way back to the sea when they reach sexual maturity and that they then start to gain weight. The capture of "monster" eels (e.g. in May 1985, we captured an eel 1430 mm long, weighing 3170 g) would seem to confirm this hypothesis.

The final remarkable feature of the fish populations in the Prespa lakes is the occurrence of five taxa that are generally considered to be rheophilous: *Cobitis meridionalis, Chondrostoma prespensis, Barbus prespensis, Alburnoides bipunctatus prespensis* and *Leuciscus cephalus.* Abundant populations of these species are rarely encountered in lakes in western Europe, but in the Balkans the genera *Barbus, Cobitis, Leuciscus* and *Chondrostoma* do occur in natural lakes (e.g. Lakes Skadar and Dojran). On the other hand, a population of spirlin (*Alburnoides*) living entirely in a lentic ecosystem such as Mikri Prespa, a lake lacking any permanent inflow streams, would seem to be unique. There are probably a number of permanent streams around Megali Prespa that would allow rheophilous species to make spring spawning migrations and spawn in the streams. We know that in the Greek part of Megali Prespa, the Agios Germanos stream plays such a role for the barbel and for other species (chub, nase and Prespa bleak).

Population trends at Lake Mikri Prespa over a ten-year period (1984–1994)

Long term monitoring of the fish population of Lake Mikri Prespa has been carried out in order to assess: (a) the fluctuations in the population of the various species present in the lake, (b) the fate of introduced species, (c) the spatial and temporal distribution of the various taxa in the lake, and (d) the interactions between fish and fish-eating birds. Only points a and b will be dealt with here.

Year-to-year changes in the spring (April, May and June) catches of all species combined are shown in Figure 2. Over a period of ten years, it would seem that there has been apparent stability or a slight increase,

Table 2. Comparison of CPUE (g/h/m^2) in 1984–1985 with CPUE (g/h/m^2) in 1990–1994 (Mann-Whitney U test). NS: not significant $P>0.05$; * up: significant increase $P<0.05$; ** up: significant increase $P<0.01$.

Taxa	Station 1	Station 2
Alburnoïdes bipunctatus prespensis	NS	NS
Barbus prespensis	NS	NS
Carassius auratus gibelio	NS	NS
Chalcalburnus belvica	NS	NS
Chondrostoma prespensis	* up	NS
Cyprinus carpio	** up	** up
Leuciscus cephalus	* up	* up
Pseudorasbora parva	NS	NS
Rutilus ohridanus prespensis	NS	NS
Tinca tinca	* up	–
Total (all species together)	NS	NS

although not significant, in the catch of all species combined at both stations 1 and 2. There has been apparent stability at station 3. It should be noted, however, that this station was only sampled once before 1990. In order to confirm this apparent trend, the CPUEs for 1984–1985 at stations 1 and 2 were compared with those for 1990 and 1994. The abundances of six taxa (*Alburnoides, Barbus, Carassius, Chalcalburnus, Pseudorasbora* and *Rutilus*) were stable at both stations (Table 2), but significant increases in abundance were recorded for two other taxa (*Leuciscus* and *Cyprinus*) at both stations 1 and 2. The abundance of *Chondrostoma prespensis* only increased at station 1 and remained stable at station 2. Catches of *Tinca tinca* increased significantly at station 1, but this species was never captured in spring at station 2.

The species composition of catches (% of biomass) at the three stations showed that roach, and to a lesser extent Prespa bleak, dominated strongly (Figure 3). Chub, and since 1992 carp, were abundant taxa in terms of biomass. Two small species, spirlin and *Pseudorasbora*, were numerically abundant but only accounted for a small proportion of biomass, whereas barbel, nase, goldfish and tench were less abundant. There were significant differences in abundance between stations 1 and 2 for four taxa: spirlin, barbel, chub and tench (Table 3). The first three were more abundant at station 2, where the substrate was their preferred spawning medium (these are lithophilous species, Balon, 1975). On the other hand, the tench, a phytophilous species, preferred the muddy substrates rich in submerged vegetation, typical of station 1. For the remaining taxa, there was no significant difference in catches between the two stations.

Only four taxa were regularly caught in the surface waters (surface to 1.5–2.0 m) at the offshore station in open water (Figure 3). These were roach, chub, nase and Prespa bleak. There was no significant difference in the abundance of the latter species in the catches made in open water and those of the inshore stations (Mann Whitney U test, $p>0.05$). The other taxa were either never captured in open water (e.g. *Barbus, Pseudorasbora*) or were rarely captured here in the spring (*Carassius, Cyprinus, Alburnoides*).

Temporal changes in abundance in catches seemed to be synchronised between stations 1 and 2 for Prespa bleak, nase, carp, *Pseudorasbora* and roach (Table 3). In contrast, there seemed to be no synchronisation between the stations in the catches of spirlin, barbel, goldfish or chub.

These results were remarkably consistent in showing the habitat differences between the two inshore stations and their utilisation by the different taxa. It should be noted that the sampling took place from April to June, which is exactly the peak spawning season for the species occurring in Mikri Prespa. In terms of their habitat preference for spawning these species can be divided into three categories: (1) lithophilous species (barbel, chub, spirlin and loach), which prefer stony or gravelly substrates with little submerged vegetation (e.g. station 2); (2) phytophilous species (carp and tench) which prefer muddy areas with abundant submerged vegetation (e.g. station 1) or, when the water level is high, flooded water meadows and (3) litho-phytophilous species (roach, Prespa bleak, nase and goldfish) which are capable of using either habitat type depending on circumstances.

Early in the century Athanassopoulos (1922) recommended introducing species into Megali Prespa, such as the 'American trout' and *Coregonus*. Apparently, fifty years had to pass before such introductions of fish species took place in the Prespa lakes (Table 1). Fortunately not all these introductions met with success. One of aims of monitoring the fish populations of Mikri Prespa was to follow the outcome of introduced species and to determine their impact on the indigenous species. Only *Pseudorasbora parva, Tinca tinca* and *Carassius auratus gibelio* have succeeded in establishing populations. The population of tench increased significantly between the two fishing periods (1984–1985 vs 1990–1994, Table 2, Figure 4). The population of goldfish also showed a tendency to increase (Figure 5), but this was not significant. As for *Pseudorasbora*, this

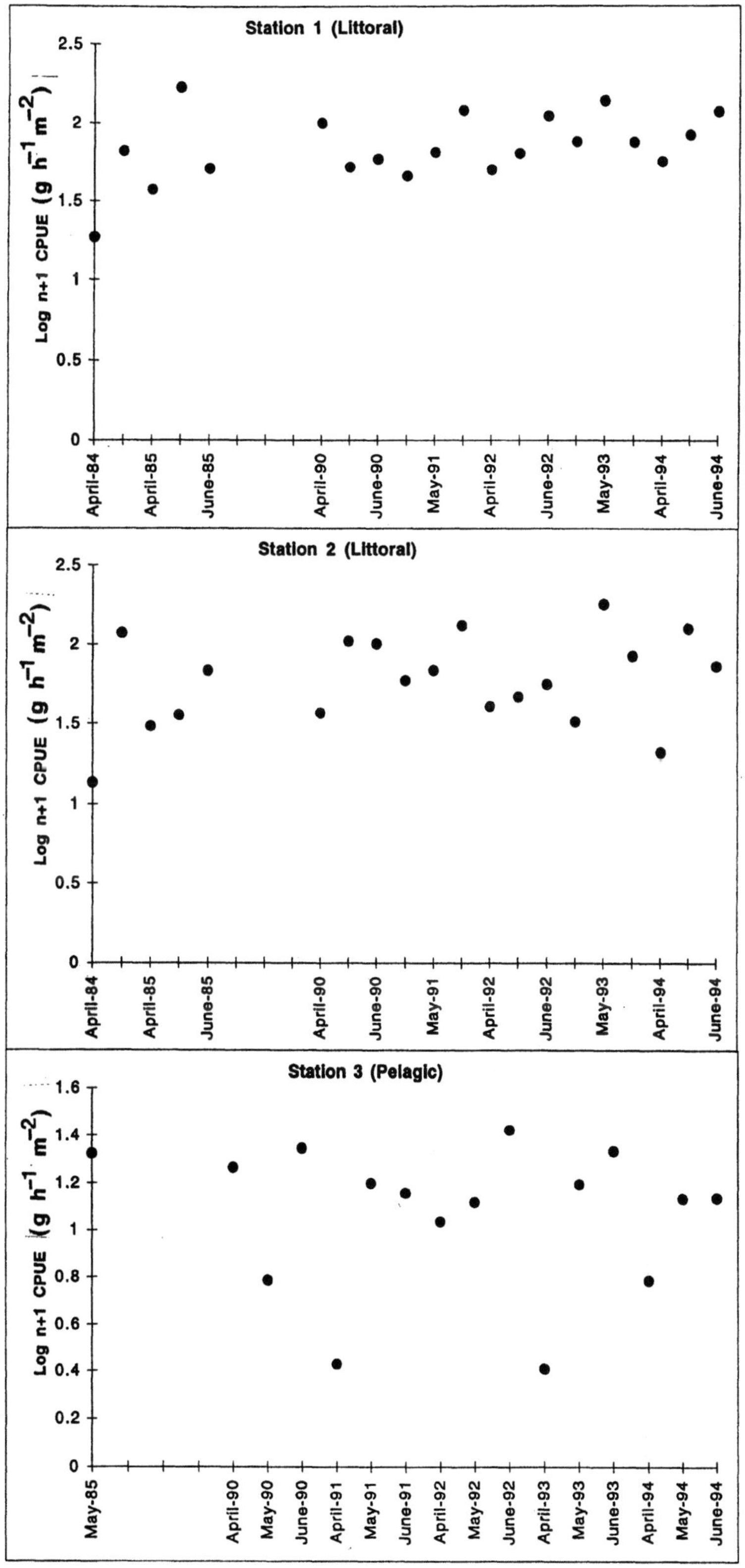

Figure 2. Catch per unit effort (g hr^{-1} m^{-2}) of all species in spring at Stations 1 to 3 from 1984 to 1994 in Lake Mikri Prespa.

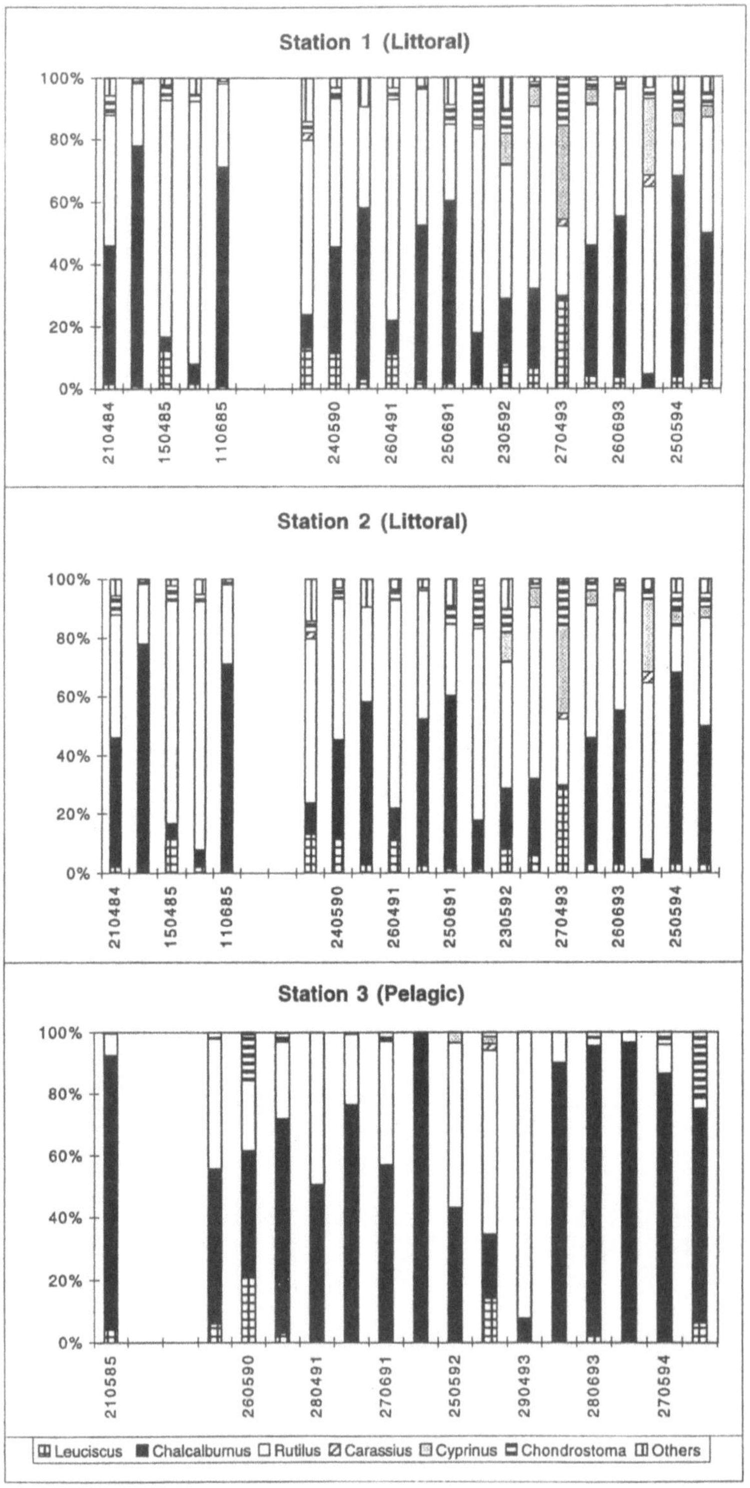

Figure 3. Percent species composition (percent of CPUE of monthly catches) at Stations 1 to 3 (Lake Mikri Prespa) from 1984 to 1994.

Table 3. Comparison of CPUE (g/h/m^2) between Stations 1 and 2 (Mann-Whitney U test) and of the time evolution of CPUE (g/h/m^2) between Stations 1 and 2 (Spearman rank correlation test). * Significant $P<0.05$; ** significant $P<0.01$.

Taxa	Mann Whitney U test	Spearman rank correlation test
Alburnoïdes bipunctatus prespensis	** higher Station 2	NS
Barbus prespensis	** higher Station 2	NS
Carassius auratus gibelio	NS	NS
Chalcalburnus belvica	NS	*
Chondrostoma prespensis	NS	*
Cyprinus carpio	NS	*
Leuciscus cephalus	* higher Station 2	NS
Pseudorasbora parva	NS	*
Rutilus ohridanus prespensis	NS	*
Tinca tinca	** higher Station 1	–
Total (all species together)	NS	*

increased in the early 1990s (Rosecchi et al., 1993), but then catches seemed to stagnate or even decrease in 1993 and 1994 (Figure 5). Three other introduced species, *Parabramis pekinensis, Ctenopharyngodon idella* and *Hypophthalmichthys molitrix*, have apparently been unable to breed successfully and only isolated old individuals are still captured in the lake. For example, in April 1990 an individual of *Parabramis pekinensis*, measuring 51.0 cm and weighing nearly 3 kg and in February 1990 a *Ctenopharyngodon idella* 105 cm long and weighing 18 kg were captured by fishermen in Mikri Prespa. Insufficiently high temperatures and the absence of running water are probably responsible for the lack of breeding success of these species.

The fisheries

Fisheries have existed on the two Prespa lakes since time immemorial (Athanassopoulos, 1922), but only the inhabitants of the village of Psarades on Megali Prespa were and still are full-time fishermen. The others are mainly farmers who fish in their spare time. Since 1945, the number of active fishermen has constantly decreased and it is difficult to know how many currently continue fishing on the Prespa lakes. Until the 1950s, the fishermen used rudimentary canoes with 4 to 8 paddles (Stankovic, 1929b); sailing boats were unknown as were boats with a tiller. Seine netting, which requires shorelines free from reeds and submerged vegetation was the most frequent means of fishing. Other methods included long lines, gill and trammel nets (13–80 mm mesh) and fish traps. From the 1960s onwards, outboard motors appeared, as did monofilament nylon nets, which allowed intensification of fishing effort, with considerable effects on the fish stocks. Even so, before the last war there was talk of decreasing catches, particularly of carp, the species most prized by Prespa fishermen.

For a long time, the only regulation concerning the fishery was the close season during the spring spawning period. But, as the state of the fishery deteriorated new attempts at regulation were made; for example the use of seine nets was prohibited in the Greek part of Mikri Prespa in 1983. Later, the length of the spring close season was lengthened and the mesh size of nets was increased (Crivelli, 1984, 1990). In 1991, for reasons that are obscure, fishery statistics stopped being collected, clearly reflecting the current precarious state of fishing on the Prespa lakes, as described below.

Traditional, so-called 'pelaizia' fishing (Figures 6 and 7)

'Pelaizia', 'Palaouzi' or 'Kedra' is the local Greek name for an ancient fishing method used by the fishermen of Lake Mikri Prespa and of Megali Prespa.

In autumn, fishermen cut branches of juniper (*Juniperus foetidissima, J. excelsa* and *J. oxycedrus*; the Greek name for juniper is 'Kedra'). They make a

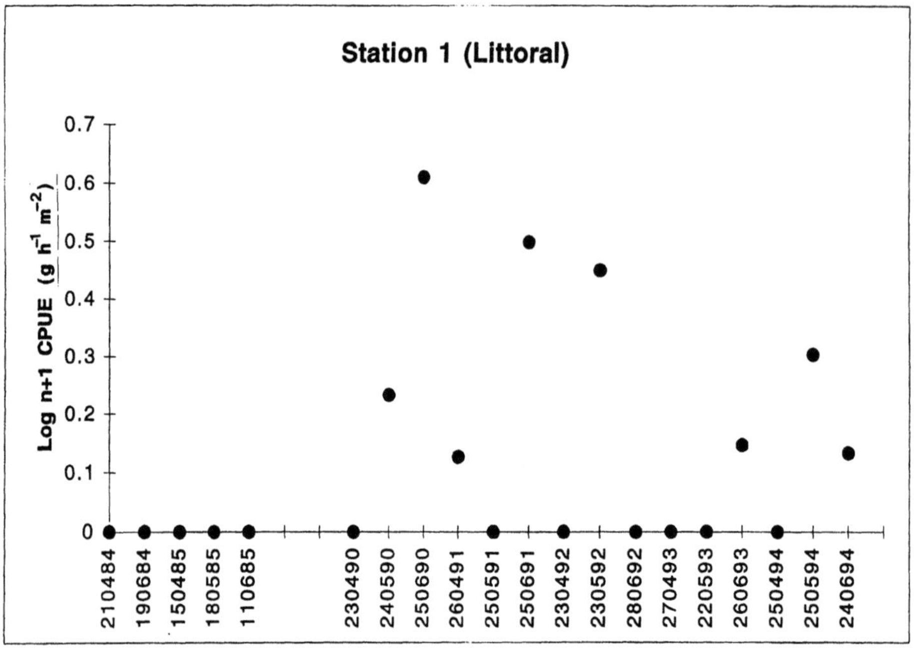

Figure 4. Catch per unit effort (g hr^{-1} m^{-2}) of *Tinca tinca* at Station 1 (Lake Mikri Prespa) from 1984 to 1994.

pile of branches in a sheltered area of the lake in a depth of 2 to 3 m, close to a reedbed. The base of the pile is 2 to 5 metres in diameter and the pelaiza sticks 20 to 50 cm above the water surface. The sites where these 'pelaizia' are built depends entirely on the experience of the fishermen who hand down the knowledge of favourable sites for this type of fishing from generation to generation. Once built, the pelaizias are used for several successive years, each year being repaired and new branches being added (Figure 6).

[20pt]

This traditional fishery only takes place in January–February, just after icemelt, the fish having spent the winter within the tangle of branches, sheltered from the frost. The fishermen encircle the pelaizia with a net 3–4 metres high, firmly fixed into the mud at the bottom and reaching to above the water surface. The net has an opening ending in a pocket. Fishing involves scaring the fish using long sticks handled from boats, those fish that try to escape going straight into the trap (Figure 7). In the best years, fishermen could thus catch up to 200 tonnes of fish, especially small fry known as 'tsironia'. This is in fact the name of *Chalcalburnus belvica*, but is also used for small fish of all species.

In January 1992, 20 samples each of one kilo were collected at random from the catch of this traditional fishery to determine the species composition. All the fish in each sample were identified to species level, measured and weighed. Not only were Prespa bleak (tsironia) caught in large quantities, but also roach, chub, nase, goldfish and *Pseudorasbora* (Table 4). According to the fishermen, in the distant past when carp were still abundant, they were also caught in large numbers. From examination of the size distribution of the fish captured (Figure 8), it can be seen that all the Prespa bleak and roach were mature individuals, whereas the majority of the chub and nase were immature. Since the largest fish captured measured 264 mm, it is reasonable to assume that the majority of the carp caught in the past by this traditional fishing method were also immature (Table 5).

What do they do with these tonnes of small fry? Most are now sold as fishfood to trout farms, but at one time tsironia from Prespa were exported to France. In the 1970s, these fish were also cooked and canned on the spot. Nowadays, they go the fish farms of Epirus and Macedonia.

This traditional fishery is now almost extinct. Before the 1940s, such pelaizias could be seen opposite the village of Mikrolimni (Mikri Prespa) and in the bay facing the village of Psarades (Megali Prespa). At present, in the Greek part of the Prespa lakes, there are only 45 remaining actively used pelaizias in the area between the island of Agios Achillios and Koula. Most

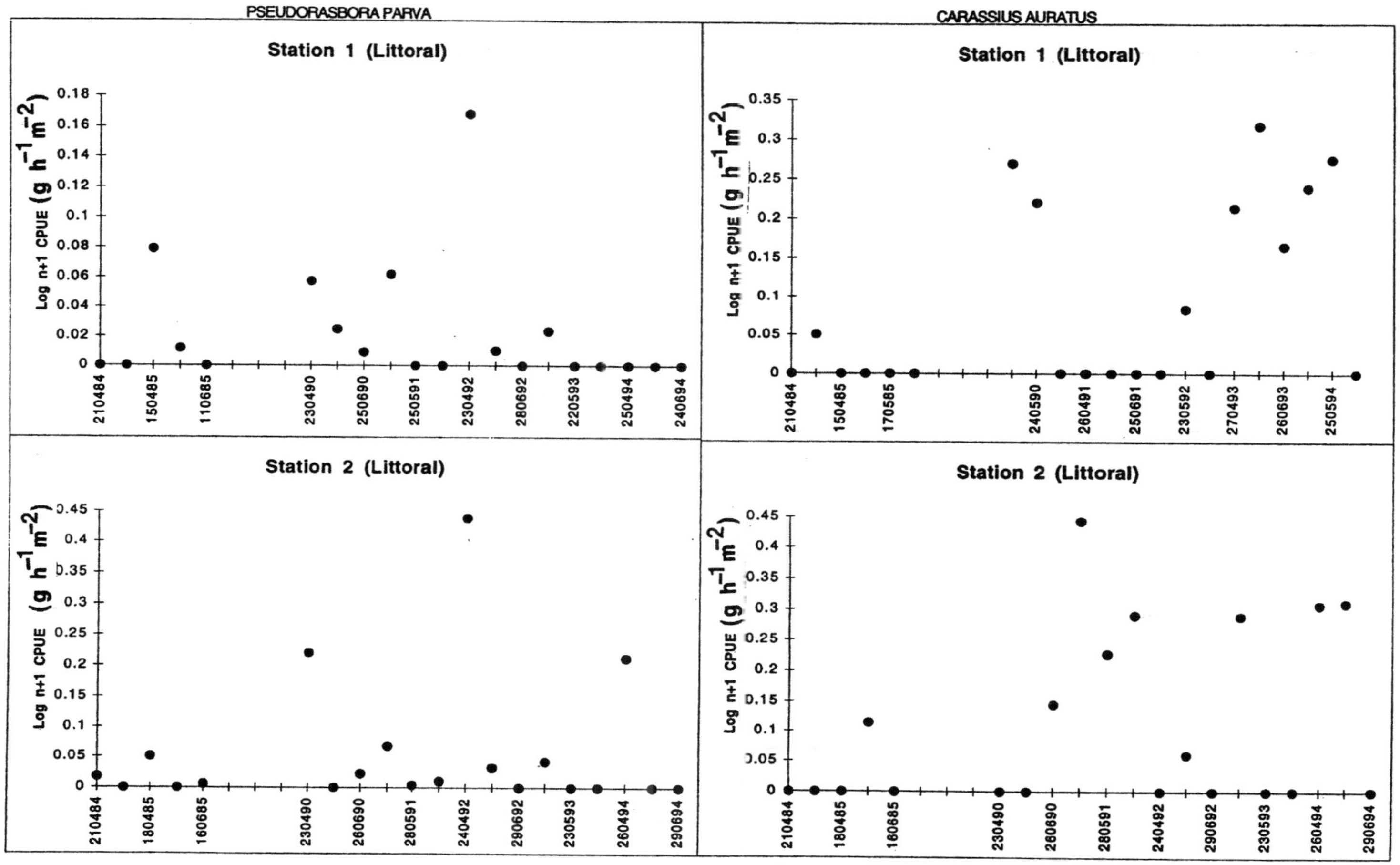

Figure 5. Catch per unit effort (g hr^{-1} m^{-2}) of *Carassius auratus gibelio* and *Pseudoraspora parva* at Stations 1 and 2 (Lake Mikri Prespa) from 1984 to 1994.

Figure 6. 'Pelaizia' near Agios Achillios (Photo: G. Catsadorakis).

Table 4. Species composition in the 1992 January subsample of fish caught in the pelaizia fishery at Mikri Prespa.

Taxa	% occurence	% of total weight	% of total number
Chalcalburnus belvica	100	48	33.9
Rutilus ohridanus prespensis	100	37.2	42.6
Pseudorasbora parva	10	0.3	0.5
Chondrostoma prespensis	90	6.3	13.2
Carassius auratus gibelio	10	1.5	0.5
Leuciscus cephalus	60	6.7	9.3

of these pelaizias belong to the fishermen from Agios Achillios, and the rest to fishermen from Psarades. It is not known if such pelaizias still exist in other parts of the Prespa lakes. In recent winters, pelaizia fishing was not very productive and during the 1994–95, there was almost no such fishing. The future of this traditional fishery seems very much in doubt. There are probably many reasons: (a) the catches have decreased, perhaps because the fish now find refuge in winter in the reedbeds, rather than in the pelaizias; (b) a disenchantment by the younger generation for this type of fishing, which is particularly arduous and (c) a decline in demand for this type of low-priced fish, probably resulting from competition with similar fisheries that have existed since 1983 in Lake Kerkini in northern Greece (Crivelli et al., 1995), and also the increasing used of artificial pelleted feed in trout farms.

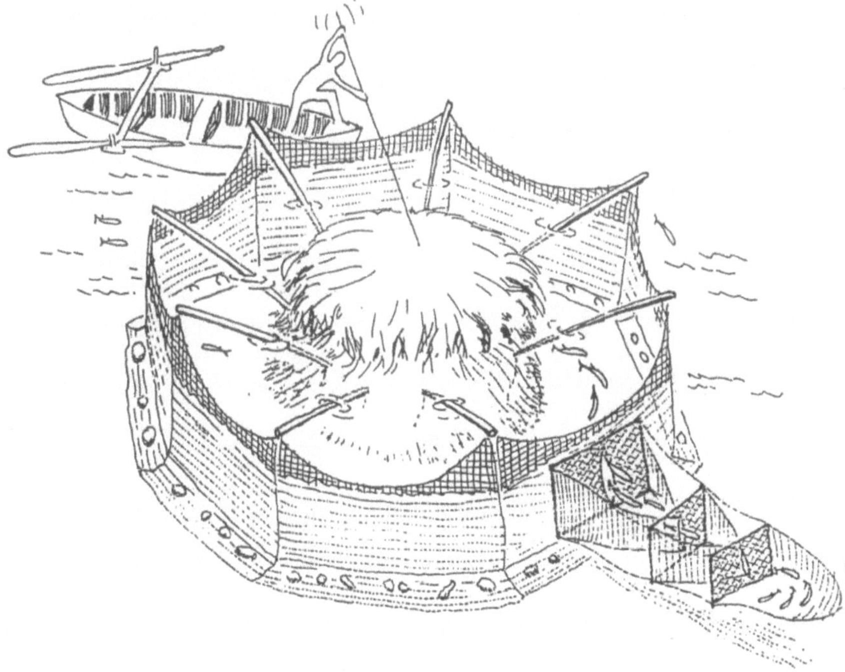

Figure 7. The 'Pelaizia' fishing method (Drawing: V. Hatzirvassanis).

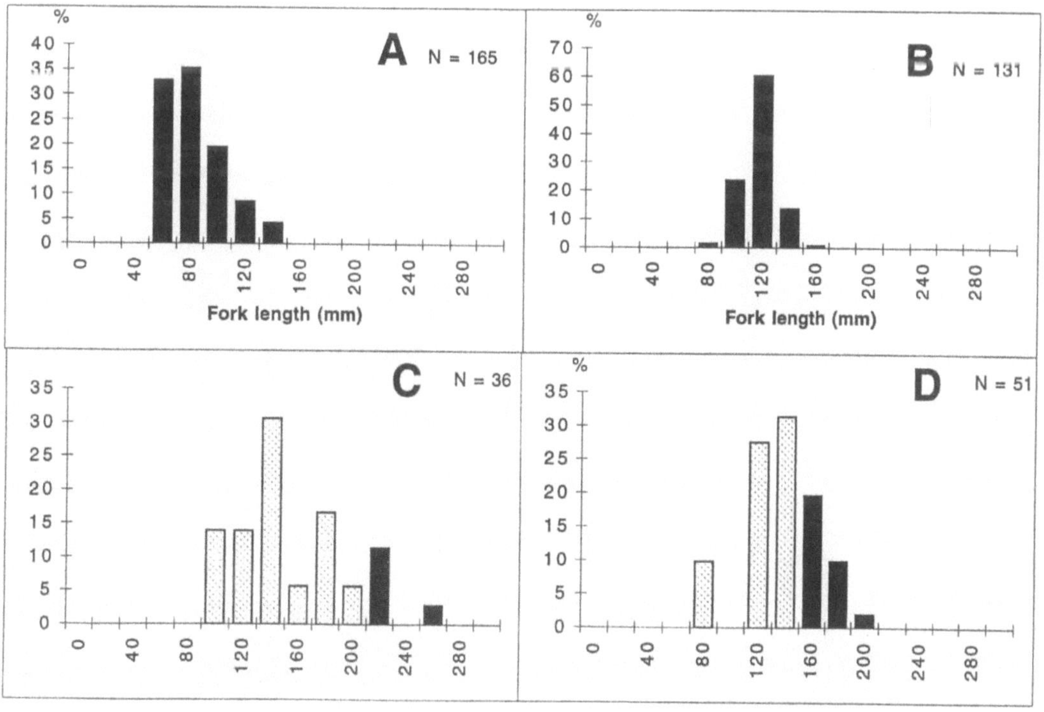

Figure 8. Size distribution of *Rutilus ohridanus prespensis* (A), *Chalcalburnus belvica* (B), *Leuciscus cephalus* (C) and *Chondrostoma prespensis* (D) caught in the pelaizia fishery (Lake Mikri Prespa). Mature fish are in black and immature fish are shaded.

Table 5. Average length ± S.D. of female fish caught by different mesh sizes (mm) during experimental gill net sampling in Lake Mikri Prespa (1990–1994 combined).

Species (Number of fish)	10	14	18	23	28	32	38	45	52	60
Rutilus ohridanus prespensis (3856)	74.1 ± 6.4	98.2 ± 6.1	122 ± 10.1	143.1 ± 10.1	171.6 ± 15.1	197.8 ± 12.6	210.4 ± 14.2	-	-	-
Chalcalburnus belvica (757)	98.1 ± 6.4	128.2 ± 7.8	147.2 ± 9.1	166.0 ± 11.4	-	-	-	-	-	-
Carassius auratus gibelio (49)	0	0	103.0 ± 6.0	116.0 ± 3.4	141.4 ± 4.8	169.2 ± 6.0	194.9 ± 9.8	217.4 ± 15.0	264.5 ± 14.8	0
Alburnoides bipunctatus (358)	74.1 ± 3.4	97.6 ± 6.9	-	-	-	-	-	-	-	-
Chondrostoma prespensis (138)	82.5 ± 0.7	117.0 ± 7.7	154.7 ± 9.3	179.2 ± 8.3	215.4 ± 21.1	255.8 ± 22.4	278.5 ± 10.3	-	-	-
Barbus prespensis (46)	100.0 ± 1.2	0	156.3 ± 6.3	197.6 ± 24.9	228.6 ± 19.2	238.6 ± 25.4	0	252.0 ± 9.9	-	-
Leuciscus cephalus (122)	85.5 ± 0.7	114.0 ± 5.5	145.9 ± 8.2	171.3 ± 8.5	234.2 ± 24.5	248.9 ± 14.2	285.3 ± 23.2	331.5 ± 20.1	362.2 ± 26.1	0
Cyprinus carpio (65)	0	0	106.0 ± 0.6	0	159.5 ± 13.4	196.4 ± 13.0	225.8 ± 17.4	245.4 ± 26.5	297.0 ± 12.1	330.0 ± 4.8

Shaded area represents length of fish which are still immature (>50% of immature fish against the total number of fish caught in a given mesh).

The fishery in Lake Mikri Prespa

The production of the Mikri Prespa fishery crashed in the 1960s (Figure 9) and has continued to decline since, reaching its nadir in 1989 (34.6 t yr^{-1}), the last year for which there are statistics. The mean production over 26 years was 183.8 t yr^{-1} (range: 34.6–488.6) equivalent to a mean yield of 38.0 kg ha^{-1} yr^{-1} (range: 7.1–100.9).

The species composition of catches, unfortunately only available since 1973, shows that the Prespa bleak, plus four other Cyprinidae (*Leuciscus, Chondrostoma, Rutilus* and *Barbus*) and carp dominated (Figure 10). One of the introduced species, the goldfish, appeared in the fishery statistics in 1984, and has continually increased since.

There is also a fishery in the Albanian part of the lake (500 ha) for which there are fishery statistics for three years: 1987, 1989 and 1990. The production was 11.8, 32.3 and 9.6 t yr^{-1}, respectively, equivalent to a yield of between 19.3 and 64.6 kg ha yr^{-1}.

In the experimental sampling with gill nets of 10 to 60 mm mesh, fish belonging to populations of different species were captured in proportion to their abundance. Only the youngest fish (young of the year) and in the case of carp and goldfish, the oldest individuals, were not captured. The fishermen's catches, on the other hand, were taken with trammel nets of 45 to 80 mm mesh or with long lines (2 km long with 1400 hooks baited with maize grains). The size distribution of two fish species (*Cyprinus carpio* and *Carassius auratus gibelio*) caught between 1990 and 1994 by our sampling and by the fishermen are compared in Figure 11. It can be seen that only mature and the oldest fish were captured by fishermen, no goldfish under 200 mm or carp under 240 mm was taken by fishermen. This illustrates perfectly the application of the regulation aimed at prohibiting fishing with nets of mesh size under 45 mm (Table 5).

The Megali Prespa fishery

Data (1973–1990) on the fishery in the Greek part of Lake Megali Prespa are less complete than those for Mikri Prespa (Figure 9). The mean production in these 17 years was 126.2 t yr^{-1} (range 12.5–276.8 t yr^{-1}) equivalent to a mean yield of 33.5 kg ha^{-1} yr^{-1} (range 3.3–73.5). Production fluctuated enormously from one year to another, and showed a slight tendency to decrease towards the end of the 1980s.

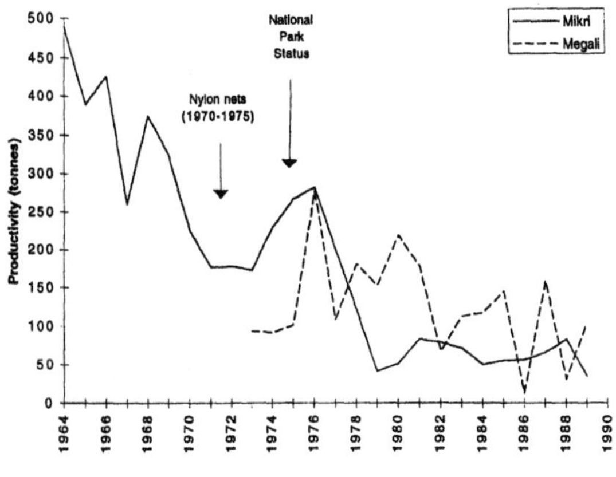

Figure 9. Productivity in tonnes of Lakes Prespa fisheries based on official fishery statistics.

Prespa bleak accounted for two-thirds of the catch (Figure 10), the remaining third consisting of four Cyprinidae (*Leuciscus, Chondrostoma, Rutilus* and *Barbus*) and carp. Prespa bleak in Megali Prespa are only caught using the 'Pezovola' fishing method during winter nights (December–March) at the deepest parts of the lake (Figure 12). The other fish species are caught using gill and trammel nets. During the last twenty years, trout have always been captured in small quantities (between 5 and 100 kg yr^{-1} in the Greek part of the lake. Catches of this species were already small at the start of the century (Athanassopoulos, 1922).

The mean production in the Greek part of Megali Prespa between 1919 to 1922, when it covered 10544 ha, was 52.2 t yr^{-1} (range 32.5–80.8), equivalent to a mean yield of 4.9 kg ha^{-1} yr^{-1} (Athanassopoulos, 1922). At the same period (1922–1925) the production of the Yugoslavian part of Megali Prespa was 48.8 t yr^{-1} (range: 16.1–73.8), equivalent to a mean yield of 2.6 kg ha^{-1} yr^{-1} (Stankovic, 1929). From 1958 to 1970 (Apostolski, 1972) the mean production for the Yugoslavian part of the lake was 138.2 t yr^{-1} (range 93.7–173.2) equivalent to a mean yield of 7.3 kg ha yr^{-1} (range: 4.9–9.2). The most fished species at this time were nase and Prespa bleak.

In 1987 and 1988, the production of the fishery in the Albanian part of Lake Megali Prespa was 246.7 and 234.5 t yr^{-1}, equivalent to yields of 36.4 and 34.6 kg ha^{-1} yr^{-1}, respectively. Prespa bleak accounted for more than 70% of the catch. These are the only available data for the Albanian part of the lake.

It is therefore evident that the yields of the fisheries on Lake Megali Prespa from 1922 to 1990 have continually increased.

The future of the fisheries on the Prespa lakes

It is difficult to analyse the long-term changes in the fisheries on the Prespa lakes for a single reason: there are no data on fishing effort. Nevertheless several comments can be made on the current state of knowledge.

With the exception of the fishermen from the village of Psarades, on Megali Prespa, all the other fishermen have other jobs, usually as farmers, and only fish part-time. Athanassopoulos (1922), even stated that in some years, there was no fishing at all, as the fishermen were too busy elsewhere. This is an important factor, since it shows that for the majority of fishermen, fishing is just a supplementary source of income, making this a marginal activity in terms of the economic development of the region.

It is undeniable that socio-economic factors play a major role in the development and changes in the fishery on the Prespa lakes. In the past, Athanassopoulos (1922) mentioned that almost the entire catch from the Prespa lakes was sold to Monastir or to Albania. But, because of political difficulties, the frontiers were closed and fishing activity at Prespa almost ceased completely. Nowadays, since Greece became a member of the European Union in 1981, the standard of living of the inhabitants of Prespa has considerably improved and the physically arduous and financially uncertain activity of fishing has become much less attractive for the younger generations. The geographically remote location of the Prespa lakes within Greece has also posed problems in terms of transport and marketing of fishery products.

The absence of fishery regulations and/or the lack of application of existing regulations (e.g. the prevention of poaching), plus the technical improvements that have taken place in the last twenty-five years have led to overfishing of the most highly prized species (e.g. carp, chub, nase, etc.). The use of small meshed nets (e.g. 22 mm), increased fishing effort due to the use of motor boats which allow the whole of the lake to be exploited and the introduction of nylon nets have all led to increased fishing pressure, a fall in stocks and evidently a decrease in catches (Apostolski, 1972; Crivelli, 1984, 1990). The long-lived species with delayed sexual maturity (e.g. carp, chub and barbel) have obviously been more affected than the short-lived species with early maturity (Prespa bleak and roach, goldfish), which have shown no effects of overfishing on their stocks (Crivelli, 1992b). Since 1987, the introduction of regulations concerning the lengthening of the spring close season and the ban on the use of nets with a mesh

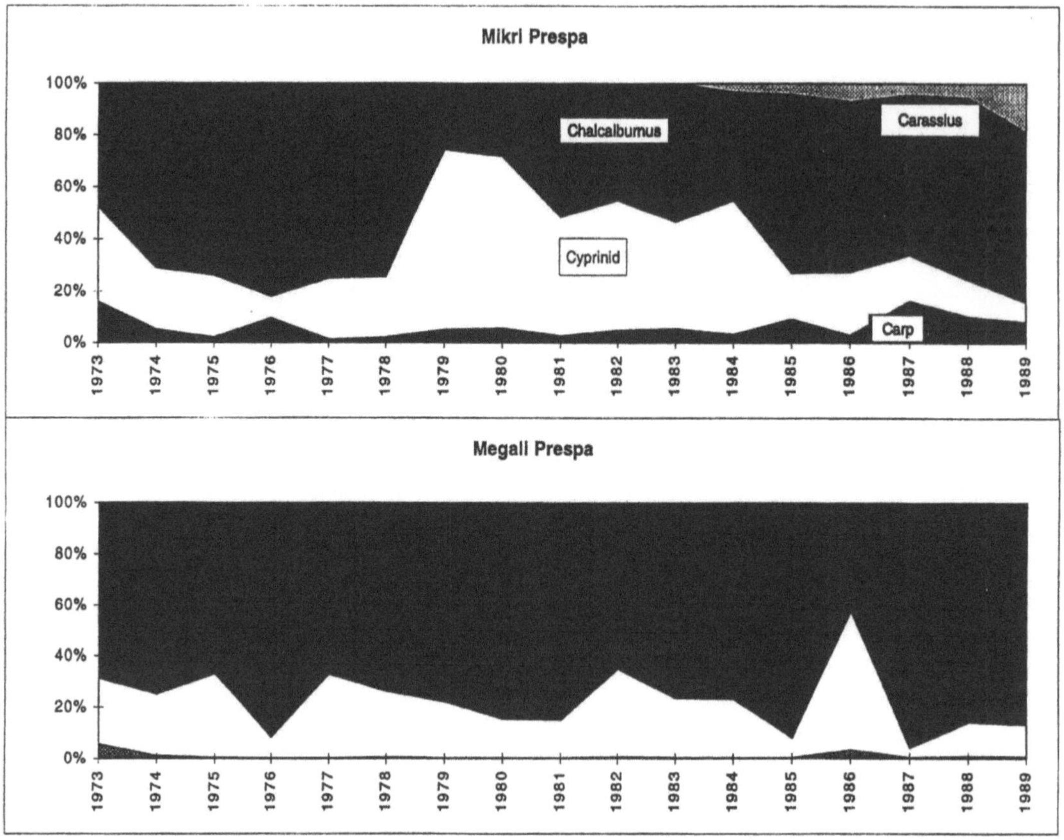

Figure 10. Percentage composition of commercial catches from Lakes Prespa. Others: *Barbus, Alburnoides, Pseudorasbora* and *Tinca.*

under 45 mm, have enabled the stocks of overfished species to recover (Table 2). It is nevertheless difficult to assess whether these regulations, or a possible decline in effort, or a decrease in the catch of immature fish by 'pelaizia' fishing have had the greatest role in the recovery of stocks.

Other important factors, such as fluctuations in water level and the state of the reedbeds, could also have played a major role in the recovery of stocks, particularly in Mikri Prespa. If the reedbeds are too dense and the spring water levels too low, fish are unable to gain access to the flooded meadows, the favourite spawning area for carp. Since 1970, the spring water level in Mikri Prespa has only exceeded 854.5 m asl on four occasions, but this is thought to be the minimum level required for flooding the meadows. Moreover, the extent of the reedbeds has steadily increased (Karteris & Pyrovetsi, 1986) and they have become denser since they are now burnt less frequently and grazing has decreased. Such a situation has obviously been unfavourable for successful annual recruitment by carp. Immediately after recording the lowest ever spring water level since 1953, in the spring of 1990 (Hollis & Stevenson, 1997; Catsadorakis et al., 1996b), the water level in 1991 was constantly above 854.6 m asl from April to mid-July. In addition, the reedbeds, which had been dry throughout 1990, were very sparse in 1991. These two factors combined allowed carp to have an exceptionally successful breeding season in the spring of 1991. From the spring of 1992, large numbers of one-year-old carp were sampled, and in 1993 and 1994 it was possible to confirm that the stock of this species had recovered. It can therefore be concluded, that not only is it essential to have strictly applied regulations, but also that the water level and the reedbeds are managed correctly if to ensure the long-term future of the carp stocks, which alone are able to ensure an acceptable and durable income for fishermen.

With the current state of knowledge of the fish populations and fisheries of Megali Prespa, it is difficult to predict what the consequences will be of the constant fall in the water level of this lake since June 1987 (Hollis & Stevenson, 1997), and particularly since 1990 when the level has been falling at the rate of 10 cm per month. It is nevertheless true that the fishery yield in Megali Prespa has increased constantly since the 1960s. There are two, non-exclusive explanations for this trend: a possible eutrophication of Megali Prespa, or an increase in fishing effort.

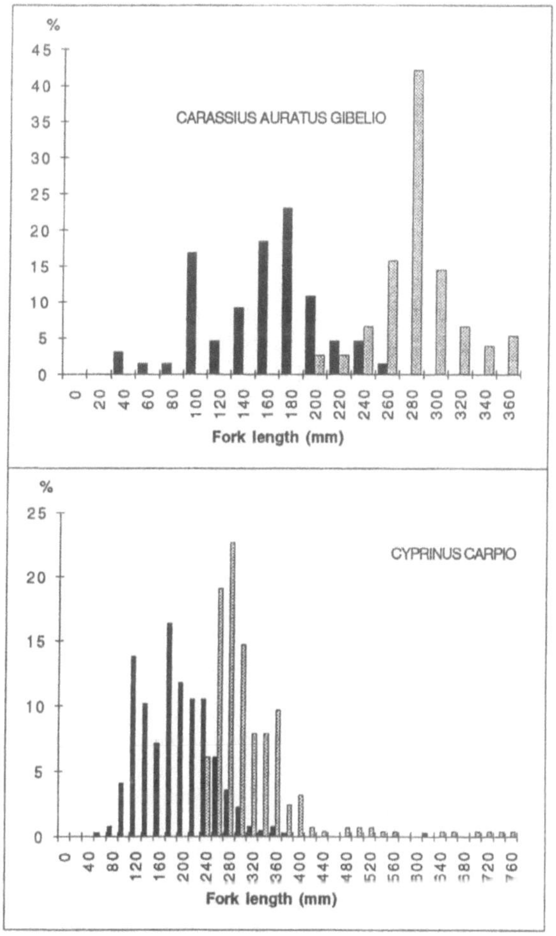

Figure 11. Size frequency distributions of carp and goldfish (1990–1994) in our experimental sampling (black; carp N = 391, goldfish N = 65) and in the fishermen's catches (shaded; carp N = 278, goldfish N = 76) at Lake Mikri Prespa.

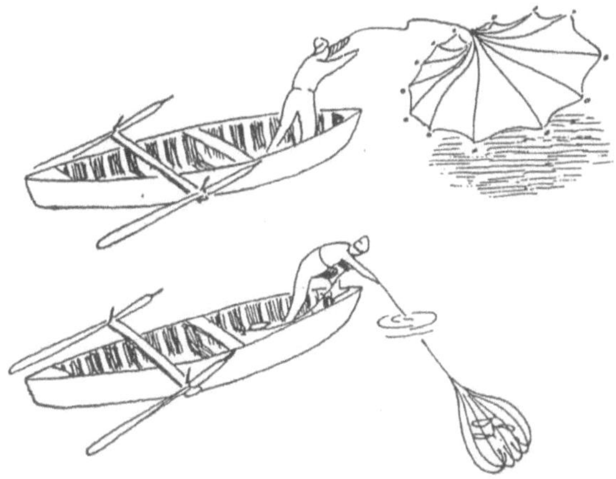

Figure 12. The 'Pezovola' fishing method (Drawing. V. Hatzirvassanis).

Conclusions

As we have seen, the fish populations of the Prespa lakes are remarkable in several respects. Unfortunately, the populations have been altered in the last thirty years, as a result of overfishing, habitat changes and the introduction of several exotic fish species.

The total biomass of the fish population in Lake Mikri Prespa is stable or has slightly increased in the last decade. The following reasons have been identified to explain this phenomenon: (a) a strong decline in fishing effort in the second half of the 1980s, for socio-economic reasons and because of the major decline in carp, the fishermen's favoured target species; (b) the coming into force of regulations prohibiting the use of nets of mesh smaller than 45 mm (Crivelli, 1984, 1990); (c) a longer fishery close season since 1983 (40–50 days instead of 30–40 days) improving breeding success; (d) a remarkably successful breeding season for carp in the spring of 1991, related to an exceptionally high water level from April to mid-July (between 854.64 and 855.04 m asl); (e) a decrease in catches in the winter 'pelaizia' fishery and consequently a decrease in the by-catch of immature fish and (f) a decrease in predation pressure by fish-eating birds on Mikri Prespa at the expense of Megali Prespa, probably resulting from increased turbidity in Mikri Prespa (Figure 1, Crivelli & Catsadorakis, unpubl.). After ten years of monitoring, the future of the species introduced into Mikri Prespa that have established populations, and their impact, still remain uncertain (Rosecchi et al., 1993; Crivelli, 1995). More time and more detailed studies are needed before any conclusions can be made on the possible impact of these introductions on the fish populations in Lake Mikri Prespa and on the whole ecosystem, including fish-eating birds. Further introductions, especially of predators such as percids, could have unimaginable consequences on the populations of the Prespa lakes (cf. Brabrand & Faafeng, 1993). We therefore recommend that awareness campaigns are launched in Greece, Albania and FYROM among decision makers, fishermen and local authorities, in order to convince them to prohibit any idea of further introductions.

It should not be forgotten that the breeding populations of many fish-eating birds depend entirely on these fish populations (Catsadorakis et al., 1996b). The same holds for the fisheries. In addition to the socio-economic aspects affecting this activity, it is clear that the maintenance of the fisheries depends above all on the presence of abundant and renewable populations of carp. To achieve this, a correct balance must be found between the harvest from fishing and recruitment, which must be high enough to compensate for both natural mortality and fishing. On Mikri Prespa, this would require a better management of water level and reedbeds which would have to be done in collaboration with the Albanians. The collection of fishery statistics also urgently needs to be restarted, fishing licences need to be introduced and possibly also fishery logbooks that would have to be filled out for each fishing trip. Our monitoring of the fish populations of Lake Mikri Prespa has shown how correctly conducted monitoring with clear objectives can be useful for assessing population trends and improving understanding of the functioning of the fish populations of these lakes. It is therefore desirable that such monitoring should continue in future years.

Acknowledgments

We would sincerely like to thank the many people who have helped us in fish-sampling over the years for their invaluable assistance. None of our studies on the Prespa lakes would have been possible without the constant financial support of the Fondation Tour du Valat. We would also like to thank V. Hatzirvassanis for preparing the drawings.

References

Anderson, N. J., 1995. Naturally eutrophic lakes: reality, myth or myopia? TREE 10: 137–138.

Anonymous, 1985. The work of the hatchery room of the Institute. Edition jubilaire consacrée à l'honneur de la 50ème anniversaire de la Fondation de l'Institut hydrobiologique, Ohrid. Institut hydrobiologique d'Ohrid, FYROM, Volume 1, 67.

Apostolski, K., 1972. Contribution to the fish population study in Lake Prespa. Godisen Zbornik na Zemjodelsko. Sumarskiot Fakultet na Universitetat, Skopje 24: 207–220. (In Serbian).

Athanassopoulos, G., 1922. Le plateau des lacs de la Macédoine occidentale. Int. Revue ges. Hydrobiol. 10: 31–39.

Athanassopoulos, G., 1923. La faune ichtyologique et la production des lacsde macédoine, Thessalie etc, durant les années 1919 à 1922. Bull. Pêche Aquicult. 10–12: 115–117.

Balon, E. K., 1975. Reproductive guilds of fishes: a proposal and definition. J. Fish. Res. Bd Can. 32: 821–864.

Banarescu, P., 1964. Pisces Osteichthyes. Fauna Rep. Pop Rom 13: 961 pp.

Berrebi, P., F. Dupont, G. Cattaneo-Berrebi & A. J. Crivelli, 1989. An isozyme study of the natural cyprinid hybrid *Alburnus alburnus* × *Rutilus rubilio* in Greece. J. Fish Biol. 34: 307–313.

Bianco, P. G., 1988. Occurence of the Asiatic gobionid *Pseudorasbora parva* (Temminck and Schlegel) in south-eastern Europe. J. Fish Biol. 32: 973–974.

Boy, V. & A. J. Crivelli, 1988. Simultaneous determination of gillnet selectivity and population age class distribution for two cyprinids. Fish. Res. 6: 337–345.

Brabrand, A. & B. Faafeng, 1993. Habitat shift in roach (*Rutilus rutilus*) induced by pikeperch (*Stizostedion lucioperca*) introduction: predation risk versus pelagic behaviour. Oecologia 95: 38–46.

Catsadorakis, G., M. Malakou & A. J. Crivelli, 1996a. The Prespa barbel, *Barbus prespensis*, Karaman 1924, in the Prespa lakes basin, north-western Greece. Tour du Valat, Arles, 79 pp.

Catsadorakis, G., M. Malakou & A. J. Crivelli, 1996b. The effects of the 1989–1990 drought on the colonial waterbirds nesting at Lake Mikri Prespa with special attention on pelicans. *Colonial Waterbirds* 19 (Special Publication 1): 207–218.

Catsadorakis, G., 1997. The importance of Prespa National Park for breeding and wintering birds. Hydrobiologia 351: 157–174

Cowx, I. G., 1991. Catch effort sampling strategies: conclusions and recommendations for management. In Cowx, I. G. (ed.), Catch Effort Sampling Strategies, Fishing News books, Oxford: 404–413.

Crivelli, A. J., 1984. Lakes – Fisheries. In Integrated Environmental Study of Prespa National Park, Greece. Final report. Commission of the European Communities DG XI, 49–86 and 144–155. Bruxelles, Belgium, 205 pp.

Crivelli, A. J., 1990. Fisheries decline in the freshwater lakes of northern Greece with special attention for Lake Mikri Prespa. In W. L. T. van Densen, B. Steinmetz & R. H. Hughes (eds), Management of Freshwater Fisheries, Proc. symp. Eur. Inland Fish. Advisory Commission, Göteborg, Sweden, 31 May–3 June 1988. PUDOC, Wageningen: 230–247.

Crivelli, A. J., 1992a. Fisheries of the Mediterranean wetlands. Will they survive beyond the year 2000? In O'Grady, T., A. J. B. Butterworth, P. B. Spillett & J. C. J. Domaniewski (eds), Fisheries in the Year 2000, Proc. 21st Anniversary Conference, the Institute of Fisheries Management, 1990, London, England: 237–252.

Crivelli, A. J., 1992b. Fisheries. In Gerakis, P. A. (ed.), Conservation and Management of Greek wetlands, Proc. Greek Wetlands Workshop, Thessaloniki, Greece, 1989, IUCN, Gland, Switzerland: 115–127.

Crivelli, A. J., 1995. Are fish introductions a threat to endemic freshwater fishes in the northern Mediterranean region? Biol. Conserv. 72: 311–319.

Crivelli, A. J., 1996. The freshwater fish endemic to the northern Mediterranean region. An Action Plan for their conservation. Tour du Valat, Arles, 171 pp.

Crivelli, A. J. & F. Dupont, 1987. Biometrical and biological features of *Alburnus alburnus* × *Rutilus rubilio* natural hybrids of Lake Mikri Prespa, northern Greece. J. Fish Biol. 31: 721–733.

Crivelli, A. J., H. Jerrentrup, T. Nazirides & P. Grillas, 1995. Effects on fisheries and waterbirds of raising the water levels at the Kerkini reservoir, a Ramsar site in Northern Greece. Envir. Mgmt 19: 431–443.

Crivelli, A. J., M. Malakou, G. Catsadorakis & E. Rosecchi, 1996. The Prespa barbel, *Barbus prespensis*, a fish species endemic to the Prespa Lakes (north-western Greece). Folia Zool. 45: 21–32.

Cvijic, J., 1911. L'ancien lac Egeen. Ann. Geogr. Paris 20: 233–259.

Daget, J., P. S. Economidis & J. Louis, 1977. Sous-espèces d'*Alburnoides bipunctatus* (Pisces, Cyprinidae) de la Grèce continentale. Cybium 2: 7–23.

Dupont, F. & A. J. Crivelli, 1988. Do parasites confer a disadvantage to hybrids? A case study of *Alburnus alburnus* × *Rutilus rubilio*, a natural hybrid of Lake Mikri Prespa, northern Greece. Oecologia (Berlin) 75: 587–592.

Dupont, F. & A. Lambert, 1986. Etude des communautés de monogènes Dactylogyridae parasites de Cyprinidae du Lac Mikri Prespa (Nord de la Grèce). Description de trois nouvelles espèces chez un *Barbus* endémique: *Barbus cyclolepis prespensis* Karaman, 1924. Ann. Parasitol. Hum. Comp. 61: 597–616.

Economidis, P. S., 1986. *Chalcalburnus belvica* (Karaman, 1924) (Pisces, Cyprinidae), nouvelle combinaison taxinomique pour la population provenant du lac Petit Prespa (Macédoine, Grèce). Cybium 10: 85–90.

Economidis, P. S., 1989. Distribution pattern of the genus *Barbus* (Pisces, Cyprinidae) in the freshwaters of Greece. Trav. Mus. Hist. nat. 'Grigore Antipa' 30: 223–229.

Economidis, P. S., 1991. Check list of freshwater fishes of Greece. Recent status of threats and protection. Hellenic Society for the Protection of Nature, Athens, 47 pp.

Economidis, P. S., 1992. Fish. In M. Karandinos & E. Legakis (eds), The Red Data Book of Threatened Vertebrates of Greece, H.Z.S. & H.O.S., Athens: 43–81.

Economidis, P. S., 1993. The uniqueness of ichtyofauna of Prespa Lakes. Proc. 15th Congr. the Hellenic Soc. Biol. Sci., Florina, 1993, 22–24 (In Greek).

Elvira, B., 1987. Taxonomic revision of the genus *Chondrostoma* Agassiz, 1835 (Pisces, Cyprinidae). Cybium 11: 111–140.

Helawell, J. M., 1971. The autoecology of the chub, *Squalus cephalus* (L.) of the river Lugg and the Afon Llynfi. III. Diet and feeding habits. Freshwat. Biol. 1: 369–387.

Hollis, G. E. & A. C. Stevenson, 1997. The physical basis of the lake Mikri Prespa systems: geology, climate, hydrology and water quality. Hydrobiologia 351: 1–19.

Hristovski, N. D., 1976. The helminth fauna of the cyprinid fish from Prespa lake. Acta Parasitologica Jugoslavia 7: 21–28.

Ivanovic, B. M., 1973. Ichthyofauna of Skadar Lake. Institution for biological and medical research in Montenegro, Titograd, 146 pp.

Jovanovic, V., 1973. The rate of individual growth in Prespa bleaks (*Alburnus* sp.) in the lakes of Ohrid, Prespa, Doyran, and Skutari. Glas. Republ. Zavoda zast Prirode Prirodnjackog Muzeja, Titograd 5: 115–127.

Karakousis, Y. & C. Triantaphyllidis, 1990. Genetic structure and differentiation among Greek brown trout (*Salmo trutta* L.) populations. Heredity 64: 297–304.

Karakousis, Y., C. Triantaphyllidis & P. S. Economidis, 1991. Morphological variability among seven populations of brown trout, *Salmo trutta* L. in Greece. J. Fish Biol. 38: 807–817.

Karaman, S., 1924. Pisces Macedoniae. Split.

Koussouris, T. & A. Diapoulis, 1983. For the development and protection of freshwater resources in Greece. I. Lake Mikri Prespa. I.O.F.R. Spec. Publ. 6, 89 pp.

Mann, R. H. K., 1976. Observations on the age, growth, reproduction and food of the chub, *Squalius cephalus* (L.) in the river Stour, Dorset. J. Fish Biol. 8: 265–288.

Mihaloudi, E., M. Zarfdjian & P. S. Economidis, P., 1997. The zooplankton of Lake Mikri Prespa. Hydrobiologia 351: 77–94.

Moss, B., S. McGowan & L. Carvalho, 1994. Determination of phytoplankton crops by top-down and bottom-up mechanisms in a group of English lakes, the West Midland meres. Limnol. Oceanogr. 39: 1020–1029.

Petridis, D. & A. Sinis, 1997. The benthic fauna of Lake Mikri Prespa. Hydrobiologia 351: 95–105.

Petrovski, H. & M. Sidorovski, 1971. Age and growth of *Alburnus alburnus belvica* Kar. from Prespa Lake. Inst. Pisc. Macédoine 4: 1–12.

Popovska-Stankovic, O., 1972. On nutrition of carp (*Cyprinus carpio* L.) from Prespa lake. Ichtyologia 4: 67–76. (In Serbian).

Pyrovetsi, M. D. & M. Karteris, 1986. Forty years of land/cover use change in Prespa National Park, Greece. J. environ. Mgmt., 23: 173–183.

Rakaj, N. & A. Flloko, A. 1995. Endemic freshwater fish of Albania. Biol. Conserv. 72: 195–199.

Rossecchi, E., A. J. Crivelli & G. Catsadorakis, 1993. The establishment and impact of *Pseudorasbora parva*, an exotic fish species introduced into Lake Mikri Prespa (north-western Greece). Aquat. Conserv. Mar. Freshwat. Ecosyst. 3: 223–231.

Stankovic, S., 1929a. La productivité piscicole des lacs balkaniques du sud. Bull. Soc. Geogr. Belgrade 25: 20–35. (In Serbian).

Stankovic, S., 1929b. Types of fishing boats used in the lakes of southern Balkans. Glasnik Etnografskog Muzeja Beogradu 4: 82–100 (In Serbian).

Stankovic, S., 1931. Les particularités limnologiques des lacs égéens. Verh. int. Ver. Limnol. 5: 158–196.

Stankovic, S., 1960. The Balkan lake Ohrid and its living world. Dr W. Junk Publishers, The Hague: 357 pp.

Stevenson, A. C. & R. J. Flower, 1991. A palaeoecological evaluation of environmental degradation in Lake Mikri Prespa, NW Greece. Biol. Conserv. 57: 89–109.

Tocko, M., 1958. Développement de *Rutilus rubilio prespensis* Karaman. Recueil des travaux de la Station hydrobiologique, Ohrid 11: 1–8.

Tryfon, E. & M. Moustaka-Gouni, 1997. Species composition and seasonal cycles of phytoplankton with special reference to the nanoplankton of Lake Mikri Prespa. Hydrobiologia 351: 61–75.

Hydrobiologia **351:** 127–133, 1997.
A. J. Crivelli & G. Catsadorakis (eds), Lake Prespa, Northwestern Greece.

The distribution and habitat preferences of the amphibians of Prespa National Park

D. Bousbouras[1] & Y. Ioannidis[2]
[1] *Hellenic Ornithological Society, Thessaloniki Office, 8 Kalapothaki Street, 54624 Thessaloniki, Greece*
[2] *Goulandris National History Museum, Levidou 13, 14562 Kifissia, Greece*

Key words: amphibians, Caudata, Anura, habitat preferences, Greece

Abstract

The Prespa National Park has a large number of amphibian species compared to other parts of Greece and the Balkan Peninsula. This is due to the high variety of habitats, the existence of relatively large areas with shallow water around the lake and also due to the existence of mountainous habitats very close to it. In this contribution, information is given on the biology of the amphibians in the area, plus data on the distribution and habitat preferences.

Introduction

A diverse amphibian fauna occurs in the Prespa National Park. Amphibians play an important role in the food chain of the area especially in spring when huge numbers of tadpoles are present. Some information on the amphibians was published by Kuhnelt (1981) but Catsadorakis (1986) produced the first list of the species present in the area and some distribution maps. In the present work a brief description is given of the distribution and habitat preferences of the amphibians.

Methods

Observations were made in spring and summer of 1986 and 1987. We recorded the same data as for reptiles (cf. Ioannidis & Bousbouras, 1997), and our data were supplemented with data offered by G. Catsadorakis (unpubl. data). The study area included the area of National Park and the Greek part of the drainage basin surrounding the National Park which was divided into 2 km × 2 km squares. The vegetation has been described by Pavlides (1985). An attempt was made to give equal weight to each square by devoting the same total amount of time to each square under various weather conditions and various times of the day. As this was impossible to put into practice, the data which were collected give a clear image of the preferences of the species, but not of their abundance in the various habitats. At the same time some measurements were made on the breeding areas (marshes, ponds, wet meadows, streams) to discover the most important reproduction sites.

Species list with notes on their biology in Prespa

A species account of the 11 amphibian species (3 Caudata and 8 Anura) found in Prespa National Park follows; with comments on their reproductive biology in the area:

Caudata

Salamandra salamandra *Fire Salamander*
A common species in mountainous Greece. It occurs in beech and oak forests, in moist places. From March to April larvae occur in slow-flowing mountain and upland streams with clear water. At this time of the year large concentrations of females are observed close to the mountainous streams. It is a nocturnal animal, but it is also active on cloudy or rainy days. It is usually inactive during summer.

Triturus cristatus *Warty Newt*
Lives in stagnant deep water at lower altitudes close to the lakes. This is the typical breeding habitat of the species also mentioned in other areas (Cooke & Frazer, 1976). In this region even mature individuals are rarely found away from water. Females lay eggs in spring from March to May.

Triturus vulgaris *Smooth Newt*
It is found mainly in permanently flowing streams in the mountain region. Whenever there is sufficient atmospheric humidity, it is found on land in damp places. It breeds in still bodies of water or slowly flowing streams, where it lays eggs between March and June.

Anura

Bombina variegata *Yellow Bellied Toad*
An exclusively aquatic species, readily adapted to all kind of shallow water without dense vegetation. It is found in ponds, marshes, rivers, streams and in temporary puddles of water in unpaved forestry or farm tracks. 60% of observations were in places with no vegetation. A very social, always aquatic species. In order to protect itself it displays the bright yellow colours of the belly as a warning. The breeding period is very extended, often lasting even after August.

Pelobates syriacus *Eastern Spadefoot*
This area is the western limit of its range in Greece and possibly in Europe. It occurs at a very low density. A nocturnal species spending its day hiding in a burrow that it digs in the soil. It occurs in cultivated areas and sandy soils in marshes close to the lakes.

Bufo bufo *Common Toad*
A common species occurring in a wide variety of habitats, even very dry ones, from the lake level to subalpine meadows. It is nocturnal and independent of water which it visits only at the breeding period. It breeds in still or slow flowing shallow water.

Bufo viridis *Green Toad*
This is much smaller than the former species with almost the same behaviour. It prefers lower altitudes and areas adjacent to lakes. As with the Common Toad it can be found kilometers away from its breeding place. In spring and especially in March, many animals of both *Bufo* species are found dead on the roads, killed by cars.

Hyla arborea *Common Tree Frog*
This is a mainly nocturnal species, the only one with well developed climbing abilities. In this region it is usually found close to water in marsh vegetation and bushes. It breeds in marshes and wet meadows. It is one of the first species to start calling for breeding advertisement, usually in mid-March.

Rana graeca *Stream Frog*
It prefers fast-flowing clear water with stones on banks or soil clear of vegetation. A typical species of streams in mountain areas, but it occurs at low densities in the area. In contrast to the other mountainous species (*Rana dalmatina*) it is always found in or close to streams.

Rana dalmatina *Agile Frog*
A species fairly independent of water but in contrast to the toads it is absent from dry habitats. It is found mainly in the damper parts of the forests and only 10% of observations were close to water. It breeds very early in spring, mainly in the wet meadows and shallow ponds with still water. In contrast to Northern Europe, in this region as in the lower altitude areas of Greece (Sofianidou & Kyriakopoulou, 1983), this species breeds immediately after hibernation.

Rana ridibunda *Marsh or Lake Frog*
The commonest frog species. It occurs in all types of water habitats from the lake to streams, but always in places with reeds and aquatic vegetation. Less common or absent from the mountain streams. Usually and especially close to the lake, it is present in high densities.

The individual species within the green frog complex are identified by bioacoustic methods and field identification is difficult as hybrids are present with intermediate morphometric characteristics. The taxonomic problems of the green frog complex are well known in the Mediterranean region (Lescure, 1992) and furthermore the species *Rana epeirotica* has been identified in N.W. Greece (Schneider et al., 1984). So, although we refer to *Rana ridibunda*, the specific status of the green frogs in Prespa has not yet been defined.

With the exception of the Alpine Newt (*Triturus alpestris*), all the European amphibian species whose

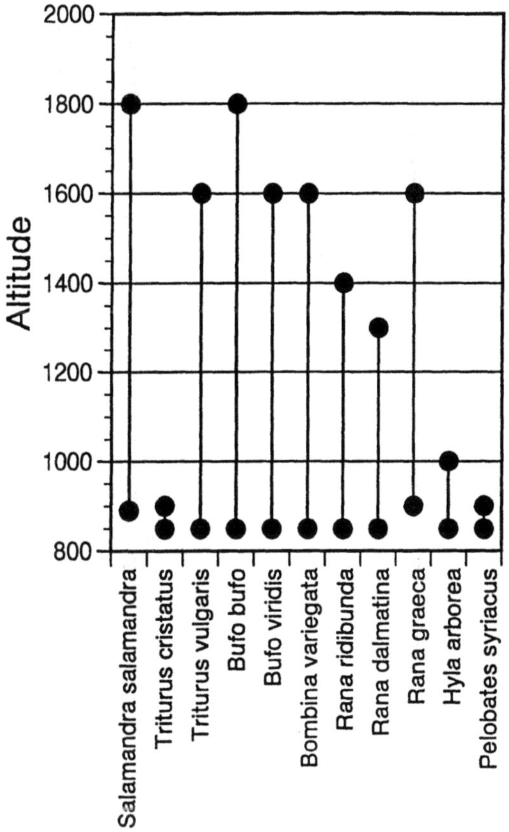

Figure 1. Altitudinal distribution of amphibiens in the study area.

geographical range include Prespa National Park, are indeed found here.

The absence of the Alpine Newt might be attributed to the absence of mountain ponds in the Varnous and Triklario mountains.

The presence of the Asian species *P. syriacus* is of special interest as the Lakes Prespa and Ohrid are the westernmost points of its distribution (Bruno, 1989) and moreover in the Prespa area it is found at relatively high altitudes (850 m, Figure 1), whereas usually it is met not higher than 500 m (Bohme, 1975; Sofianidou, 1977). This is possibly attributable to the mild climate of the area compared to the surrounding mountains, due to the presence of the lakes. 79% of the amphibian species of Continental Greece and 42% of those of the Balkan peninsula are found within the Prespa National Park, according to the catalogue of Societas Europaea Herpetologica (Corbett, 1989).

Distribution and habitat preferences

The distributions of the amphibian species in the area are shown in Figures 2 and 3. The small circles denote the presence of the species in the area. The large circles denote an important breeding site for the area. The criterion for an important breeding site is the presence of the following densities at the peak of the breeding season:

S. salamandra: >30 females/100 m. river bank
T. cristatus: >200 larvae/10 m^2
B. variegata, R. ridibunda: >200 ind./100 m^2
H. arborea, Bufo bufo, B. viridis: >50 ind./100 m^2

These criteria for each species are only applicable to the National Park area because, unfortunately there are no general population studies from other areas of Greece and the southern Balkans.

The various types of habitats are used in different ways by the amphibians. Some species are closely related to one type of habitat while others have much wider distribution. The main factors that define the distribution of amphibian species (see also Figure 1) in the area are:

– The dependence on water of the mature individuals. The species *Triturus vulgaris, T. cristatus, Bombina variegata, Rana graeca* and *R. ridibunda* are highly dependent on water and they are always found in or close to it. In contrast, the species *Bufo bufo* and *B. viridis* are entirely independent of water and can even occur in very dry habitats, while *Salamandra salamandra, Pelobates syriacus, Hyla arborea,* and *Rana dalmatina* are found in places away from water but always with a high humidity.

– The type of aquatic habitat for the exclusively aquatic species. *Triturus vulgaris* and *Rana graeca* prefer clear, cold, flowing water, while *Triturus cristatus* prefers still water. In contrast, for *Rana ridibunda* the main factor is the presence of aquatic vegetation while this is not the case for *Bombina variegata.*

– The type of substrate and vegetation. Of the terrestrial species, the *Salamandra salamandra* is an exclusive forest species while *Pelobates syriacus* prefers open sandy soils and *Hyla arborea* areas with aquatic vegetation.

Most of the species are found near the lakes due to the variety of habitats associated with water. The number of species decreases as one moves up from the lake level to the alpine meadows. (Figure 1)

The shallow shores of lake Mikri Prespa are mostly covered by reedbeds. Other areas are rocky and steep but in some places of low inclination wet meadows

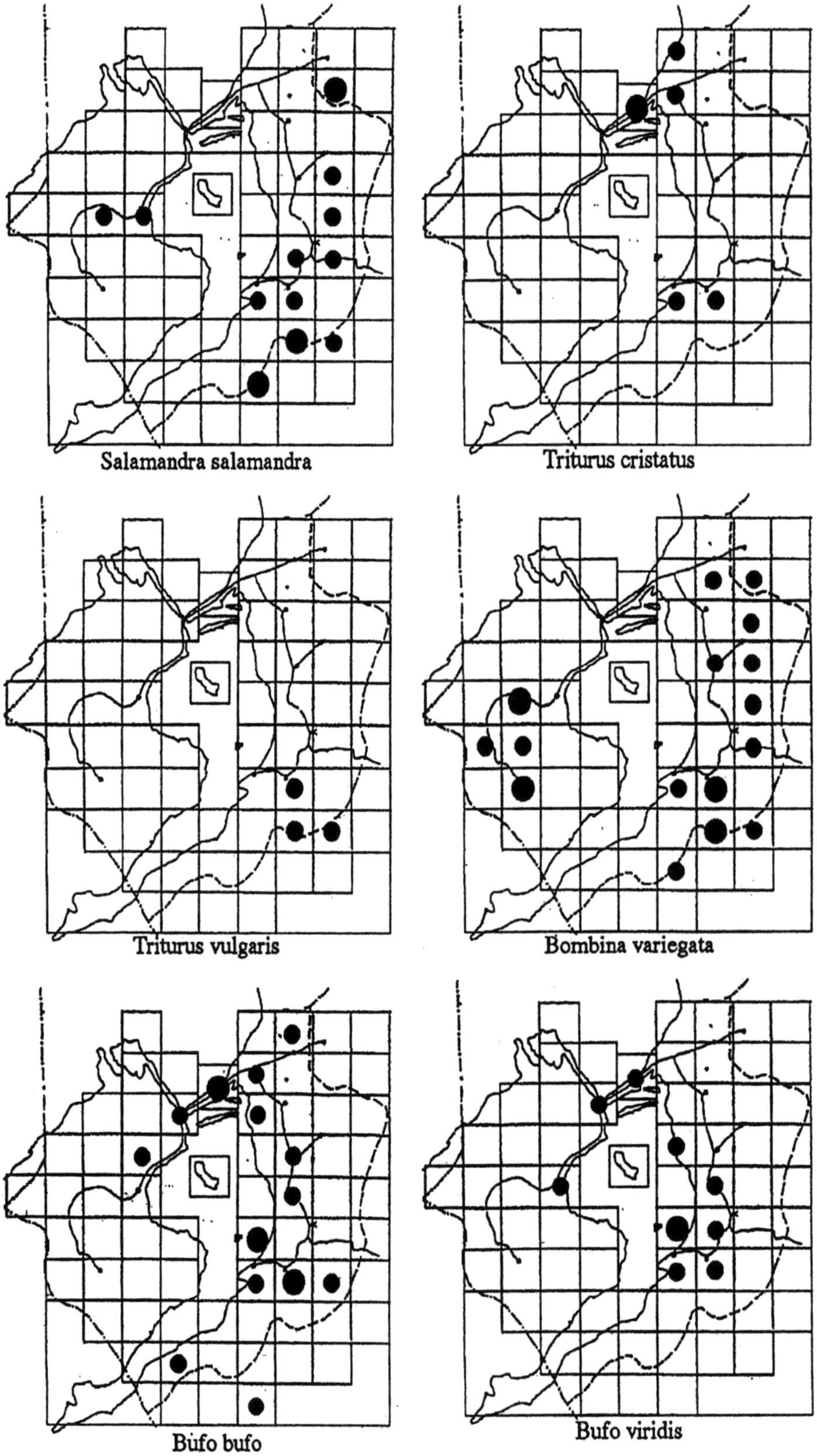

Figure 2. Distribution of the amphibien species in the study area. The small circles denote the presence of the species in the area. The large circles denote an important breeding site for the area.

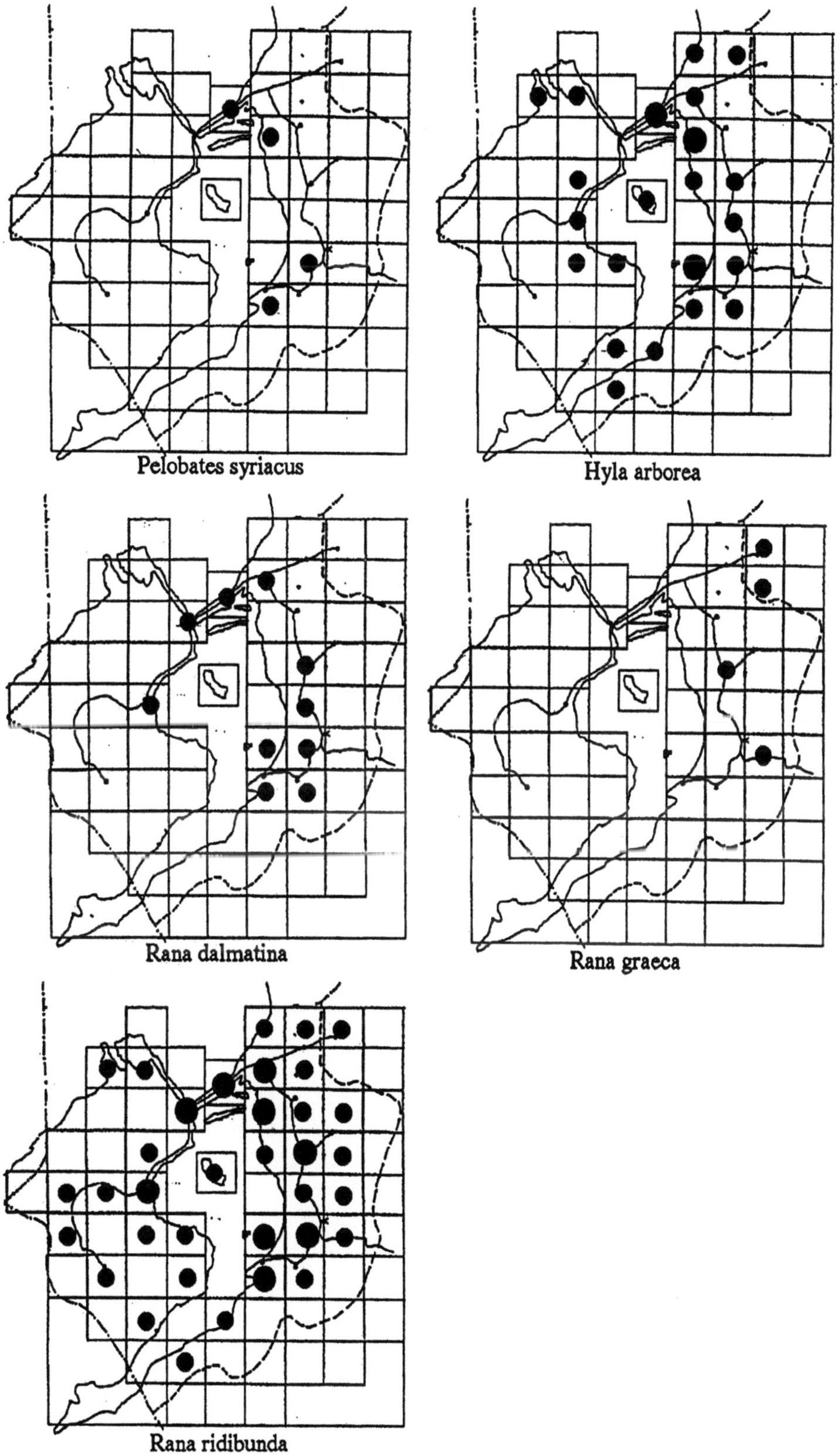

Figure 3. Distribution of the amphibien species in the study area. The small circles denote the presence of the species in the area. The large circles denote an important breeding site for the area.

exist. These wet meadows are flooded with water early in the spring and become dry at the end of the summer when the water level of the lake falls. Hence, during the amphibian reprodution period they maintain shallow water.

Most of the species of amphibians reproduce in this area, with the exception of *S. salamandra, T. vulgaris, R. graeca* and *B. variegata.* The two species of *Bufo* toads, that live in other areas for most of the year, move during the spring particularly to these areas (i.e. wet meadows and other temporary water areas) to reproduce. *R. ridibunda* and *H. arborea* are especially abundant in the reedbeds. Wherever reedbeds are too dense, there is no space for egg-laying and the amphibians have to move to nearby areas, such as the wet meadows and swamps.

R. ridibunda, H. arborea and hundreds of young *T. cristatus* were observed in summer in an area with small temporary ponds in the alluvial strip of land between the two lakes. Many of these small lakes have been produced by former sand quarrying for building purposes. The continued excavation of sand that is taking place in the area could destroy these water bodies. If new ponds are not created nearby, there could be a negative impact on the population of *T. cristatus* because of their high preference for these areas for reproduction. The same area is also used for breeding by *R. ridibunda* and possibly by *P. syriacus*. The population of the latter species is low and their breeding sites have not been found.

The area of cultivated land and wet meadows is traversed by streams and drainage ditches which are the principal breeding and living sites of amphibians. The artificially embanked streams in this zone have a more or less constant flow but unfortunately in 1985 their bankside trees and bushes were cleared. Despite this, they still constitute important breeding sites for the *R. ridibunda, H. arborea, R. dalmatina* and for the three toads *Bufo* and *Bombina*. The two *Bufo* toads and the Agile Frog reproduce at the slower flowing parts of rivers and in temporary pools created in the river bed by falling water levels.

In the drainage ditches and the marshes near Lake Mikri Prespa there are dense populations of *R. ridibunda* and *H. arborea*. Many individuals of most species can become trapped in concrete irrigation channels but the maintenance of some water or of increased humidity during summer in the unused water channels, enables most of the trapped animals to survive.

The forest zone between 850 and 1800 m is crossed by small streams of little flow with clean, oxygenated water. In these small streams, within the oak woodland zone (at altitudes of less than 1200 m) the following species are present: *R. ridibunda, R. dalmatina, R. graeca, B. variegata, S. salamandra* and *T. vulgaris*.

Within the beech woodland zone (between 1200 and 1800 m), the species richness and abundance decreases: *S. salamandra, T. vulgaris, B. variegata* and *R. dalmatina* are present.

The *H. arborea* is recorded only in the meadows and forests at lower altitudes near the lake. Only *Bufo bufo* has been observed in the subalpine meadows.

Conclusion

The high biodiversity of the amphibians and the presence of two rather rare species for Greece, *Triturus cristatus* and *Pelobates syriacus*, (Karandinos & Legakis, 1992) make it important to take measures for the conservation of the species and their populations. The conservation of large populations, mainly of the *Anura* around the lake, must be a basic aim as they are considered to be a basic food source for many species of aquatic birds and especially for the Ciconiiformes (Cramp & Simmons, 1977). The feeding sites of the aquatic birds in Prespa National Park coincide with those where amphibians are active and breed (Catsadorakis, 1986).

Unfortunately the populations are dwindling as their breeding habitats are threatened. The fast draining of wet meadows in spring because of the pumping of water for irrigation, the excavation of sand in the area between the two lakes and the cutting of the vegetation in the artificial streams and drainage ditches in the cultivated area, are considered to be the most important threats. Also, the pesticides which are used in the cultivated area are flushed out by intensive irrigation, and soon end up in the drainage ditches where they may cause deaths or even extinction of species (Honegger, 1978; Berger, 1987).

These problems must be faced through a global framework of protection and improvement of the wetlands, having as a goal the conservation of the population of the amphibians as well as the aquatic birds.

References

Berger, L., 1987. Impact of agricultural intensification on Amphibia. Proc. Fourth Ord. Gen. Meet. S.E.H. Nijmengen, 79–82.

Bohme, W., 1975. Zur Vorkommen von *Pelobates syriacus* in Griechenland. Senckenbergiana Biol. 56: 199–202.

Bruno, S., 1989. Introduction to a study of the herpetofauna of Albania. Brit. Herp. Soc. Bul. 29: 16–41.

Catsadorakis, G., 1986. Biotopes and vertebrates in Prespa National Park. Univ. of Athens, 161 pp. (in Greek).

Cooke, A. & J. Frazer, 1976. Characteristics of newt breeding sites. J. Zool. Lond. 178: 223–236.

Corbet, K., 1988. Conservation of European Reptiles and Amphibians. Croom Helm, London.

Cramp, S. & K. E. L. Simmons (eds), 1977. Handbook of the birds of Europe, the Middle East and North Africa. The birds of the Western Palearctic. Oxford University Press.

Karandinos, M. & A. Legakis (eds), 1992. The Red Data Book of the threatened vertebrates of Greece. Hellenic Zoological Society and Hellenic Ornithological Society. Athens, 356 pp.

Honegger, R., 1978. Amphibiens et reptiles menacés en Europe. Conseil de l' Europe, Strasbourg, 127 pp.

Ioannidis, Y. & D. Bousbouras, 1997. The space utilization by the reptiles in Prespa National Park. Hydrobiologia 351: 135–142.

Kuhnelt, W., 1981. Vorlaufig uebersicht ueber die wechselwarme (Poikiloptherma) Land und ufertierwelt der umgebung der biologischen station bei Mikrolimni am kleinen Prespasee. Bulletin of the Hellenic Soc. Prot. Nature 26: 32–39.

Lescure, J., 1992. Les Amphibiens du pourtour Méditerranéen. Bull.Soc. Herp. Fr. 64: 1–14.

Schneider, H., T. Sofianidou & P. Kyriakopoulou, 1984. Bioacoustic and morphometric studies in water frogs of Lake Ioannina in Greece, and description of a new species. Sond. aus z. f. zool. syst. u. evol. 22: 349–366.

Pavlides, G., 1985. Geobotanical study of the National Park of Lakes Prespa. Part A: Ecology, Flora, Phytogeography, Vegetation. Aristotelian Univ., Thessaloniki, 308 pp (in Greek with English summary).

Sofianidou, T., 1977. Study on the biology and ecology of *Pelobates syriacus* Boettger (Anura: Pelobatidae). Doct. Diss. Univ. of Thessaloniki, 181 pp. (in Greek)

Sofianidou, T. & P. Kyriakopoulou, 1983. Studies on the biology of the frog *Rana dalmatina* Bonaparte during the breeding season in Greece. Amphibia-Reptilia 4: 124–136

Hydrobiologia **351:** 135–142, 1997.
A. J. Crivelli & G. Catsadorakis (eds), Lake Prespa, Northwestern Greece.

The space utilization by the reptiles in Prespa National Park

Y. Ioannidis[1] & D. Bousbouras[2]
[1] *The Goulandris Natural History Museum, Levidou 13, GR-14562 Kifissia, Greece*
[2] *31 Pergamou Street, GR-60100 Katerini, Greece*

Key words: lizards, snakes, habitat segregation, activity patterns, Greece

Abstract

In the first part of this paper data are presented on the habitat segregation of the reptile species of Lake Prespa National Park. Twenty species of reptiles are present in the study area. Most of them are of East-Mediterranean origin, some are Balkan endemics and a few are of northern origin. Systematic surveys and counts of reptiles were conducted in each of the following habitats: marshes and reedbeds, wet meadows, rocky coastal areas and sandy beaches, cultivated land, grazing lands, mixed deciduous, oak and beech forests, sub-alpine and alpine meadows. A list is provided of the typical reptile species occurring in each of these habitats, with notes on abundance and habits. The highest reptilian diversity was found in man-made/modified habitats, namely the farmland zone, which structurally is the more diversified. There was a dramatic fall of species richness with increasing altitude. The second part of the paper describes qualitative differences in habitat utilization among the members of the two major reptilian groups, lizards and snakes. The Sand Lizard *Lacerta agilis* is found exclusively in the alpine and subalpine grasslands above 1600 m. asl., the Common Wall Lizard *Podarcis muralis* is the more euryecous while all other Lacertids show a clear specialization in habitat use though retaining various degrees of overlap. Among snakes the Adder *Vipera berus* is restricted to high mountain grasslands, the two *Natrix* species are common and more or less confined to areas close to the lakes while the Nose-horned Viper *Vipera anmmodytes* the third commonest species of the area prefers rocky/stony slopes.The notes on habitat utilization by the remaining species of snakes are only of indicative value because of small number of observations.

Introduction

Although Prespa National Park is one of the best studied areas of wildlife interest in Greece, few publications refer to its herpetofauna. Scarce information on the reptiles of this area can be found in Karaman (1928), Kuhnelt (1981), Chondropoulos (1989) and Clark (1992). Catsadorakis (1986) presented an almost complete list of species with some notes on their habitat preferences. A checklist for all the mountainous area of Florina and Prespa has been prepared (Bousbouras & Ioannidis, 1994). A typical division of resources among species is along three main dimensions: Habitat, time and food type (Pianka, 1969). This study is mainly concentrated on habitat segregation, but we also give data on the abundances, distribution and activity patterns of the reptile species present in the area.

Methods

This study covers the area of the National Park and the surrounding mountains. The vegetation of the area is extensively described by Pavlides (1985).

Field work was carried out in April–May 1986 and 1987, July 1986 and August 1987. Additional data were collected from June to August 1988 and in June 1990.

During the summer months, we mainly conducted a general herpetological search and for each individual a number of variables were recorded: time of day, weather conditions, altitude, habitat type, microhabitat type and behavioural observations. Whenever possible, sex was also recorded and external body measurements were taken. Due to reasons of consistency the data, on which most of the analysis is based, was not combined

Table 1. List of species observed in the area and number of observations recorded.

Species	English name	n
Testudo hermanni boetgerii	Hermann's Tortoise	54
Emys orbicularis	European Pond Terrapin	33
Algyroides nigropunctatus	Dalmatian Algyroides	45
Lacerta viridis viridis	Green Lizard	192
Lacerta trilineata trilineata	Balkan Green lizard	2
Lacerta agilis bosnica	Sand Lizard	6
Podarcis erhardii liveti	Erhard's Wall Lizard	140
Podarcis muralis albanica	Common Wall Lizard	213
Podarcis taurica taurica	Balkan Wall Lizard	183
Ablepharus kitaibelii kitaibelii	Snake-eyed Skink	42
Anguis fragilis colchicus	Slow Worm	16
Malpolon monspessulanus insignitus	Montpellier Snake	3
Coluber caspius	Large Whip Snake	15
Elaphe quatuorlineata quatuorlineata	Four-lined Snake	5
Elaphe longissima longissima	Aesculapian Snake	7
Natrix natrix persa	Grass Snake	74
Natrix tessellata tessellata	Dice Snake	63
Coronella austriaca austriaca	Smooth Snake	7
Vipera ammodytes meridionalis	Nose-horned Viper	30
Vipera berus bosniensis	Adder	1

with the results of more specialized methods, such as density measurement or species-specific observations. Abundance was estimated from transects through a habitat, along predetermined paths of known distance, recording the numbers of individuals observed (Pianka, 1970). The results are expressed as numbers of individuals observed ha^{-1}.

We use the following indices:

Margalef's species richness index: $D_m = S - 1/\log N$ (Margalef, 1951)

Shannon-Wiener's diversity index (Pielou, 1966)

$$D_s = -\sum_{i=1}^{S} \text{pi} \, (\log_e \text{pi}),$$

where pi = proportional abundance of the ith species, S = number of species, N = total number of individuals.

Niche Breadth was calculated using the reciprocal of Simpson's (1949) measure (Pianka, 1986):

$$B = 1/\sum_{i=1}^{S} \text{pi}^2,$$

where pi = the proportion of the ith category, S = total number of categories.

The reptile community

A total of 20 species of reptiles was found to occur in our study area (Table 1), including 2 Chelonia, 9 Sauria and 9 Ophidia. The absence of the thermophilous members of the Gekkonidae family and the presence of two species with a far more northerly distribution centre *(Lacerta agilis, Vipera berus)*, clearly separates this area from the mediterranean lowlands.

The majority of the species present in the area are of East Meditterranean origin. There are also some Balkan endemics (*Algyroides nigropunctatus, Podarcis erhardii*). A few species have reached this area from the north. Such species with a Pontic origin are *Lacerta agilis* and *Coronella austriaca.* It is also worth mentioning the presence in the area of intergrades between *Emys orbicularis hellenica* and *E. o. orbicularis*, the Greek and the Pontic subspecies of the European Pond Terrapin (Fritz, 1992). No endemic forms are present in the area, which was to be expected as no real geographical barriers for reptiles are present and the majority of the species have a continuous distribution encompassing the National Park area.

We have not included *Elaphe situla* in this list of species, although a road casualty specimen (from Prespa) was given to us. As no other animal has been

Table 2. Distribution of the species in the four major habitat groups.

	Testudo hermanni	*Emys orbicularis*	*Algyroides nigropunctatus*	*Lacerta viridis*	*Lacerta trilineata*	*Lacerta agilis*	*Podarcis erhardii*	*Podarcis muralis*	*Podarcis taurica*	*Ablepharus kitaibelii*	*Anguis fragilis*	*Malpolon monspessulanus*	*Coluber caspius*	*Elaphe quatuorlineata*	*Elaphe longissima*	*Natrix natrix*	*Natrix tessellata*	*Coronella austriaca*	*Vipera ammodytes*	*Vipera berus*	Margalef's species richness index	Shannon-Wiener's diversity index
Lake zone	+	+					+	+	+						+	+	+	+	+		4.3	1.6
Man-influenced zone	+	+		+	+		+	+	+	+	+	+	+	+	+	+	+	+	+		5.5	2.0
Forest	+		+	+			+	+		+	+	+		+	+			+	+		4.1	1.8
Subalpine meadows						+		+												+	2.1	0.8

observed by us, we decided to exclude it as not fully documented. The presence of *Coluber najadum* is also possible, as we have observed this species less than 30 km. E.SE. from the lakes, in an area with similar climatic conditions. Finally G. Catsadorakis has informed us that he has repeatedly found road-killed individuals of *Coluber gemonensis* from the National Park area.

Habitats and species composition

The lake area

The two *Natrix* spp. and *E. orbicularis* are quite common in the lakes. Apart from these three, a few other species were observed on the coastal area. The coastal area is not homogeneous and can, in general terms, be divided as follows:

– marshes and reed beds, where only the three above mentioned species are present;

– wet-meadows. *Podarcis muralis, Vipera ammodytes* and *Elaphe longissima* were observed in these areas, whenever the water had withdrawn. Despite this it seems that no reptiles, except the three closely associated with water, permanently use these areas;

– rocky coastal areas. These areas are largely used by the two *Natrix* species as basking places, in contrast to *Emys orbicularis* which usually prefers less exposed and safer locations. *Podarcis muralis* and especially *P. erhardii* are also often encountered in these places. Occasionally we have recorded the temporary presence of a few other snakes (*V. ammodytes, C. austriaca)* near the water;

– Coastal areas with low gradients, usually with sandy soil. The main difference of this area is the presence of *Podarcis taurica.*

The zone influenced by humans

Apart from the human settlements this area can be divided into two categories:

– the cultivated zone, with irrigation canals, drainage ditches and hedges. *Lacerta viridis* and *Testudo hermanni* are common along the hedges and canal borders as long as there are adequate hiding places. *Podarcis muralis* and *P. taurica* are also encountered in this area. The cultivated zone is also used as feeding grounds for a number of snake species such as *Malpolon monspessulanus, Elaphe quatuorlineata, Coluber caspius* and *Vipera ammodytes.* The drainage ditches are used by *N. natrix* and *E. orbicularis* which are found in high densities but only if the edges are well vegetated and can provide a good refuge;

– grazing pastures, these are situated either in the lowlands or in the mountains, sometimes up to the alpine meadows. There is a variety of pasture types and many species are present here. In the low altitude vegetation areas, on flat ground *P. taurica* and *A. kitaibelli* are very common as are *L. viridis* and *T. hermanni* if shrubs are present. *P. erhardii, P. muralis,*

V. ammodytes and *C. austriaca* are often present on the rocky overgrazed slopes. *N. natrix* is sometimes found in the rivers that run throught this area. Some other species are found mainly on the borders of this zone.

The forests

These can be separated into three main categories:

– mixed forests, often situated on rocky ground. *P. muralis, P. erhardii* and *L. viridis* are very common in the open parts of these forests but also of great importance is the presence of *A. nigropunctatus,* which is almost exclusively found in these areas. A number of snake species is also present, with *V. ammodytes* being the most commonly encountered, at some distance from the lakes;

– oak forests. The main differences from the previous category in the lizard fauna is the absence of *A. nigropunctatus* and the frequent presence of *A. fragilis* and *A. kitaibelii. P. muralis* is more common along the roads. *T. hermanni* is locally abundant here as well as in the mixed forests. Among the snakes, *C. austriaca* and mainly *E. longissima* are present;

– beech forests. These are situated at higher altitudes. Few species are present generally at low densities. *P. muralis, A. fragilis, A. kitaibelii* and occasionally *L. viridis* are the main reptile species of this area.

Subalpine-alpine meadows

This area is used for livestock grazing during summer and it is characterized by *L. agilis. V. berus* has also been found in the vicinity of surface water (Ioannidis & Bousbouras, 1989). *P. muralis* is also encountered in a few localities.

In Table 2, two species diversity indices for the four main 'groups' of habitat are given. Margalef's index is used as a simple expression of species richness but it does not take into account the relative abundance of each species and is affected by the sample size. Shannon-Wiener's index is used to show the diversity of species within a habitat-group. The effects of sample size are low and apart from the number of species, it is also influenced by the evenness with which individuals are distributed among the species (Spellerberg, 1991). Both indices have lowest values for the alpine meadows. This habitat type is the least diverse of all four and as it is situated at high altitudes, unfavourable for the reptiles, the low values are expected. These values can therefore be used as a comparative measure showing a habitat type to be more or less unsuitable for reptiles. For the other three habitat type groups, the results can be misleading. The highest values are for the man-influenced zone. Although this may seem strange, it is quite expected, as this is the most diverse habitat group. A clearer picture can be given if we calculate the two indices for each habitat-type within each group (Table 3), as these have been presented earlier.

Table 3. Marlagef's and Shannon-Wiener's indices for the more important habitats.

	Marlgalef's richness index	Shannon-Wiener's diversity index
Marshes	1.2	1.1
Beaches	3.6	1.5
Wet meadows	3.8	1.3
Grazing pastures	5.1	1.9
Cultivation	5.3	2.0
Mixed Forest	4.1	1.7
Oak Forest	4.0	1.5
Beech Forest	2.3	1.3
Subalpine meadows	2.1	0.8

Some explanations in Table 3 are necessary. Margalef's index for marshes is lower than for the alpine meadows, although the number of species is the same. This is due to the smaller sample size from the latter habitat. At the same time Shannon-Wiener's index is higher for the former habitat showing a higher evenness. Both indices still attain their highest values in the two man influenced habitat types. Apart from conditions being suitable for the reptiles in these areas, we can still attribute this to the variability of microhabitats present in these two habitat types. So, if we exclude the individuals present in or at the vicinity of drainage ditches, the values for cultivated areas ($D_m = 3.4$, $D_s = 1.4$) are lower than those from the various forest types. Similarly the values we get for the overgrazed rocky slopes ($D_m = 2.5$, $D_s = 1.0$) are much closer to the 'unsuitable' alpine meadows.

The general conclusion is that the variability of habitats is a major reason for the diverse herpetofauna being present in this mountainous region.

Altitudinal influences

The altitudinal range of each species in the Prespa area is presented in Figure 1. Eight species are restricted

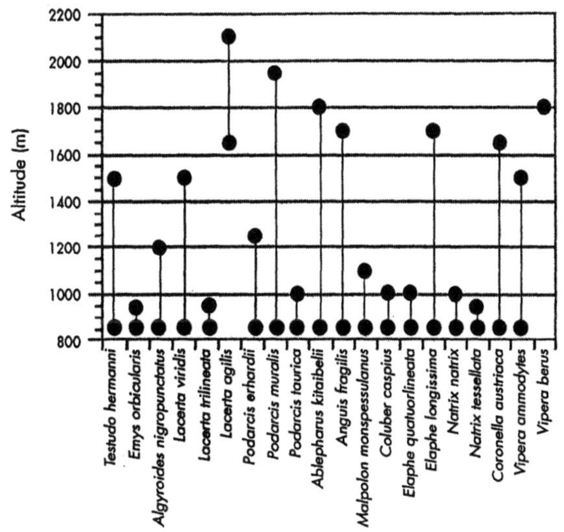

Figure 1. Altitudinal distribution of the reptile species in the study area.

to the lower parts of the area, that is around the lakes, while two others are restricted to the higher mountains. There is a significant decrease in the number of species with increasing altitude (Bousbouras & Ioannidis, 1994). This can be attributed to the decrease of spatial heterogeneity (fewer habitats at higher altitudes) and to the limitations imposed by climatic factors. The active period at higher altitudes is generally retarded because of the snow cover and the lower temperatures. Strijbosch et al. (1989) have demonstrated this for *A. kitaibelli* from the Evros province. In our area the peak activity for this species, at an altitude of around 900 m, was between mid April to mid May and some individuals where active even in summer, mainly in early morning or late evening. This is in agreement with the data from the higher mountains (>500 m) of Evros province (north-eastern Greece).

Habitat utilization

A major ecological differentiation between the lizard and the snake community is their food preferences (invertebrates vs vertebrates). In addition, the amount of data is much less for snakes than for lizards, we will therefore discuss the two communities separately.

Lizard community

One way of examining spatial partitioning among species is to look at the habitats and their relative use by the various species. We can even try to calculate the niche breadth of each species, but as the amount of field effort was not equal for all habitats this can only be used as an indication. There are some limits in using of spatial niche breadth from general data like ours, because of the difficulties of setting exact limits on the various habitats and also because of the difficulty in differentiating between the extent of the habitat and of the microhabitat components which may be present in various habitats, and have the same value for the occurrence of a species. For this more precise data are necessary.

For two members of the Lacertidae in our area the niche breadth is very small (close to 1), which means that mainly only one habitat type was used. *Lacerta agilis*, by using the subalpine and alpine meadows, has avoided interspecific competition, at least from other lizards. *Algyroides nigropunctatus* on the other hand shares the mixed forests with two other small lacertids, *P. muralis* and *P. erhardii. A. nigropunctatus* seems to be restricted to the shaded, rocky parts of the wood (73% of the observations) while the two other species are more abundant in the open parts (67%). If we try to correlate the species with the various components within each habitat (microhabitat preferences) we can have a much clearer picture of niche separation.

Podarcis taurica is restricted to sunny areas with low vegetation (90% of the observations), away from trees and shrubs (79% over 2 m away), seeking cover in the herb layer (Chondropoulos & Lykakis, 1983). It is more abundant in sandy areas (with a herb layer) in the northern part of the small lake, reaching densities of 80 ind. ha^{-1}. *Lacerta viridis,* the young of which are almost equal in size to *P. taurica,* also favours areas with low vegetation but this species is strongly connected to areas with trees and shrubs (84% less than 2 m away).

In contrast with the previous species *Podarcis erhardii* shows a strong preference to sunny rocky areas with 88% of the observations near, on, or under rocks or stones. The greatest density measured for this species was 57 ind. ha^{-1}, i.e. far lower in relation to the densities that it can reach in the Aegean islands (Catsadorakis, 1984; Valakos, 1990) or in some mainland lowlands (Ioannidis & Bousbouras, unpubl. data).

Podarcis muralis has the widest niche breadth of the four small Lacertids (3.68 in an analysis based on Table 3). It can be found almost anywhere (Clark, 1992) but in the strongholds of the other three species it has a rather marginal distribution and relatively low densities.

Ablepharus kitaibelii is usually found among leaf litter (55% of the observations) but also in the herba-

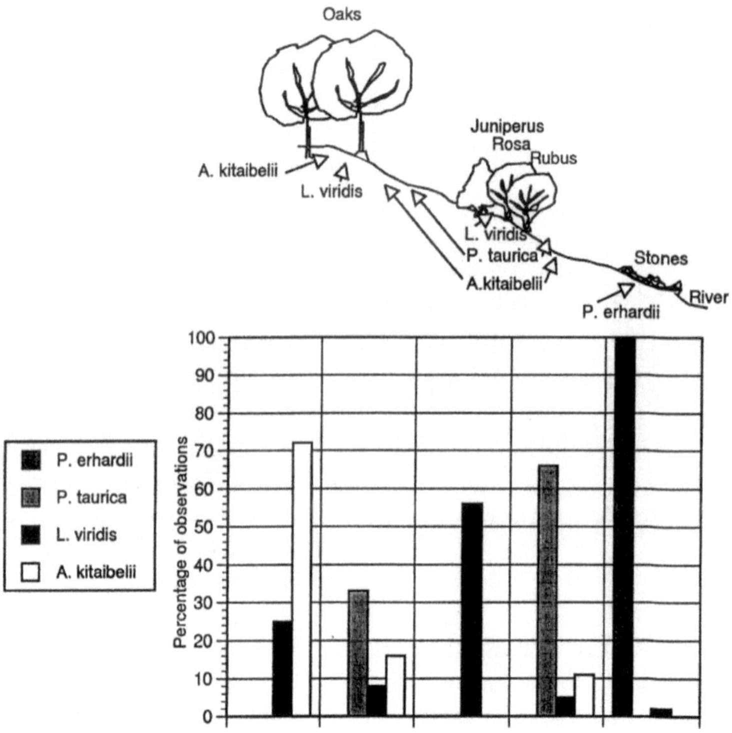

Figure 2. Percentages of captures of four lizard species in an 1.4 ha study area.

ceous cover of grazing pastures (45%). *A. kitaibelii* differs from *P. taurica*, which also favours this area, as it is much smaller and has a different temporal pattern of activity.

The spatial separation among the species found in a 1.4 ha area is shown schematically in Figure 2. Most reptiles were removed from this site over a two day period and released the day after (Bury, 1977). The positions of the animals recorded are the positions where the captured animals were first sighted.

Anguis fragilis shows a preference for humid areas near or in the forests. Most animals were found under stones at sunset but the amount of data for this species is insufficient.

With *L. trilineata* we faced a particular problem. We didn't find any adults that could be cleary identified as belonging to this species and the only specimens we definitely identified were two characteristic juveniles.

Snake community

For most of the species the results can only be considered as an indication of their habitat preferences in the area because of the low number of observations considered here.

Vipera berus is clearly separated from the rest of the species being restricted to the high mountain areas where no other snake has been observed.

Natrix natrix and *N. tessellata* are connected with water although they sometimes wander fairly large distances from it. That is especially the case with *N. natrix*. Of the two species, *N. tessellata* is by far more common in or near the lakes, while *N. natrix* is more often found away from the lakes, in drainage ditches or streams (Bousbouras & Ioannidis, 1994).

Of the other six species, three are present only in lower altitude areas and most of the observations come from cultivated areas or open grassy fields. *Coluber caspius* is the commonnest of the three. It has the widest habitat range and it is also present in wet areas. It is more often active in the high summer temperatures (Figure 3).

Malpolon monspessulanus and *Elaphe quatuorlineata* both prefer the open woods and fields but *E. quatuorlineata* is more common in cultivated zone while *M. monspessulanus* in stony places with shrubs.

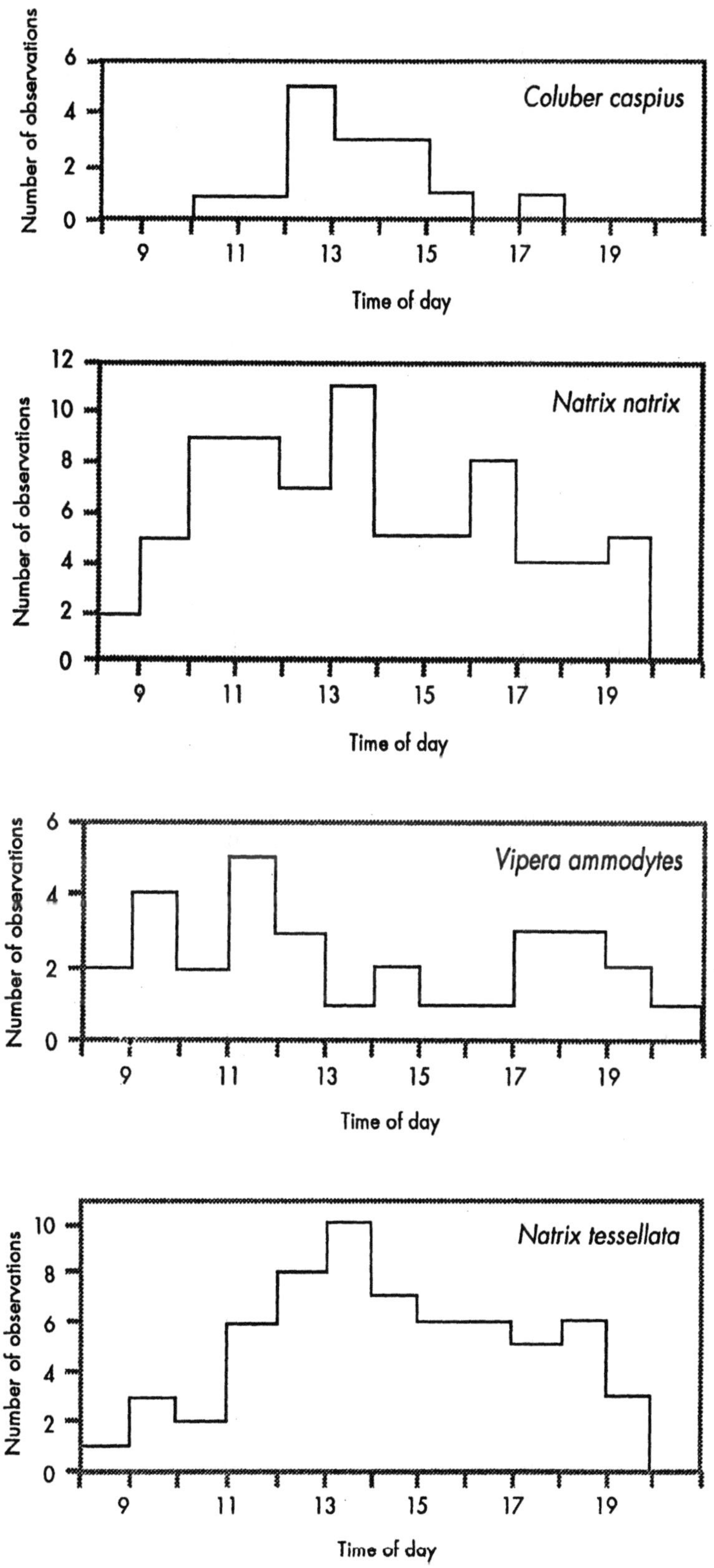

Figure 3. Activity periods for several snake species present in the area.

Of the three species that can also be found higher up on the mountains, *Vipera ammodytes* is the more common, in fact the most frequently met snake after the watersnakes. It has a wide habitat selection but seems more common in rocky or stony ground (63% of the observations). It is active in early morning or evening as well as at midday in spring (Figure 3).

Elaphe longissima is found in wooded areas and even in the open fields it is close to shrubs or trees. *Coronella austriaca* favours the same areas but it is also found in bare ground such as overgrazed slopes. While both species are active mainly in the morning or evening hours, both have also been observed at midday. While *E. longissima* has been observed basking, mainly on low temperature days, *C. austriaca* was never seen exposed under direct sunlight. It was only observed active at midday on cloudy days.

Conclusions

A very diverse herpetofaune occurs at the Prespa National Park mainly due to the variability of habitats present.

The reptile fauna of the area is composed of East Mediterranean and of Central European elements. The species of a clearly Mediterranean origin are present mainly around the lakes. Those of Central European origin are more evenly distributed occurring also among the mountains, while two are restricted to the mountains. The human influence in the area doesn't have any clear serious negative impact upon the reptiles, on the contrary some species may have benefited from the open (non-forested) areas. This is true as long as the land use is not yet highly intensive. The negative effects of intensive land use are shown on the overgrazed slopes.

None of the reptile species seems to be really threatened in the area. The only possible exceptions are *Lacerta agilis* and *Vipera berus* due to their limited range in the subalpine-alpine meadows. This area is used for grazing, in places intensively, during the spring and summer which is also the lizard's active period. Apart from this, only the human atittude, mainly towards snakes, and traffic have some negative effects but, at least until now, these are not severe enough to threaten any reptile species.

References

Bousbouras, D. & Y. Ioannidis, 1994. Amphibien und Reptilien des Prespa-Nationalparks und der Gebirgeregion um Florina (Mazedonien, Griechenland). Salamandra 30: 209–220.

Bury, R. B., 1977. Structure and composition of Mojave Desert reptile communities determined with a removal method. Symposium S.S.A.R. and Herp. League, Wildlife Res. Report 13: 135–142.

Catsadorakis, G., 1984. Aspects of Ecology and Behaviour in the lizard *Podarcis erhardii*. Biologia Gallo-Hellenica 11: 99–110.

Catsadorakis, G., 1986. Biotopes and vertebrates in Prespa National Park. Univ. of Athens, 161 pp. (in Greek).

Chondropoulos, B. P., 1989. A checklist of Greek reptiles. II. The snakes. Herpetozoa 2: 3–36.

Chondropoulos, B. P. & J. J. Lykakis, 1983. Ecology of the Balkan Wall Lizard, *Podarcis taurica ionica* (Sauria: Lacertidae) from Greece. Copeia: 991–1001.

Clark, R., 1992. An account of herpetological observations in Macedonia, The Pindos Mountains and Prespa Lake region, Greece. Herptile 17: 49–63.

Fritz, U., 1992. Zur innerartlichen Variabilitat von *Emys orbicularis* (Linnaeus, 1758). 2. Variabilitat in Osteuropa und Redefinition von *Emys orbicularis orbicularis* (Linnaeus, 1758) und *E. o. hellenica* (Valenciennes, 1832) Reptilia, Testudines: Emydidae). Zool. Abh. Mus. Tierkd. 47: 37–77.

Ioannidis, Y. & D. Bousbouras, 1989. Erster Bericht über die Balkan Kreuzotter *Vipera berus bosniensis* Boettger, 1889 aus Griechenland. Salamandra 25: 77–80.

Karaman, S., 1928. Prilog Herpetologiji Jugoslavije. Glasn. skopsk. nauc. dr., Skoplje 4: 129–143.

Kuhnelt, W., 1981. Vorlaufige Übersicht über die wechselwarme (Poikilotherma) Land und Ufertierwelt der Umgebung der Biologischen Station bei Mikrolimni am kleinen Prespasee. Physis, Athinai 26: 32–39.

Margalef, R., 1951. Diversidad de especies en las comunidades naturales. Publnes. inst. Biol. apl., Barcelona 6: 59–72.

Pavlides, G., 1985. Geobotanical Study of the Prespa Lakes National Park (NW. Greece). Part A: Ecology, Flora, Phytogeography, Vegetation. Aristotelian Univ., Thessaloniki, 308 pp. (in Greek).

Pianka, E. R., 1969. Sympatry of desert lizards (*Ctenotus*) in western Australia. Ecology 50: 1012–1030.

Pianka, E. R., 1970. Comparative autecology of the lizard *Cnemidophorus tigris* in different parts of its geografic range. Ecology 51: 703–720.

Pianka, E. R., 1986. Ecology and Natural History of Desert Lizards. Princeton Univ. Press, Princeton, New Jersey, 208 pp.

Pielou, E. C., 1966. The measurement of diversity in different types of biological collections. J. Theor. Ecol. 13: 131–144.

Simpson, E. H., 1949. Measurement of diversity. Nature 163: 688.

Spellerberg, I. F., 1991. Monitoring ecological change. Cambridge University Press, 334 pp.

Strijbosch, H., W. Helmer & P. T. Scholte, 1989. Distribution and ecology of lizards in the Greek province of Evros. Amphibia-Reptilia 10: 151–174.

Valakos, E. D., 1990. The ecology of the lizard *Podarcis erhardii* (Bedriaga, 1882) (Sauria: Lacertidae) in a typical insular ecosystem on Naxos Isl. Ph.D.Thesis, University of Athens, 213 pp. (in Greek with English summary).

Hydrobiologia **351:** 143–155, 1997.
A. J. Crivelli & G. Catsadorakis (eds), Lake Prespa, Northwestern Greece.

Breeding birds from reedbeds to alpine meadows

G. Catsadorakis
Society for the Protection of Prespa, 530 77 Agios Germanos, Greece

Key words: alpine birds, bird–habitat relationships, farmland birds, Greece, passerines, Prespa, reedbed birds, woodland birds

Abstract

The fauna of passerines (Passeriformes) and other bird families (Galliformes, Columbiformes, Cuculiformes, Coraciiformes and Piciformes) censused with similar methods, was examined; 132 species of the above families have been observed in the Prespa area. Of these, 19 non-passerines and 77 passerines breed. The composition of avian communities during the breeding season and the relative abundance of their species were studied in nine habitats, namely: reedbeds, croplands, sparse woodland on abandoned fields, hop-hornbeam woodland, oak-juniper shrubland, oak forest, beech forest, mixed deciduous forest and alpine grasslands. Six species made up the avifauna of reedbeds, the Great Reed Warbler being the most widespread. There were nine typical farmland species but Corn Bunting, Black-headed Bunting and Red-backed Shrike comprised more than 65% of the total. The most widespread and numerous species in the National Park were the typical species of temperate forests such as Blue Tit, Jay, Chaffinch, Blackcap, Chiffchaff, Marsh Tit and Great Tit. Skylark, Whinchat and Ortolan Bunting were the commonest species above the tree line. Very few similarities existed between the avifaunas of croplands, alpine grasslands and reedbeds. Correspondence Analysis was used to analyse the avifaunistic composition of each one of the six woodland habitats along with the floristic composition, seven abiotic features and 21 measures of vegetation structure. The most diverse avifauna was that of the mixed deciduous forest which had the highest horizontal and vertical structural heterogeneity. The avifauna of Juniper shrubland differed from those of all other types of woodland studied. The degree of human influence upon the vegetational features of this habitat is not easy to define, but the existence of endemic plant and animal taxa and relict species of plants and birds suggests that the differences can be attributed not only to its vegetational characteristics but may also have a paleoecological background.

Introduction

One of the main values of Prespa National Park is its habitat variety. This is due to its geographical position at the limits of the Mediterranean area, its geomorphology, its topography, the great altitudinal differences within the area and its geologic history (Pavlides, 1985). The activities of man for hundreds of years have had a differential impact upon each of these habitats, sometimes leading to an increase in biodiversity and sometimes to a decrease. Because of their high mobility and colonisation abilities birds are not the best indicators of habitat changes and usually it is not easy to draw firm conclusions on the relation between the features of a habitat and its birds from analysing their patterns of distribution and abundance, in places even slightly modified by man. On the other hand, in Greece very few studies of the habitat relations of birds have been made. Even for areas that are relatively well known ornithologically, only lists of species exist with indications of breeding and migration status and such lists often contain errors. Until recently this was also the situation for Prespa. The present paper summarizes part of the results from the only relatively long-term study of a passerine community in Greece, namely that of Prespa National Park (Catsadorakis, 1989). This study was carried out from 1983 to 1989. The present paper tries to approach the biodiversity of Prespa via its diverse fauna of birds. Its aim is to contribute to our understanding of the distribution and abundance patterns of these birds across the area's habitats as the latter have been shaped by both natural factors and human influence.

The study areas

The main REEDBED study area was situated on the 200 ha alluvial strip beween the two lakes (Figure 1). This is the main marsh area of Lake Mikri Prespa, with rich and diverse aquatic vegetation. Scattered among reedbeds are areas of dry land, open water and wet meadows flooded occasionally according to season and rainfall. Water depths in the reedbed area vary between 0 and 60 cm. *Phragmites australis, Scirpus lacustris* and *Typha angustifolia* dominate (Figure 2.1). A reedbed belt of variable width and complexity is found around the periphery of the lake. In order to make sampling more representative for the whole reedbed system of the lake, censuses were also made in six other reedbed areas (Figure 1) and along the draining ditches of the irrigation system. In these areas *Phragmites australis* dominated completely.

FARMLAND bird censuses took place within the southern sub-area of the irrigation system (Figure 1). The area consists of irrigated and non-irrigated croplands, abandoned and fallow agricultural land, pastures, access dirt-roads running through the fields, along the irrigation channels and drainage ditches (Figure 2.2). The dominant crop was winter cereals, followed by beans. The vegetation in the drainage ditches, which are 2–4 m wide and 1–1.8 m deep, was mainly *Phragmites australis, Typha angustifolia, Salix triandra* and *S. officinalis, Scirpus* spp. and *Juncus* spp. Dominant shrubs on road sides and the ditch banks were: *Rubus* spp. and *Rosa* spp. bushes, *Prunus spinosa* and *Pyrus amygdaliformis*. Ditch vegetation suffered uncontrolled cutting and burning each year by the farmers. The main draining ditches discharge into the lake and merge with the existing aquatic vegetation there. On its western side the study area was in contact with dry and wet meadows.

The OSTRIA sampling area is located in deciduous coppiced forests at the lower altitudes (around 880 m asl) of the eastern part of the lake on a limestone substrate (Figure 1). Dominant tree species are *Ostrya carpinifolia* and *Carpinus orientalis*. Accompanying species are *Juniperus oxycedrus, Quercus pubescens, Q. frainneto, Fraxinus ornus* and *Acer campestre*.

The JUNIPERUS sampling area is a shrubland with *Juniperus foetidissima* and *Quercus trojana* (or *Q. macedonica*). It occupies the limestones of the western part of Lake Mikri Prespa up to ca 1000 m (Figure 2.3). Regular accompanying species are *Carpinus orientalis, Buxus sempervirens* and *Juniperus oxycedrus.*

Oak woodlands are the commonest type of woodland in the Prespa area (Figure 1) and seven species of oak are found there (Pavlides, 1985). Widespread species are *Quercus frainneto, Q. cerris* and *Q. petraea.* The QUERCUS sampling area (Figure 1, Figure 2.5) consisted of forest of several oak species as well as *Sorbus* spp., *Acer* spp., *Corylus* spp., *Fraxinus* spp. and a few *Tilia vulgaris* and *Ulmus glabra.* The KARIES sampling area (Figure 1, Figure 2.4) consisted of areas adjacent to oak forests at low altitudes, which were cultivated until about 1950 but have since been abandoned. Woodland is in the process of re-establishing itself, forming sparse woodland with several oak species, *Juniperus oxycedrus* and *Rosa* and *Rubus* bushes. These areas are found close to the villages and between the forest and the cultivation zone and were lightly grazed by sheep (Figure 2.4).

Above the oak forests lies a narrow belt with beech forest. The dominant species is *Fagus sylvatica* and, in places, *Fagus moesiaca.* The FAGUS sampling area (Figure 1, Figure 2.6) was in an almost pure beech forest with no understore of herbs and shrubs and a thick leafy litter. The censused stands were moderately managed for timber in patches and contained mostly trees of an average DBH of 17 cm. The MIXED sampling area was located in a rich, multilayered beech forest mixed with other deciduous trees including, *Quercus* spp., *Ostrya carpinifolia, Sorbus torminalis, Acer* spp., *Fraxinus* spp. and *Corylus colurna* (Figure 1, Figure 2.7). *Juniperus oxycedrus* and, less abundantly, *J. communis* were found in the openings. All the above forests had been managed for timber in the past by the Forest Service, but the sampling plots were located in as uniformly managed areas as possible.

The area above the tree line of Mt. Varnous was also censused for birds (Figure 1, Figure 2.8). On Greek territory, the mountain has twelve peaks above 2000 m, the highest being 2334 m, but the bulk of the mountain and its summit, which reaches 2610 m (Mt Pelister), lies in the Former Yugoslavian Republic of Macedonia (FYROM). Two vegetation formations are found above the tree limit: the sub-alpine and the alpine meadows. The sub-alpine meadows consist of a ca 50 cm high layer of dense and compact shrubs forming a narrow zone between 1800 and 2000 m. Dominant species are *Juniperus communis nana* and *Vaccinium myrtillus* and

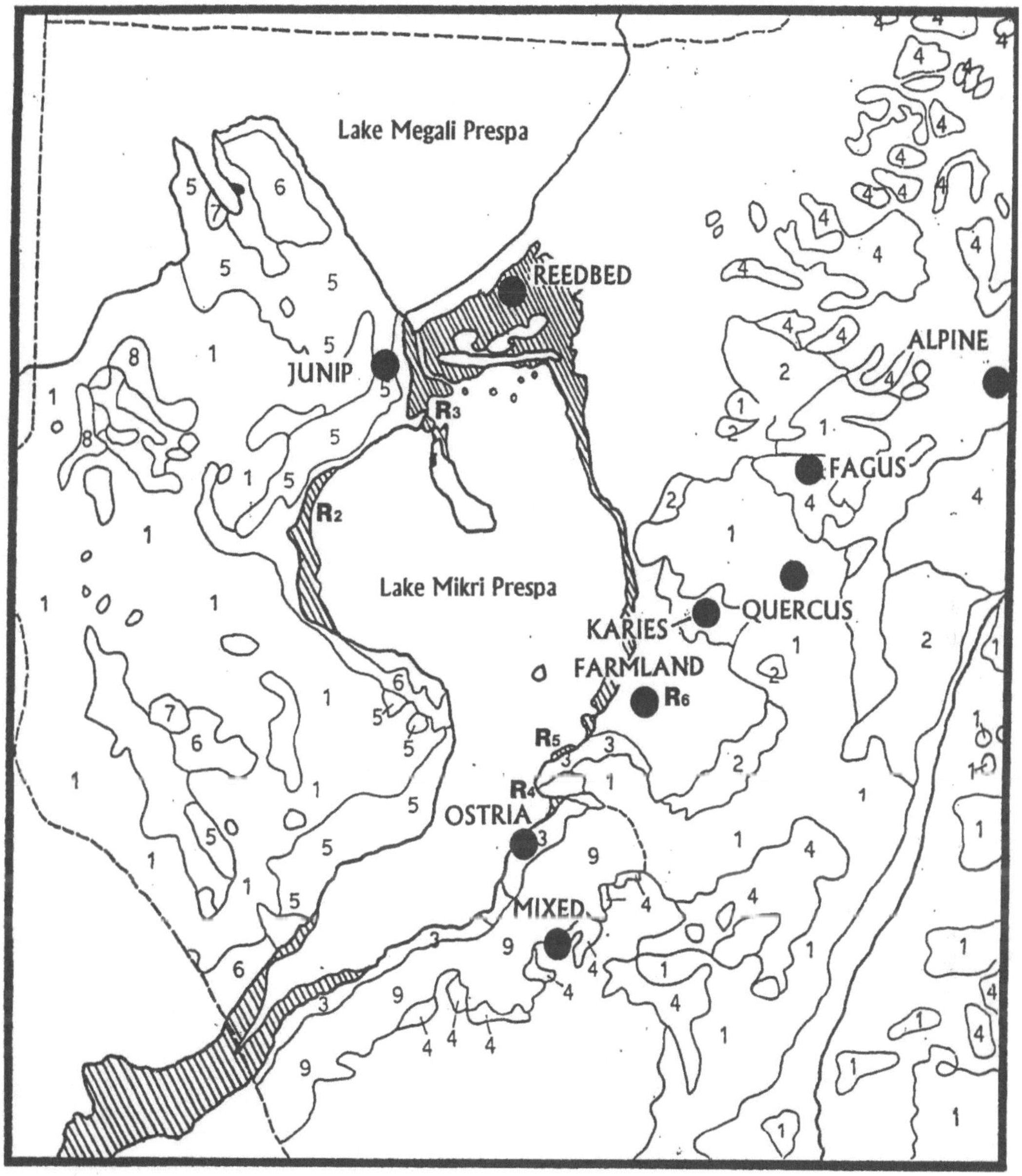

Figure 1. The Prespa area with the distribution of the main types of woodland and the sampling locations for birds. (Modified after Pavlides 1985). 1 = Oak, 2 = degraded coppiced oak, 3 = degraded Hop-Hornbeam, 4 = Beech, 5 = Juniper-Oak, 6 = degraded Juniper-Oak, 7 = Box shrubland, 8 = Macedonian Oak, 9 = Mixed deciduous. Hatched area is reedbeds. R2-R6 are sampling areas for reedbed birds.

many graminaceous species. The alpine meadows are found from ca 1800 m right up to the peaks and consist entirely of non-woody plants. Dominant species are *Arabis alpina, Alchemilla alpina, Aster alpinus* and three species of *Festuca*. The relief is generally smooth. Much of the area has deep soil, few rocks and many moist areas and brooks. The ALPINE census transects were located partly in the sub-alpine and partly in the alpine zone.

Methods

The avian orders included in this study were the Passeriformes, Galliformes, Columbiformes, Cuculiformes, Coraciiformes and Piciformes. This is usual in literature because these families are censused using the same methods. Data on presence/absence are given also for the Apodiformes.

Figure 2. (Left column): 1. Main REEDBED sampling area at Viro-Koula; 2. FARMLAND sampling area close to the village of Mikrolimni; 3. JUNIPERUS shrubland sampling area; 4. KARIES open oak woodland; (right column): 5. View of the interior of the QUERCUS sampling site in Oak woodland; 6. View under the canopy of FAGUS beech forest; 7. Mixed deciduous forest at MIXED sampling site; 8. Alpine area at 2100 m on Mt. Varnous.

Table 1. Percent composition of the breeding passerines in several reedbed sampling areas

	Viro-Koula (R1)	Pili (R2)	Krina (R3)	Latsista (R3)	Biol. station (R5)	Draining ditches (R6)
Great Reed Warbler (*Acrocephalus arundinaceus*)	14	50.5	81.6	100	100	60
Reed Warbler (*A. scirpaceus*)	49	28.2	13.3			
Sedge Warbler (*A. schoenobaenus*)	10					
Marsh Warbler (*A. palustris*)						40
Moustached Warbler (*Panurus biarmicus*)	22	7.3	2.5			
Reed Bunting (*Emberiza schoeniclus*)	3	6	2.5			
Cetti's Waubler (*Cettia cetti*)	2	8				
Surface of each sampling area (ha)	150	70	1	10	<1	25

Birds in the main REEDBED study area were censused by using mist nets (Karr, 1981) since usual censusing methods fail to estimate reedbed bird numbers with acceptable accuracy (Haukioja, 1968; Bell et al., 1968; Bell et al., 1973). Although the use of mist nets for estimating bird numbers has certain limitations (Karr, 1981) this method can be used successfully given that vegetation height does not exceed the height of the nets (Odum & Height, 1957; Terborgh, 1977); in the sampling area the vegetation was lower than the nets used.

The specific structure of the cropland/irrigation system which precluded access to every part of it, lead to the adoption of a specific census method for FARMLAND birds, namely a combination of two well known methods: the territory mapping (Kendeigh, 1944; Pinowski & Williamson, 1974) and the line or strip transect method (Merikallio, 1958; Bibby, 1978; Osborne & Osborne, 1983).

The line transect method described by Emlen (1971) was used for censusing birds in all types of woodlands. Due to various problems of access and bad weather conditions (especially strong winds) in the ALPINE areas, the Emlen line transect method was also used for censusing birds in these areas, but the exact distances covered were not measured. Random transects were made covering at least 2 km. Thus the results are given as relative abundances of birds.

The bird fauna of Prespa

The total number of bird species of the above orders observed in the Prespa area from 1983 to 1989 is 132. Of these, 24 are non-passerines and 108 are passerines (a complete list of species with indications on breeding and migration status is given in Catsadorakis (this volume). In total, 19 non-passerines and 77 passerine species of the studied orders breed in Prespa. For 7 species (Kingfisher, Wryneck, Shore Lark, Alpine Accentor, Dunnock, Willow Tit and Raven/for scientific names see Appendix 1 of Catsadorakis, 1997) it is doubtful whether they breed or not. Some of them might breed in neighbouring areas in FYROM. The Stock Dove, Moustached Warbler and Savi's Warbler formerly bred in the area but have not bred since 1989.

The habitats and their bird species

The Great Reed Warbler was the most widespread and abundant reed bird (Table 1) and was often the only representative in linear reedbeds of limited area such as R4, R5 and R6 (Figure 1). However in the more extensive and diversified marsh area at Viro-Koula (R1) the Great Reed Warbler was less common than the Reed Warbler. Only the more extensive and diverse reedbeds (R1, R2 and R3) supported more than two breeding passerine species.

The species present in $\geq 50\%$ of censuses in the FARMLAND habitat (typical species, according to Fuller 1982 in O'Connor & Shrub, 1986) were nine, with Corn Bunting, Black-headed Bunting and Red-backed Shrike together making up more than 65% of the total community (Table 2). Great Reed and Marsh Warblers were confined to the aquatic vegetation of the draining ditches. Less common breeders were: Partridge, Short-toed Lark, Whitethroat, Magpie, Blackbird, Goldfinch, Carrion Crow, Barred Warbler and Cetti's Warbler. Species breeding only occasionally

Table 2. Percentage composition of the typical avian species censused in the farmland zone of Prespa in the years 1983–1986.

Corn Bunting (*Miliaria calandra*)	31.7
Red-backed Shrike (*Lanius collurio*)	19.7
Black-headed Bunting (*Emberiza melanocephala*)	13.7
Balck-headed Yellow Wagtail (*Motacila flava feldlegg*)	8
Great Reed Warbler (*Acrocephalus arundinaceus*)	7.5
Skylark (*Alauda arvensis*)	7
Marsh Warbler (*Acrocephalus palustris*)	5.2
Quail (*Coturnix coturnix*)	4
Turtle Dove (*Streptopelia decaocto*)	2.9

included Whinchat, Stonechat, Spanish Sparrow and Tawny Pipit. Some species visited the farmland from nearby habitats only to feed or roost, viz. Turtle Dove, Starling, Hoopoe, Reed Warbler, Tree Sparrow and Jackdaw.

Only typical species (present in $\geq$ 50% of censuses) are included in Table 3 which lists bird species censused in each of the woodland habitats. The species of woodpeckers (Picidae) have been lumped together because I was frequently unable to distinguish with certainty the audible cues of presence of the four large *Dendrocopos* species: *D. medius, D. leucotos, D. major* and *D. syriacus*. However, some woodpecker species were identified with certainty in each one of the following habitats; MIXED: *Dryocopus martius, Dendrocopos medius*. QUERCUS: *Picus viridis*. KARIES: *Dendrocopos minor.*

The most widespread but often also the more numerous species within the National Park were the typical ubiquitous woodland species of European deciduous temperate forests: Blue Tit, Jay, Chaffinch, Blackcap, Chiffchaff, Marsh Tit and Great Tit. This is not surprising as these forests occupy the majority of the area of the Park (Figure 1). Also common were the Cirl Bunting, Lesser Whitethroat, and the Blackbird which are associated with shrublands and understore, as well as the Turtle Dove which is a typical species of the forest edge, nesting in trees but feeding in openings. The Chaffinch not only occurred in six habitats with trees but was among the dominant species in five of them. Other species predominating in more than one habitat type were Robin (3), Blue Tit (3), Lesser Whitethroat (2), Nightingale (2), and Red-backed Shrike (2).

Table 4 presents the relative abundances of the different species censused in the alpine grasslands. The ALPINE area can be split into three subareas with respect to its avifauna. In the tree limit-subalpine area, Tree Pipit, Yellowhammer and Ortolan Bunting strongly dominate and make up 40–50% of the total number of individuals. The second zone is the purely alpine area very far away from trees and extending up to 2200 m. There, Skylark, Whinchat and to a lesser extent Water Pipit predominate, together constituting 45-50% of the community, while accompanying species are Tawny Pipit, Wheatear, Black Redstart, Rock Thrush and Rock Partridge which together make up 15–20% of the individuals. Skylarks nest on gently sloping or level ground with deep soil and dense grass. Whinchats nest only within *Juniperus communis nana* stands and Water Pipits are scattered throughout the area. The third alpine sub-zone are the areas above 2200 m which were not censused in this study. There, the predominant species are the Shore Lark and the Alpine Accentor (Bauer & Hodge, 1970; Matvejev, 1976). These species were met rarely during censuses at lower altitudes.

Inter-habitat similarities of the avifaunas

Catsadorakis (1989) used hierarchical agglomerative clustering to classify the nine habitat types based on the relative frequencies of the species constituting their avifaunas (Figure 3). The woodland habitats MIXED, FAGUS, QUERCUS, OSTRIA and KARIES make up the most similar group, while the bird communities of the JUNIPERUS shrubland/woodland formation clearly differs from those of these five habitats. All other types of habitats are clearly separated both from each other and from the woodland group.

Within the woodland group there is strong similarity between MIXED and FAGUS and a weaker one between these two and QUERCUS, while KARIES and OSTRIA form another group of great similarity. The group of the first three forests is characterised by a strong predominance of species such as Chaffinch, Great Tit, Robin, Blue Tit and Nuthatch. In the KARIES/OSTRIA group, species such as Blackbird, Nightingale and Turtle Dove predominate. The dominant species in the MIXED/FAGUS/QUERCUS group, which also covers most of the National Park, often dominate also in the other three habitat types. By contrast the species which predominate in the KARIES/OSTRIA/ JUNIPERUS habitats have a distribution limited to these habitats only.

The avifaunas of REEDBEDS and ALPINE differ completely from each other but also from those of all

Table 3. Typical bird species censused in each woodland type and the % contribution of each one. Woodpeckers (Picidae) are included as one species.

	MIXED	FACUS	QUERCUS	OSTRIA	KARIES	JUNIPERUS
Certhia brachydactyla	2.5					
Cethia familiaris	2.3					
Parus ater	2.7					
Columba palumbus	3.3					
Anthus trivialis	1.2					
Turdus philomelos	0.8	2.2				
Turdus viscivorous	2.6	3.1				
Cuculus canorus		2.2				
Troglodytes troglodytes	6.3	7.7				
Parus palustris	5.7	5.2	3.6			
Sitta europea	10.9		2.3			
Erithacus rubecula	19.1	21.6	15	6.1		
Sylvia atricapilla	1.9	4.6	3	7.8		
Parus caeruleus	10.4	8.5	14.5	3.4		
Emberiza cia	0.8			2.2		
Parus major	7.4	1.9	28.9	3.7	4.5	
Woodpeckers [a]	1.4		1.2		2.5	
Phylloscopus collybita	2.5	14.6		2.4	3.1	
Garrulus glandarius	3.5	4.6	2.1	0.5	0.9	2.4
Turdus merula	2.2	2.7	5.6	22.6	30.7	2.3
Fringilla coelebs	12.6	13.8	17.5	10.6		7.6
Sylvia curruca		6.9	6.2	3.6	9.5	20.8
Luscinia megarhynchos				11.6	25.6	
Streptopelia turtur				15.6	3.1	2.3
Emberiza cirlus				5.1	4.6	6.6
Sylvia cantillans				4.8		21.8
Lanius collurio					6.9	7.2
Carduelis chloris					3.3	9.9
Lullula arborea					1.4	
Parus lugubris					4	5.9
Carduelis carduelis						4.7
Oriolus oriolus						3
Oenanthe hispanica						5.7
Number of pairs/40 ha	373.3	164.4	114.9	229.5	234.7	89.8
Typical species	20*	14*	11*	14	13*	13
Totally observed species	29	25	21	21	25	23

a= for species of woodpeckers (Picidae) see in text.

other habitats. Seventy three per cent of reedbed birds are exclusive to this type of habitat. In the drainage ditches which transect the entire area of the croplands, a dense reed vegetation has developed and this linear type of reedbed is inhabited only by Great Reed Warbers and Marsh Warblers. The former is the most common reed bird around the lake, whereas the latter is an ecotonal species which at Prespa is confined to the diffuse zone of reeds and tall herbs along the drainage ditches that separate these from the arable land. The Marsh Warbler illustrates how difficult it is to draw the line between arable land and wetlands in situations where the former has mainly been acquired through wetland reclamation and where there still remains much aquatic vegetation along the margins of the fields (cf. Gerakis, 1992).

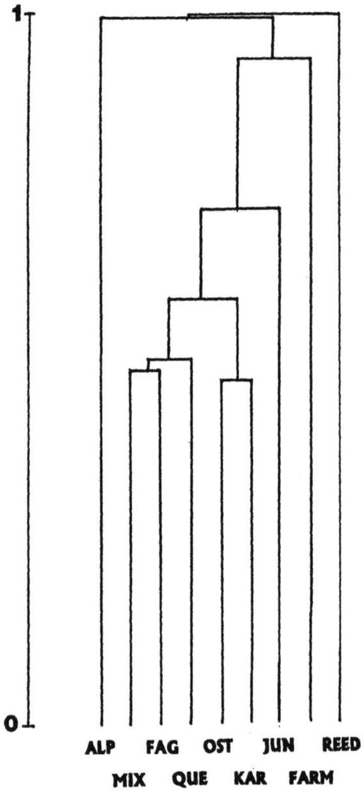

Figure 3. Dendrogram depicting the classification of the nine sampling sites according to the relative abundances of their 'typical' bird species.

Table 4. Percent contribution of each bird species to the composition of the avifauna of the subalpine and alpine zones of mount Varnous.

Alauda arvensis	24.6
Emberiza hortulana	17.4
Saxicola rubetra	17.4
Anthus trivialis	16.5
Emberiza citrinella	7.2
Phoenicurus ochruros	6.5
Oenanthe oenanthe	2.3
Anthus spinoletta	2.0
Anthus campestris	0.9
Alectoris graeca	0.9
Acanthis cannabina	0.8
Monticola saratilis	0.7
Prunella modularis	0.6
Coturnix coturnix	0.5
Prunella collaris	0.5
Corvus corone	0.5
Lullula arborea	0.5
Lanius collurio	0.5

Over 60% of the species counted in the ALPINE zone were met in no other habitat type within the area. However, six of the species found above the tree limit can be found also in the FARMLAND habitat, namely Quail, Skylark, Whinchat, Linnet, Cirl Bunting and Corn Bunting. A straight line distance of ca 10 km and an altitudinal difference of ca 1150 m separate these two habitat types. It is unlikely that the populations of the species inhabiting these two habitats are completely separated. Of the above six species, only Skylark populations are much greater in the ALPINE than in the FARMLAND zone. It is probable that changes in the crops grown and especially the strong decline of cereals, will mean that in the future species associated with cereals, such as Quail and Corn Bunting, will mainly be found in grassland areas above the tree limit.

Almost half (44.8%) of the species censused in FARMLAND occur exclusively there. Of the three dominant species in FARMLAND, the Corn Bunting is found also in the ALPINE, while the Red-backed Shrike is one of the eight farmland species also found in open woodland with bushes and shrubs (KARIES, JUNIPERUS). Only two typical farmland species occur also in the denser woodland habitats (Red-backed Shrike, Turtle Dove). This is expected, since the overall physiognomy of the FARMLAND is one of open grassland (cereals and other crops) transected by lines of natural vegetation of trees and reeds.

The forested habitat MIXED had the highest bird species diversity and species richness (Figure 4). The total number of species observed in FARMLAND was even higher, but this is due to the fact that croplands border to other habitats such as reedbeds, forests and shrublands. Thus, many species nesting in other habitats travel to feed there.

The role of vegetation structure, floristic composition and human impact upon woodland avifaunas

Catsadorakis (1989) analysed the relationships of the six woodland habitat types based upon their avifaunistic composition and seven abiotic features, floristic composition and 21 measures of vegetation structure. He concluded that the floristic composition was not correlated with the composition of their bird community except in special cases, e.g. the Lesser Whitethroat, whose abundance was positively correlated with the

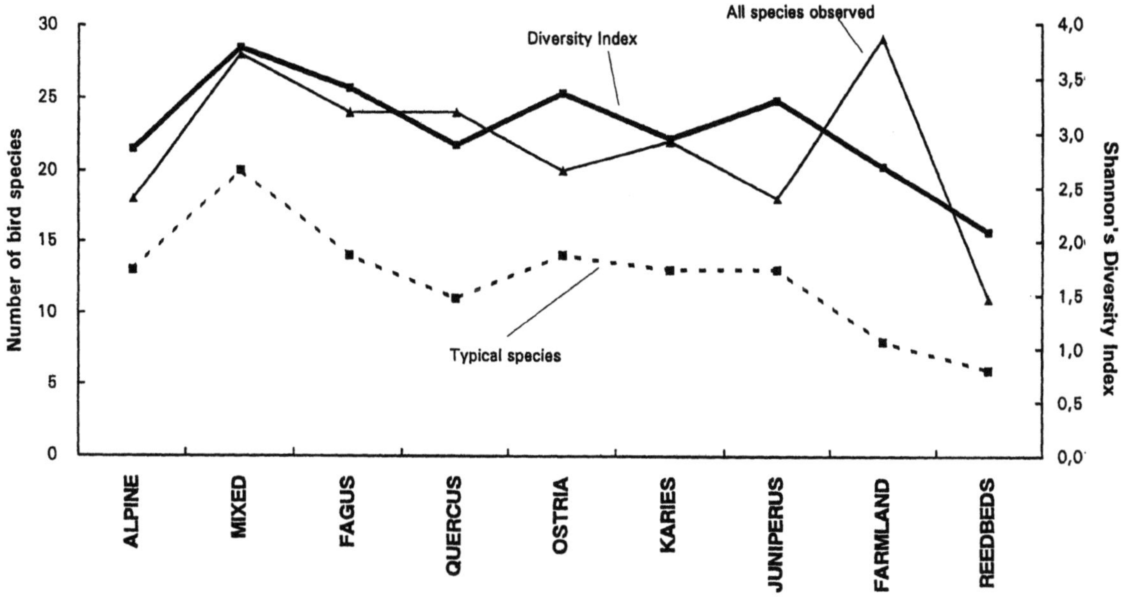

Figure 4. For each habitat type censused for birds is plotted: The number of species recorded in ≥ 50% of all censuses (typical species), the total number of bird species observed in all sampling efforts, and the Shannon-Weaver Diversity Index, based upon relative abundances of 'typical' species only.

abundance of *Juniperus oxycedrus*. This shrub is common in JUNIPERUS shrubland but in other forest types is also one of the first colonisers of forest clearings.

He also concluded that height above sea level per se did not seem to affect the bird species composition over a gradient of 860–1350 m asl. However in a historical perspective, the human impact has been more pronounced at lower altitudes and close to the villages (timber gathering for fuel, grazing, fires etc.) than higher up. Recently the opening of roads, providing easy access even to high remote patches of woodland has extended the human influence in previously untouched portions of the forest.

The analysis showed that the bird species composition was more strongly correlated with the vegetational features than with any of the other factors. Figures 5 and 6 show the similar – though inverse – way the six woodland habitats are arranged along the first two axes of a Correspondence Analysis based upon their bird composition (Figure 5) and the values of the 21 vegetational parameters (Figure 6).

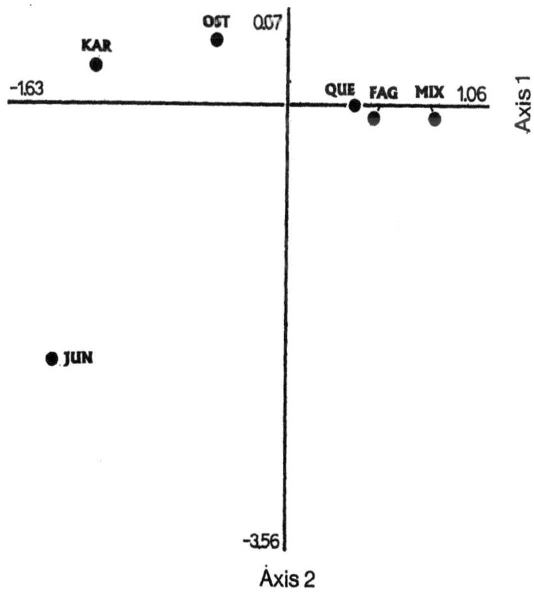

Figure 5. Arrangement of the six woodland habitats along the two first axes of a Correspondence Analysis using the abundances of the 'typical' bird species in each of them (After Catsadorakis 1989).

The MIXED sampling area shows the highest vertical and horizontal heterogeneity, and also has the most diverse avifauna. It is situated in an area where a rather varied deciduous forest growing on poor soils on sunny, south-oriented slopes is traversed by shady damp beech-forested gullies with deep soils. This habitat comes closest to the 'primeval' situation for the woodlands of the area and contains the most complete temperate forest avifauna (cf. Blondel, 1984).

FAGUS is equally multi-layered but is dominated by much more homogeneous stands of beech, with few openings; these have been recently colonised by

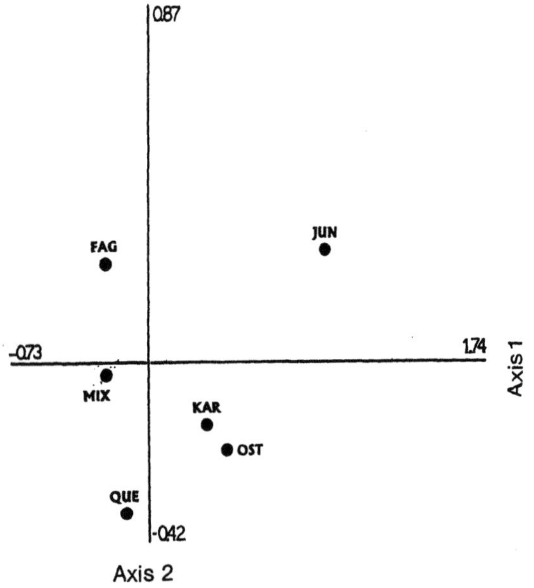

Figure 6. Arrangement of the six woodland habitats along the two first axes of a Correspondence Analysis using the average scores of each of 21 vegetational parametres measured in each of the habitats studied (After Catsadorakis 1989).

a few other broadleaves and junipers. It was disturbed by logging less than 10 years before the census. Its avifauna was less diverse and rich because the bark-gleaning species (Short-toed and Common Treecreeper and Nuthatch) were totally missing, possibly because the smooth, hole-less trunks of young beeches are unsuitable for them. The Robin, a bird nesting in a wide variety of trees and at different heights, was the commonest bird in FAGUS and MIXED. The high numbers of Chiffchaff are probably explained by the larger canopy foliage volume above 10 m (Catsadorakis, 1989) found in FAGUS and MIXED compared to all oher habitats.

QUERCUS is also a habitat heavily disturbed by logging. It is very homogeneous, dominated by even-aged (15–20 years old) oak trees. Older trees are few and scattered. A few shadier gullies with different species of broadleaves traverse the area. The commonest bird in this woodland was the Great Tit which seems to use the holes available in the few old trees left by the forest managers. As in the other two woodlands, the Chaffinch is one of the commonest species. Generally, among the forests that have trees more than 8 m high, QUERCUS has the lowest foliage height diversity (sensu MacArthur et al., 1962), the least horizontal heterogeneity (patchiness) and thus the poorest avifauna, although it constitutes the most extensive type of forest in the National Park. The clear-cutting, especially for fuelwood and the leaving of only a few old trees produce a rather poor environment for birds.

OSTRIA is also a highly disturbed habitat. It is found only in a limited area of the southern part of the National Park close to the lake. It is a rather low coppiced woodland, intensively cut by the local people in the past to obtain bean sticks. It is also grazed. Openings are mainly occupied by junipers. This habitat type, which in its primeval form would have been a mixed deciduous forest, is strongly dominated by bird species of the dense understore such as Nightingale and Blackbird, while the openings have been invaded by shrubland species such as Subalpine Warbler and Cirl Bunting. The habitats retain a substantial number of widely distributed forest bird species such as Robin, Chaffinch, etc. The second commonest species is the Turtle Dove, a species nesting at forest edges but feeding in openings.

The KARIES sampling area is situated at the edges of the oak forest at low altitude and is composed partly of abandoned fields and strips of dense oak woodland in small linear depressions. Isolated trees and Juniper bushes have started growing in the abandoned fields. The two commonest species, Blackbird and Nightingale, mainly use the strips of woodland for nesting and the adjacent openings for feeding. Typical open shrubland species, such as Cirl Bunting, Red-backed Shrike and Lesser Whitethroat, have invaded gradually along with the forest encroachment on former fields.

The JUNIPERUS shrubland differs in a number of features from all other habitats examined. Its unique vegetational features correlate well with its different avifauna. Its exposure is eastern and the substrate is exposed limestone rocks with very few pockets of terra rossa soil. Although in the past it must have suffered from pruning (by livestock keepers) and grazing, the quality of the substrate indicates that the type of vegetation must have been the same for long periods of time. As expected from the overall low vegetation complexity in comparison to the other habitats, it has the lowest densities of breeding birds (Catsadorakis, 1989) and the dominant species are those typical of shrubland, viz. Lesser Whitethroat, Subalpine Warbler, Greenfinch and Red-backed Shrike.

Biogeographic history and origins

The human influence on the natural habitats disturbs succession and natural changes of bird communities and thus conclusions about bird-habitat relationships based upon analyses of heavily human-influenced situations may be erroneous and misleading (Lack, 1965; Terborgh, 1977; Blondel, 1981; Blondel & Farre, 1988). Similarly misleading conclusions about the bird-habitat relationships can be drawn when studying birds in a patchy environment. The likeliness of wrong conclusions is higher in a rather small area such as Prespa, where the bird populations of each habitat are in a continuous interaction with each other. In addition to the above difficulties there are also others relating to variations from one year to another and the effects of winter periods upon populations of birds. However real these limitations are, it is worth trying, at this local scale, to interpret the distribution patterns of the passerine communities in the area of Prespa in relation to the various habitats.

Figure 7 shows the composition of the avifauna of each habitat type according to the five zoogeographic faunal types of Voous (1960) as proposed by Blondel (1978). As expected, all woodland habitats are strongly dominated by Palaearctic and European elements, a well established situation for South Europe (Blondel, 1982; Blondel, 1987; Blondel & Farre, 1988). Among habitats with trees and shrubs, the proportion of Mediterranean elements is from nil (MIXED, FAGUS, QUERCUS) to less than 10% (KARIES, OSTRIA) but increases dramatically to 40% in JUNIPERUS which is thus the only habitat dominated by Mediterranean and Euroturkestanian elements. Does this apparent differentiation reflect a real difference caused by the zoogeographical history of this area or is it a pure artifact of man's impact upon the natural vegetation of that habitat? Matvejev (1976) in his work on the bird zoogeography of the Balkans, suggests that in this intermediate zone of the Balkans which was only moderately affected by the glacial and interglacial periods, some parts of the mountains, especially those at low altitudes and facing S, E and SE, acted as 'refugia' for the Mediterranean and the sub-Mediterranean fauna. The low eastern slopes near the lakes on which JUNIPERUS is situated show all the features of a 'refugium' (Matvejev, 1976). Additionally, the area hosts the endemic plant *Centaurea prespana* (Rechinger, 1974) and is the only site in Greece for a completely isolated population of *Phelypaea coccinea* (*Orobanchaceae*, Thorning-Lund & Crivelli, pers.com.). The same area's cliffs at the coast of Lake Megali Prespa host the southernmost European population of Goosanders (*Mergus merganser*, Aves, Anatidae). This small and very isolated population exists some 300 km away from the nearest breeding population in Europe (Bauer et al., 1969; Cramp & Simmons, 1977). Also, a subspecies of a land snail (*Montenegrina dofleini prespaensis*) was described from this same area (Nordsieck, 1988). All these floral and faunal elements which might be considered as relicts, are perhaps indications that this particular area in Prespa has remained relatively isolated since the last glaciation and indeed has been a kind of 'refugium'.

However, these are speculations and whether the JUNIPERUS avifauna with its highly Mediterranean composition is purely an artifact of human influence (and hence an effect of its present day vegetation structure) or also has something to do with the particular biogeographic history of this piece of land remains an open question. The existence of the old Tertiary lakes (Cvijic, 1911; Jakovljevic, 1935) surrounded by high mountains, the large altitudinal differences within a restricted area and the human impact upon its habitats, make Prespa extremely interesting from a paleoecological point of view, not least with respect to the origin and composition of the bird faunas of its habitats.

Conclusion

Perhaps the most important feature of Prespa regarding the examined bird groups is the fact that within a straight-line distance of only 11 km a wide array of typical communities of reedbeds, farmland, temperate forests and alpine grasslands is found. The existence of ecotone situations make the bird diversity even higher. Except for one species that is widespread, all other reedbed species occur mainly at the marsh/reedbed area north of Lake Mikri Prespa, whose importance for the preservation of bird diversity is crucial. The farmland avifauna is dominated by three species relying heavily upon the vegetation of road sides and drainage ditches for feeding and nesting. An intensification of agricultural practices, most probably accompanied by crop changes, which might influence these habitats, would have a dramatic effect on farmland avifaunas as has happened elsewhere in Europe. The farmland avifauna of Prespa, living in a totally man-influenced habitat, is thus the one most directly vulnerable to human caused changes. The forests of the area are of a central European type and are thus dominated by typical resident temperate forest bird species, which are common even in those habitats heavily affected by human activities. The physiognomic features and the

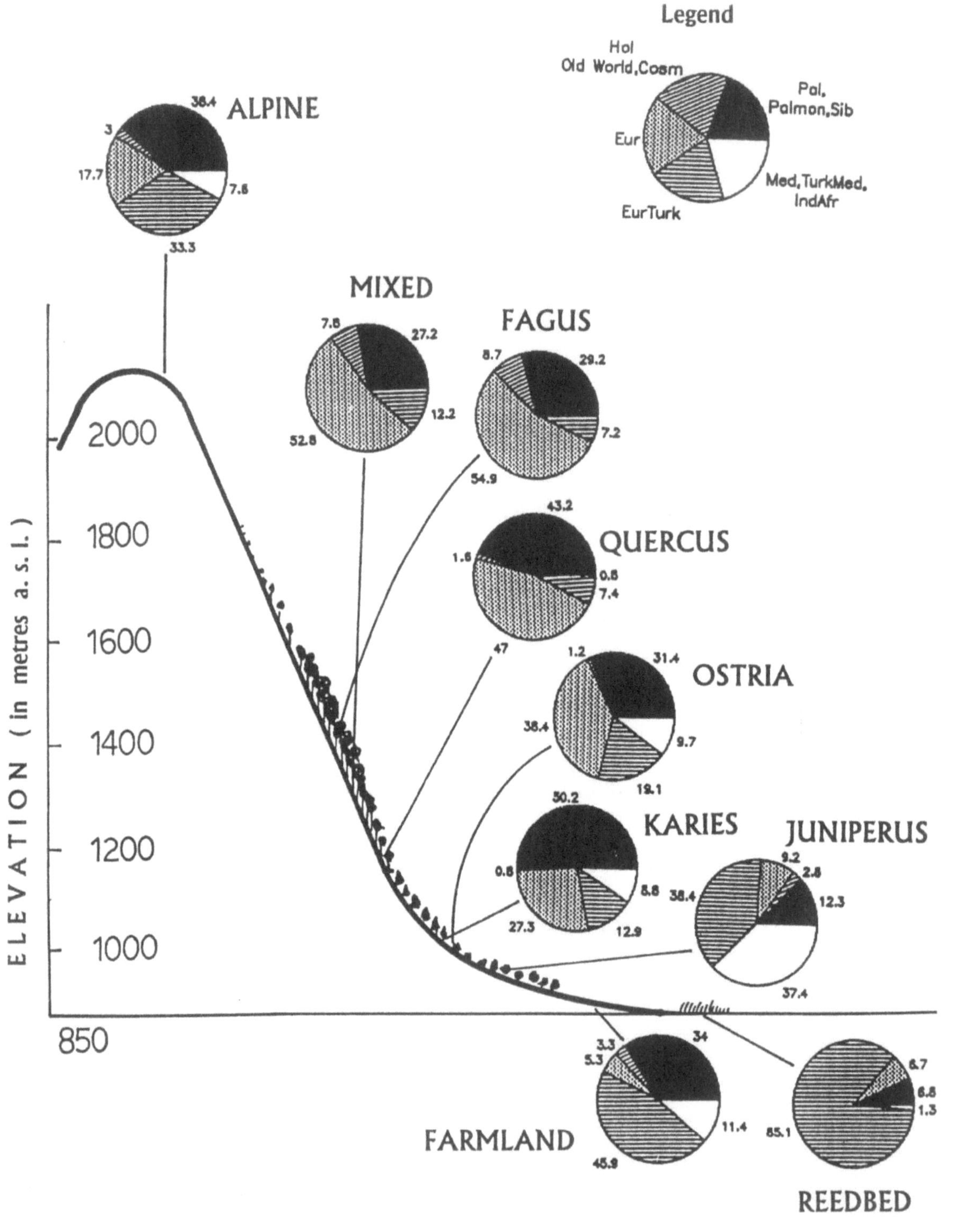

Figure 7. Schematic representation of the proportion of different avifaunal elements according to Voous (1960) and as modified by Blondel (1978), in each of the habitats studied at Prespa. Hol = Holarctic, Old World = Old World, Cosm = Cosmopolitan, Eur = European, EurTurk = EuroTurkestanian, Med = Mediterranean, TurkMed = Turkestano-Mediterranean, IndAfr = Indo-African, Pal = Palearctic, Palmon = Paleomontane, Sib = Siberian. (After Catsadorakis, 1989.)

highly Mediterranean bird composition of the Juniper shrubland indicate not only a long human influence but, combined with other biogeographic particularities of the surrounding area, may suggest that this part of Prespa might have been a refugium during glacial periods.

This suggests that more systematic studies should be devoted to that special north-western part of Prespa National Park which in terms of biodiversity and biogeography seems to be the most important.

References

Bauer, W., O. V. Helversen, M. Hodge & J. Martens, 1969. Catalogus Faunae Graeciae; Pars II. AVES. Cura A. Kanellis. Thessaloniki.

Bauer, W. & M. Hodge, 1970. 1. Nachtrag zum 'Catalogus Faunae Graeciae; Pars Aves'. Vogelwelt 91: 96–106.

Bell, B. D., C. K. Catchpole & K. J. Corbett, 1968. Problems of censusing Reed Buntings, Sedge Warblers & Reed Warblers. Bird Study 15: 16–21.

Bell, B. D., C. K. Catchpole, K. J. Corbett & R. J. Hornby, 1973. The relationship between census results and breeding populations of some marshland passerines. Bird Study 20: 127–140.

Bibby, C. J., 1978. A heathland Bird Census. Bird Study 25: 87–96.

Blondel, J., 1978. L'avifaune du Mont Ventoux, essai de synthese biogéographique et écologique. Terre et Vie 32: 111–145.

Blondel, J., 1981. Practical and theoretical problems of bird censusing in a mosaic of Mediterranean habitats. Proc. VII. Int. Conf. Bd Census and IBCC.: 121–126. Leon.

Blondel, J., 1982. Caracterisation et mis en place des avifaunes dans le bassin méditerranéen. Ecol. Med. VIII: 253–271.

Blondel, J., 1984. Avifaunes forestieres méditerranéens; histoire des peuplements. Aves 21: 209–226.

Blondel, J., 1987. From biogeography to life theory: a multithematic approach illustrated by the biogeography of vertebrates. J. Biogeogr. 14: 405–422.

Blondel, J. & H. Farre, 1988. The convergent trajectories of bird communities along ecological successions in European forests. Oecologia 75: 83–93.

Catsadorakis, G., 1989. Bird communities in Prespa, Florina, Greece: Elements of ecology and biogeography. Ph. D. Diss. University of Athens. Department of Biology. (In Greek)

Cramp, S. & K. E. L. Simmons, 1977. The Birds of the Western Palearctic. Vol. I. Oxford University Press.

Cvijic, J., 1911. L'ancient Lac Égéen. Ann. Geogr. 20: 233–259.

Dawson , D. G., 1981. Counting birds for a relative measure (index) of density. In Ralph, C. J. & J. M. Scott (eds), Estimating numbers of terrestrial birds. Stud. Avian Biol. 6: 12–16.

Emlen, J. T., 1971. Population densities of birds derived from transect counts. Auk 88: 323–342.

Gerakis, P. A., 1992: Wetland and Cropland interactions In Gerakis, P. A. (ed.), Conservation and Management of Greek Wetlands, Proceedings of a Greek Wetlands Workshop held in Thessaloniki, Greece. The IUCN Wetlands Programme. Gland, Switzerland: 129–148.

Fuller, R. J., 1984. The distribution and feeding behaviour of breeding songbirds on cereal farmland at Manydown Farm, Hampshire, in 1984. Report to the Game Conservancy. Tring. British Trust for Ornithology.

Handrinos, G., 1992. Birds. In Karandinos, M. & A. Legakis (eds), The Red Data Book of Threatened Vertebrates of Greece. Hellenic Zool. Soc. & Hellenic Ornithol. Soc. Athens.

Haukioja, E., 1968. Reliability of the Line Survey Method in bird census with reference to Reed Bunting and Sedge Warbler. Ornis Fenn. 45: 105–113.

Jakovljevic, S., 1935. Recherches limnologiques sur le lac de Prespa. Verh. int. Verh. Limnol. 7: 187–206.

Karr, J. R., 1981. Surveying Birds with Mist Nets. In Ralph, C. J. & J. M. Scott (eds), Estimating numbers of terrestrial birds. Stud. Avian Biol. 6: 62–68.

Kendeigh, S. C., 1944. Measurement of Bird Populations. Ecol. Monogr. 14: 67–106.

Lack, D., 1965: Evolutionary Ecology. J. anim.Ecol. 34: 223–231.

Mac Arthur, R. H., J. W. Mac Arthur & J. Preer, 1962. On bird species diversity II. Prediction of bird censuses from habitat measurements. Am. Nat. 96: 167–174.

Matvejev, S. D., 1976. Survey of the avifauna of the Balkan Peninsula. The Serbian Academy of Sciences and Arts Monog. Vol. CDXCI. Beograd. 321 pp.

Merikallio, E., 1958. Finnish birds, their distribution and numbers. Fauna Fenn. 5: 1–181.

Nordsieck, H., 1988. Neue Clausilien aus NW-Griechenland. Ann. Nat. Mus. Wien. 90: 197–201.

Odum, E. P. & G. L. Height, 1957. The use of mist nets in population studies of winter fringillids on the AEC Savannah River area. Bird banding 28: 203–213.

Osborne, P. & L. Osborne, 1985. A winter bird census on farmland. InTaylor, K., R. J. Fuller & P. C. Lack (eds), Bird Census and Atlas Studies. Proc. VIII Int.Conf. Bd Census and Atlas Work, 287–302. BTO. Tring.

Pavlides, G., 1985. Geobotanical Study of Prespa National Park in Florina. Part A. University of Thessaloniki, 308 pp. (In Greek).

Pinowski, J. & K. Williamson, 1974. Proceedings of the Second Meeting of the European Ornithological Atlas Committee. Acta Ornithol. 14: 32–37, 152–164.

Rechinger, K. H., 1974. *Centaurea prespana*, eine neue gelbluchende Art am der Section *Acrolophus*. Ann. Musei Goulandris 2: 55–58.

Robbins, C. S., 1981. Effect of the time of day on bird activity. In Ralph C. J. & J. M. Scott (eds): Estimating numbers of terrestrial birds. Stud. Avian Biol. 6: 275–286.

Terborgh, J., 1971. Distribution of environmental gradients: theory and preliminary interpretation of distributional patterns in the avifauna of the Cordillera Vilcabamba, Peru. Ecology 52: 23–40.

Terborgh, J., 1977. Bird species diversity on an Andean elevational gradient. Ecology 58: 1007–1019.

Voous, K. H., 1960. Atlas of European Birds. Nelson. Edinburgh.

Willson, M. F. & D. J. Moriarty, 1976. Bird species diversity in forest understory. Analysis of mist net samples. Oecologia 25: 373–379.

Hydrobiologia **351**: 157–174, 1997.
A. J. Crivelli & G. Catsadorakis (eds), Lake Prespa, Northwestern Greece.

The importance of Prespa National Park for breeding and wintering birds

G. Catsadorakis
Society for the Protection of Prespa, 530 77 Agios Germanos, Greece

Key words: avifauna, avian diversity, breeding birds, Greece, Prespa, wintering birds

Abstract

The breeding avifauna of Prespa National Park is of national and international importance both due to its high richness and the internationally important populations of rare species. The latter include the Dalmatian (*Pelecanus crispus*) and Great White (*Pelecanus onocrotalus*) pelicans and the Pygmy Cormorant (*Phalacrocorax pygmaeus*). Populations of national interest include Cormorant (*Phalacrocorax carbo*), Greylag Goose (*Anser anser*), Goosander (*Mergus merganser*), Great White Egret (*Egretta alba*) and six other species of ardeids. Among the 261 bird species observed in the area of ca 250 km^2 since the late 60s, 164 breed and 41 are very rarely observed. Passage migrants and winter visitors include 91 species. Twelve breeding species became extinct in the National Park during the last 25 years and others are endangered. Raptors and aquatic birds suffered more due to habitat and land use alterations brought about by the construction of an irrigation scheme. Eighty-one species observed in the area are listed as in need of special protection in Annex I of the 79/409 EEC Directive and 32 of them breed. Lake Mikri Prespa is of virtually no importance for wintering waterfowl since it is frozen every winter, but the Greek part of Lake Megali Prespa, which never freezes, concentrates among other species the 20% and 35% respectively of the Great Crested Grebes and Tufted Ducks wintering in Greece.

Introduction

Despite the existence of a number of relatively recent publications on the avifauna of Prespa National Park (Catsadorakis, 1986; Pyrovetsi & Crivelli, 1988; Catsadorakis, 1994) and also of many unpublished data, Prespa is still known mainly for its pelicans while the status of the other species are not well known.

The present paper aims at presenting a global account of the avifauna of Prespa, by providing a full list of species, data on breeding and seasonal occurrence, and for the most important species and groups of species, identifying population trends and evaluating this avifauna in relation to the breeding and wintering of birds for Greece and Europe. Additionally it presents the most updated review of extinct and threatened species of birds in the area.

Breeding species of national and international interest

Dalmatian Pelican *Pelecanus crispus:* this world 'vulnerable' species (Groombridge, 1994) has bred in Prespa at least since before World War II, but probably very much earlier. In the period 1983–1989 the total number of breeding attempts fluctuated between 114 and 163 pairs (Figure 1), but after 1990 they showed a continuous increase and in 1995 reached a maximum of 505 which is the highest breeding concentration of this species on earth and its largest single colony (Figure 2), if compared to numbers mentioned in Crivelli et al. (1994) for the former USSR countries which host the majority of the world population. The reasons for the increase are not very clear. It could result of natural improvement of conditions at feeding or nesting sites or both, but it is certainly not unrelated to the reduction of disturbance due to increased wardening in recent years.

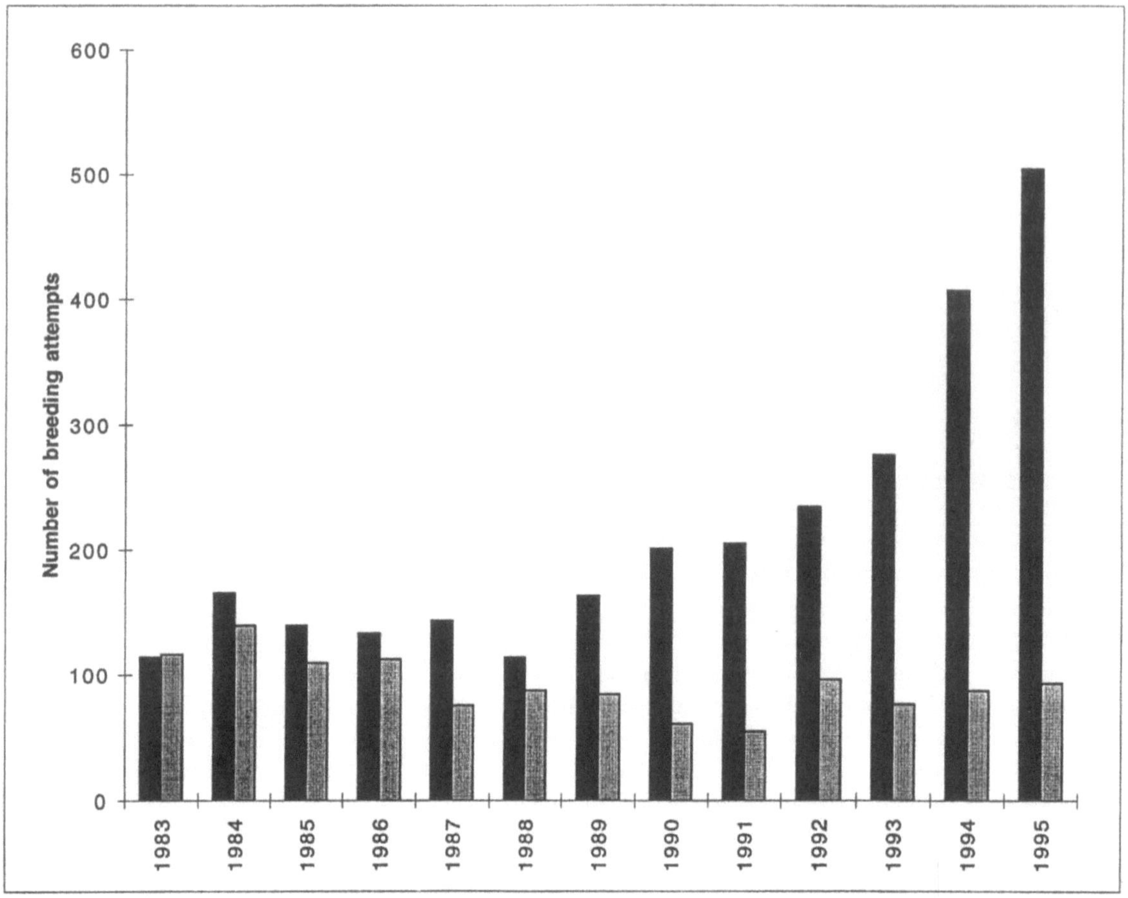

Figure 1. Number of breeding attempts of *Pelecanus crispus* (black) and of *Pelecanus onocrotalus* (shaded; 1983–1985: from Hatzilacou, 1992) at Lake Mikri Prespa from 1983 to 1995.

Great White Pelican *Pelecanus onocrotalus*: This pelican species also nests at Prespa alongside Dalmatian Pelicans, probably also from old times. From 1983 to 1995 the number of breeding attempts ranged between 139 and 54 (Catsadorakis et al., 1996, 1997) and they show a declining trend but the reasons for this are not clearly understood (Figure 1). This is the only breeding colony of the species in Europe outside Rumania (Danube delta) and the countries of the former USSR. The breeding and feeding ecology of Dalmatian and Great White pelicans in Prespa are the subject of a long term study (Crivelli, 1987; Crivelli et al., 1991; Catsadorakis et al., 1996).

Pygmy Cormorant *Phalacrocorax pygmaeus*: the data of breeding pairs since 1968 show that the population is fluctuating but the overall trend is to remain steady or even increasing at around 200–300 pairs (Figure 3a). Since 1988 the colony has been situated in the reedbed of Mikrolimni (Catsadorakis et al., 1996, for changes in the last few years) with the mixed colony of herons. The Prespa colony is one of the two large colonies remaining in Greece and is the second in size after that of Lake Kerkini (Crivelli et al., 1995).

Cormorant *Phalacrocorax carbo*: the Cormorant has in Prespa the second largest breeding colony in Greece after the colony of Lake Kerkini (Crivelli et al., 1995). Since the late 60s when there are the first data on numbers of breeding pairs, the total number of pairs has fluctuated with an increasing trend (Figure 3b). Thus, in Prespa the species, as in many other areas in Europe seems to be increasing (Th. Nazirides, A. H. Joensen, N. O. Preuss, pers. comm.). Up to 1989 Cormorants bred at three sites: the Albanian border area in willows, on coastal cliffs at Megali Prespa and on the breeding islands of pelicans in Vromolimni area (Catsadorakis et al., 1996). Specifically Megali Prespa was

Figure 2. A pelican breeding colony at Lake Mikri Prespa (Photo: G. Catsadorakis).

Table 1. Bird species that have ceased breeding in Prespa area.

	Last reported or suspected breeding year	Reference
Platalea leucorodia	1992	This study.
Plegadis falcinellus	1970	Bauer et al. 1973.
Aythya nyroca	1989	This study.
Neophron percnopterus	1990	This study.
Haliaeetus albicilla	1972	Kempf & Wersinger 1973.
Aquila heliaca	1969	Bauer & Hodge 1970.
Hieraaetus pennatus	1967–1969	Terrasse et al. 1969.
Falco peregrinus	1969	Terrasse et al. 1969.
Phasianus colchicus	1984	This study. (Introduced species).
Columba oenas	1969	Bauer & Hodge 1970, Terrasse et al. 1969.
Locustella luscinioides	1989	This study.

the only place in Europe where they bred on inland cliffs (W. Suter, pers. comm.). After 1989 the situation changed and they now all breed on old Junipers at Vidrinetsi island (Catsadorakis et al., 1996). Cormorants fish mainly in the open waters of Lake Megali Prespa and Dalmatian pelicans exploit this behaviour to catch bleak (*Chalcalburnus belvica*), their common food species (Crivelli & Vizi, 1981; Crivelli, 1987). In this unusual commensal relation, cormorants may also benefit by using pelicans as visual marks to locate the existence of fishing activity (Crivelli & Vizi, 1981; Catsadorakis, unpubl. data.)

Greylag Goose *Anser anser rubrirostris*: the largest population of only two left in Greece. The other is at

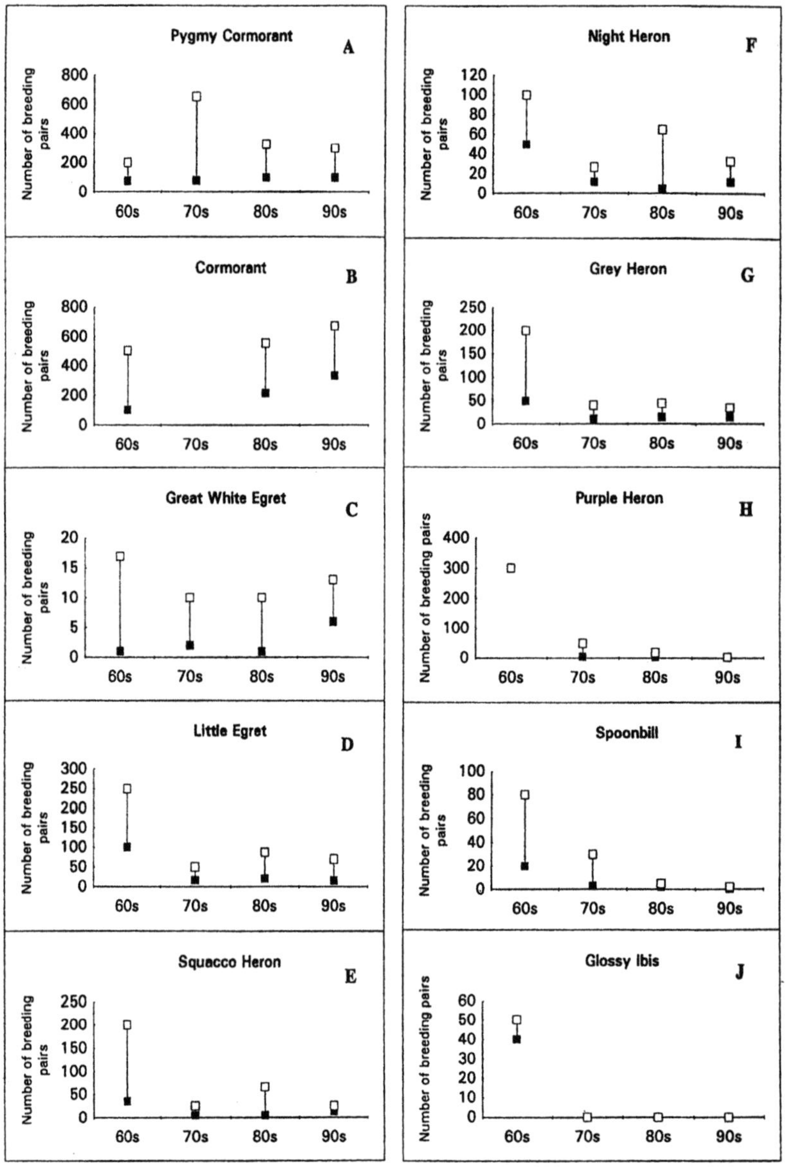

Figure 3. Absolute minimum and maximum numbers of breeding pairs of large waterbirds for each decade from 1967 onwards, for those years data were available. Numbers were taken after scanning all available data in a large number of published and unpublished papers and personal unpublished data.

Lake Kerkini (Crivelli et al., 1995). Always suffering from poaching in Prespa, they increased in mid 80s (Figure 4) but later on they suffered a low breeding success due to a late cold spell in March 1987 which destroyed many clutches and the dramatic shrinkage of suitable breeding grounds during the drought of 1987–1990, when very few pairs bred. In 1993–1994 the population shows signs of recovery. This distinct population is resident except in cases of prolonged harsh weather in winter.

Goosander *Mergus merganser*: the only nesting site in Greece and the furthest south-east in the Mediterranean area. A relict and very small population, almost 300 km south from the southern limit of its range in Northern Europe (Cramp & Simmons, 1977). Known to breed in the Yugoslavian part of Megali Prespa since 1939 (Makatch, 1950). It was discovered breeding in the Greek part in 1968 (Brosselin & Molinier, 1968), in crevices of the rocky coast. The population in Greece does not exceed 30 breeding pairs. Entirely resident,

Table 2. Occasional, probable or/and very sparse breeders in Prespa area.

	Status	Year of breeding, reference and number of pairs
Anas clypeata	Has bred only once.	1971. Bauer et al. 1973, 1 p
Aythya ferina	Probable breeder	1989. 1 p. (This study)
A. fuligula	Probable breeder	1989. 1 p. (This study)
Podiceps nigricollis	Occasional breeder	1972. (Kempf & Wersinger 1974). 1986. 1993 (This study).
Vanellus vanellus	Occasional breeder	1969. (Terrasse et al. 1969), 5p. 1987, 1 p. (This study).
Cygnus olor	Occasional breeder	1982–1985. After, has ceased breeding (This study).
Porzana parva	Uncertain breeding.	1971, Bauer et al. 1973.
Chlidonias hybrida	Occasional breeder	1984, 1987 (This study).
Otus scops	Occasional breeder	1987, 1989 (This study).
Saxicola torquata	Occasional breeder	Some years in relatively high numbers. (This study).
Passer hispaniolensis	Occasional breeder	1984, 1988, 1990 (This study).
Acrocephalus melanopogon	Extinct ?	1969. (Bauer & Hodge 1970, Terrasse et al. 1969). 1986, (This study).
Serinus serinus	Occasional breeder	1988, (This study).

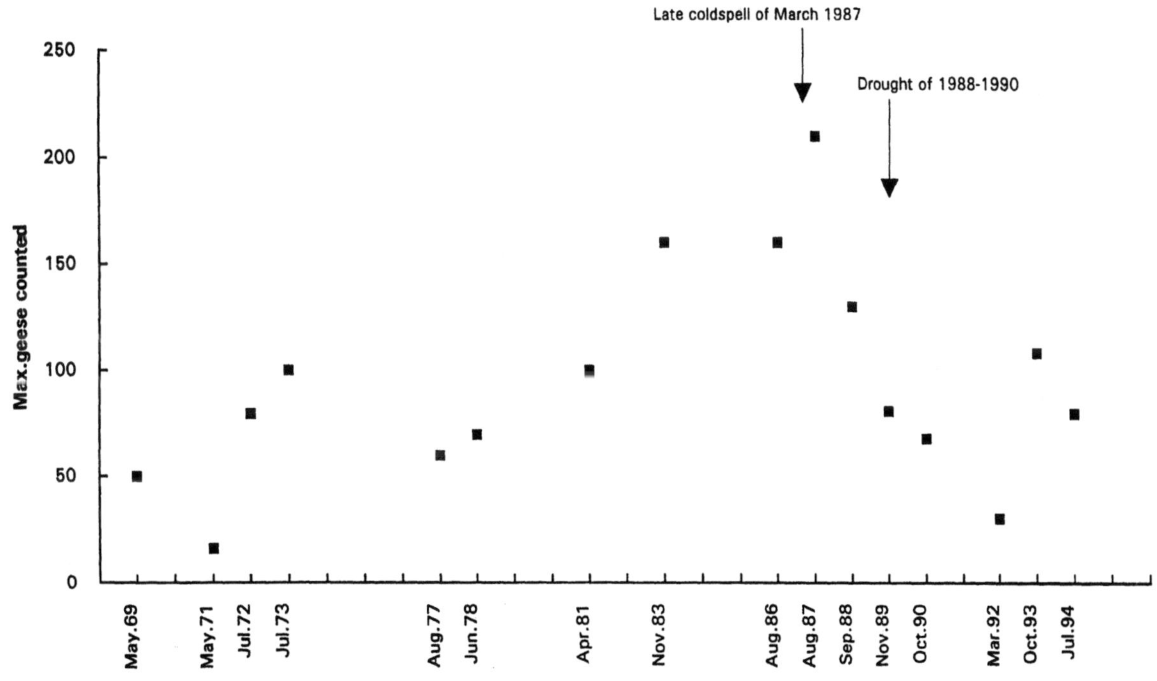

Figure 4. Maximum available counts of Greylag Geese (*Anser anser rubrirostris*) in Prespa National Park, in each year since 1969. Counts in the winter months (December, January, February) are not included since they might include migrant birds.

may move some kilometres in cases of extreme cold in winter. Normally they use the waters of Megali Prespa in winter which never freeze. In early spring they fish in Lake Mikri Prespa. Mendel (1988) mentioned breeding of the Red-breasted Merganser *Mergus serrator* but he was probably mistaken, as this record has not been confirmed by any other researcher even after intense research.

White Stork *Ciconia ciconia:* one or two or (rarely as in 1990) 3 pairs of White Storks have bred in the villages Plati, Lefkonas and Lemos, at least since 1986. These villages are situated at an altitude of ca 900 m asl and as nest records in Europe above 500 m are exceptional (Cramp & Simmons, 1977), these nests certainly are among the highest in Europe (Broussalis, 1974).

Great White Egret *Egretta alba*: the largest nesting population in Greece is at Prespa. The population has fluctuated roughly between 6 and 12 pairs, since 1968, though there may have been years with up to 17 pairs. The population is entirely resident. Some pairs always nest within the mixed heron colony, but most nests are found in loose concentration within the northern reedbed. Although, compared to all other heron species it had been always the less numerous, it is the only one that did not decrease after 1968 (Figure 3c). Is this connected with its residency? to a higher plasticity in foraging tactics? or to recruitment from an increasing population elsewhere?

Mute Swan *Cygnus olor*: every year from 1982 to 1985 one pair of Mute Swans nested succesfully in Vromolimni of Lake Mikri Prespa (H. Jerrentrup, pers. comm.; Catsadorakis, unpubl. data), rearing 2, 6, 5 and 7 juveniles respectively. Although present in winter in small numbers this species has never nested again. In Greece, there are currently only two feral populations of introduced birds, in Lake Kastoria and Agras reservoir. In the wild the species bred in many places in Greece at the beginning of the century but the last nesting in the wild happened in the Evros Delta in the late 70s (Cramp & Simmons, 1977). Thus, the nesting in Prespa has been the first of really wild birds recently in Greece.

Yellow-legged Gull *Larus cachinnans*: a recent addition to the list of breeding birds of the Prespa area, this species established a breeding population in 1991 with one pair and in 1994 counted up to 12–15 nests and a maximum of 60 individuals. They breed on Vidrinetsi island and this is the only inland colony of this species in Greece.

Herons, Spoonbills (*Platalea leucorodia*) and Glossy Ibises (*Plegadis falcinellus*): out of the seven heron species still breeding in Prespa, five, namely Squacco (*Ardeola ralloides*), Night (*Nycticorax nycticorax*), Grey (*Ardea cinerea*) and Purple (*Ardea purpurea*) herons and the Little Egret (*Egretta garzetta*) have suffered a dramatic decline after the major habitat changes that occurred with the construction of the irrigation system in the 60s (Figures 3d–h), (but see Great White Egret above). There are few data for the Little Bittern *Ixobrychus minutus* (5 pairs in Biber & Crivelli, 1978), but there are certainly more than 30 pairs (this study). The Spoonbill *Platalea leucorodia* and the Glossy Ibis *Plegadis falcinellus* became extinct (Figure 3i, k). The decline was attributed to the shrinkage of feeding areas (e.g. wet meadows) (Crivelli, pers. comm.; Pyrovetsi & Karteris, 1986).

Table 3. Bird species breeding in Prespa area which are vulnerable to extinction since they breed in very limited numbers.

	Number of pairs in 1992–1994
Ciconia ciconia	1
Ardea purpurea	2–3
Circus aeruginosus	0–1
Sterna hirundo	1–5
Perdix perdix	<20
Bubo bubo	1 bird seen repeatedly
Asio otus	1 ?
Caprimulgus europaeus	<10
Merops apiaster	0–3
Dendrocopos leucotos	?
Calandrella brachydactyla	? (extinct ?)
Remiz pendulinus	0–3

Raptors: the raptors of Prespa have been greatly impoverished over the last 20 years (Catsadorakis, 1994). At least five species became extinct and three more are threatened with extinction (Tables 1 and 3). Their decline can be attributed mainly to land use and habitat changes that happened during the 60s, (Pyrovetsi & Karteris, 1986) not only in Prespa but in many rural areas in Greece too, and to the fact they were breeding in very low numbers as top predators.

The diversity of the avifauna of Prespa

The first ornithological reference to the Prespa area was made in the 30s by Thorpe et al. (1936). Since that time and up to 1994, in the area encompassing the Greek part of the two lakes, up to the peaks of mountains delineating their drainage basin, 261 species of birds have been observed (Appendix 1). Three species which are not included in the above list are considered as wrongly identified, namely: The Sardinian Warbler *Sylvia melanocephala* by Kempf & Wersinger (1974), the Shag *Phalacrocorax aristotelis* by Berlic et al. (1983) and the Red-Breasted Merganser by Mendel (1988). Out of the 261 species, 151 are non-passerines and 110 are passerines. A large number of species (41) has been observed only once or twice at irregular periods of the year and are considered as accidentals or vagrants. A total of 164 species (81 non-passerines and 83 passerines) breed, have bred in the past, are occasional breeders, or it is very likely that they breed in Prespa National Park but there are no proofs of breeding at present. In terms of seasonal occurrence a total of 91 species (at least 63 non-passerines and 28 passerines) are, or have been, passage migrants and/or winter visitors in the area. This number must be considered as a minimum because for many species it is not possible to discriminate between individuals that migrate or stay in the area to breed, or whether the same individuals stay to overwinter. No comments are made in this work on partial migrants since this issue also remains largely undocumented.

Extinct and highly threatened bird species

Catsadorakis (1991) made the latest review of the extinct and threatened species for the period 1967–1990, but since that time and up to 1994, unfortunately there are additions to the list. Twelve species which bred within or at the near vicinity of Prespa National Park have become extinct (Table 1). The Spoonbill was the last to cease breeding in 1992 and the Egyptian Vulture in 1990. The following nine species were reported as breeders in the same area in a list by Broussalis (1974), but they certainly do not breed any more: Lesser spotted Eagle *Aquilla pomarina*, Griffon vulture *Gyps fulvus*, Long-legged Buzzard *Buteo rufinus*, Lesser Kestrel *Falco naumanni*, Roller *Coracias garrulus*, Woodchat Shrike *Lanius senator*, Orphean Warbler *Sylvia hortensis*, Bonelli's Warbler *Phylloscopus bonelli* and Rock Sparrow *Petronia petronia*. Since no other details were available in that work and no other references mentioned proofs of breeding, the breeding of these species must be treated with caution. Twelve more species (Table 2) are species that breed occasionally and irregularly, or they breed in such small numbers and are elusive, that it is not always possible to certify their nesting. Among the latter are species that were certainly recorded to breed at least once but with no subsequent records.

Very important from the conservation point of view are the highly endangered species (Table 3). Most of them nest in very low numbers or/and have shown strong declining trend in recent years. A typical example is the Common Tern *Sterna hirundo* of which 27 pairs nested in 1984, 20 pairs in 1987, but less than 5 pairs in recent years. The White Stork and the Marsh Harrier *Circus aeruginosus* are most vulnerable, with 1 or at the most, 2 breeding pairs. The Marsh Harrier was reported to be a 'very common' nester in the past (Thorpe et al., 1936), and in 1968–1969 had a population of 5–10 pairs (Brosselin & Molinier, 1968, Terrasse et al., 1969). The recent years it has not even bred every year. One nest of the Eagle Owl *Bubo bubo* found in 1987 was deserted and since then only one bird has been observed. The Bee-eater *Merops apiaster* population has dwindled to zero to 3 nests, while the last nest of the Long-eared Owl *Asio otus* was discovered in 1990. The Partridge *Perdix perdix* and the Short-toed Lark *Calandrella brachydactyla* were influenced negatively by the reduction of cereals and the intensification of farming practices and they both face danger of extinction. Data on the White-backed Woodpecker *Dendrocopos leucotos* are insufficient to evaluate the threat, but the nests of Penduline Tits *Remiz pendulinus* are too obvious to elude observation. Other species such as Golden Eagle *Aquilla chrysaetos*, Night Heron, Squacco Heron, Gadwall *Anas strepera*, Goosander, Short-toed Eagle *Circaetus gallicus*, Honey Buzzard *Pernis apivorous*, Hobby *Falco subbuteo*, Little Ringed Plover *Charadrius dubius*, Woodlark *Lullula arborea*, Dipper *Cinclus cinclus*, Black-eared Wheatear *Oenanthe hispanica* and the Rock Thrush *Monticola saxatilis* nest in very low numbers and are thus vulnerable but no data exist to document possible threats. In the list of Appendix 1, all species included in the Red Data Book of the Vertebrates of Greece (Handrinos, 1992) are shown with indications on their threat category.

Eighty one out of the 261 bird species observed in the area are listed as in need of protection in Annex 1 of EEC Directive 79/409 as amended by Directive 85/411, for the Conservation of Wild Birds (Appendix 1). Of these, 32 are breeding birds. The populations of Syrian Woodpecker *Dendrocopos syriacus*, Ortolan Bunting *Emberiza hortulana* and Barred Warbler *Sylvia nisoria* do not face direct problems under the current conditions of habitat management in Prespa, but a further intensification of agricultural practices might affect these species in the future. The populations of Black Woodpecker *Dryocopus martius*, Middle Spotted Woodpecker *Dendrocopos medius* and White Backed Woodpecker are indeed small (Catsadorakis, 1989) and are perhaps threatened given the scarcity of old mature trees under current forest management practices. The Wryneck *Jynx torquilla* might have already become extinct but the data available are insufficient. The populations of Woodlark and the Tawny Pipit *Anthus campestris* are observed in rather low but seemingly stable. They nest in areas of low grass kept open by grazing and perhaps in the long run they will be threatened by a further reduction of grazing and the encroachment of woodland. The population of Lesser Grey Shrike *Lanius minor* is sparse but does not face a direct danger. The Red-backed Shrike *Lanius collurio* has shown a strong decline the last years (Catsadorakis unpublished data) related seemingly to the intensification and changes in farming practices.

In general the impoverishment of croplands which are denuded of their natural vegetation patches is nowadays the single most important factor affecting at present and in the future the populations of passerines and related birds in Prespa National Park.

The importance of the Lakes Mikri Prespa and Megali Prespa for wintering water-birds

Almost all the research on the avifauna of Prespa up to now has considered the breeding season only. Only Catsadorakis (1986) has provided a list of bird species with indications on seasonal occurrence for two years (1983–1984) and Catsadorakis (1989) gave a similar updated list for 1983–1989 as well as data on the composition of passerine communities in three woodland habitats in winter. Nevertheless the Midwinter Waterfowl Census has been conducted on the Prespa lakes systematically since 1987. Tables 4 and 5 contain the numbers of a total of 26 species of waterfowl counted in mid-January in these two lakes in the period 1988–1994.

Lakes Mikri and Megali Prespa are the two lakes situated at the highest altitude in Greece (853 m and 850 m asl respectively) and they lie within the coldest area of Greece. Mikri Prespa is a shallow lake which freezes almost every winter and remains frozen from a few days (e.g. 1988) to more than 2 months (e.g. 1992). It is of little importance for wintering birds since virtually no birds occur on the lake in severe winters (Table 4: 1989, 1990, 1992, 1993). However the birds have the opportunity to shift easily to the nearby Lake Megali Prespa due to its proximity.

The large lake never freezes and thus concentrates the main bulk of the wintering water-birds of Prespa area (Table 5). Data by Visser & Koning (1968) who counted waterbirds in the Greek part of Megali Prespa in January 1968 corroborate the opinion that the composition and frequencies of the wintering species of waterbirds have remained unaltered. Due to the severe climatic conditions in such a mountain plateau, few birds generally overwinter here in comparison to those in other coastal wetlands of northern Greece (Athanassiou, 1987; Handrinos, 1989). Lake Megali Prespa is important only for its numbers of Great Crested Grebes *Podiceps cristatus* and Tufted Ducks *Aythya fuligula* hosting on the average a 20% and 35% respectively of the total wintering numbers of these two species in Greece.

Generally, the numbers of Pygmy Cormorants wintering at Lake Megali Prespa are much underestimated because they stay spread all along the coast of the lake and are very difficult to count.

A factor that diminishes the value of the winter counts in drawing conclusions about trends of bird numbers and habitat quality in Lakes Prespa in Greece, is that a large part of Lake Megali Prespa belongs to Albania and the Former Yugoslavian Republic of Macedonia (FYROM). Birds can move both within the Lake Megali Prespa but also can shuttle between Lake Mikri and Lake Megali Prespa. As data of each country are not available to the other and there is no coordination of censuses, counts yield results of limited utility. For example, Mikuska (1992) reported the following numbers of birds on the FYROM parts of Megali Prespa: 1988 >14 800, 1989 >8851 and 1990 >11 501 birds. These numbers do not show any similar trends with those in the Greek part (Table 5). A measure of the different environments within this large lake which can attract different species is given by comparing the numbers of Red-crested Pochards *Netta rufina* on the FYROM part: 1987: 150, 1988: 100, 1989: 46 (Schneider-Jacoby & Vasic, 1989) with those on the Greek part (Table 5).

The same problems exist from the conservation point of view regarding the overall management of these shared wetlands (Catsadorakis & Malakou, this volume). In winter, hunting is allowed in all parts of Lake Megali Prespa, but hunting by boat is forbidden in the Greek part only, thus large numbers of birds concentrate on the offshore open waters of the lake away from the coast to roost during daylight (Figure 5). Additionally, issues of pollution, water use and management and fisheries are not tackled in common and the FYROM and Albanian parts are not protected at all.

Conclusion

Though after the late 1960s it has suffered impoverishment and changes and has lost many important species, the avifauna of Prespa occupies a distinct place in Greece and in Europe as well. In the relatively small area (257 km^2) of the National Park, a large part of which (ca 80 km^2) is occupied by open water surfaces, 261 species of birds can be observed and an even more impressive 164 of these are regular or irregular breed-

Table 4. Numbers of wintering waterfowl in the Greek part of Lake Mikri Prespa as counted during IWRB's Mid-winter Waterfowl Censuses.

	1988	1989	1990	1991	1992	1993	1994
Podiceps cristatus	110						
P. nigricollis	60						
Pelecanus crispus	10						67
Phalacrocorax carbo	37						60
P. pygmaeus	415	16		371			54
Egretta alba	18	1		2			5
Ardea cinerea							4
Cygnus olor	1						
Anser anser	76			56			87
Anas platyrhynchos	present			90			20
A. penelope				300			12
A. strepera							35
A. crecca				40			
A. acuta				20			
Aythya fuligula	176						
A. ferina	1350	45		750			20
A. nyroca		3					
Mergus merganser	13						
Fulica atra	1250	70		630			180
Larus ridibundus	present			330			
Other ducks	272						
TOTAL	>3788	135	0	2589	0	0	544
Percent frozen	5%	75%	100%	50%	100%	100%	5%

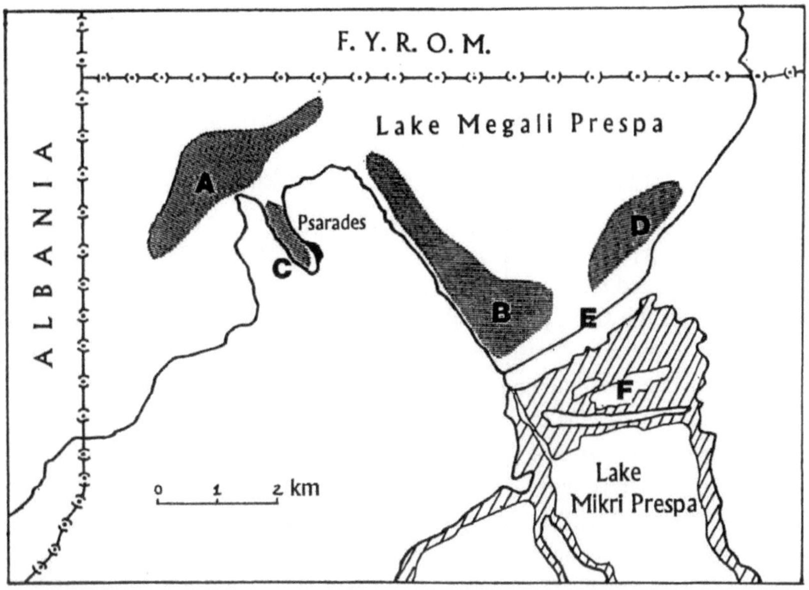

Figure 5. Areas of open water and the coastline of lakes Megali and Mikri Prespa in which certain species of waterfowl are regularly observed during the Mid-winter Waterfowl Censuses conducted in mid January. Data from 1988–1994. A: Great-crested Grebes, gulls, few Black-necked Grebes. B: Tufted Ducks, Coots, Pochards. C: Pygmy Cormorants, gulls, Coots and Black-necked Grebes. D: Coots and diving ducks. E, F: Various dubbling ducks, herons and geese.

Table 5. Numbers of wintering waterfowl in the Greek part of Lake Megali Prespa as counted during IWRB's Mid Winter Waterfowl Census.

	1988	1989	1990	1991	1992	1993	1994
Gavia arctica				3		3	5
Tachybaptus ruficollis	205	20		33		63	
Podiceps cristatus	160	1615	1250	2755	4175	9830	1700
P. nigricollis	130	118	1035	712	155	715	100
Pelecanus crispus			4				
Phalacrocorax carbo			19		2	4	41
P. pygmaeus	20	59	50	128	140	5	100
Egretta alba				18		2	4
Ardea cinerea						5	
Cygnus olor		2			3		
Tadorna tadorna				3			
Anas platyrhynchos	present			205	15	31	15
A. penelope				159			80
A. crecca				81			
A. acuta				6			
Netta rufina						1	
Aythya fuligula	180	638	2400	3600	150	4125	8850
A. ferina	110	90	150		300	388	180
A. nyroca		2					
Bucephala clangula		10			3		
Mergus merganser	13	6	2		7	2	1
Fulica atra	760	720	1280	1155	810	760	1800
Larus ridibundus	30	260	150		60	238	530
L. cachinnans		1			11	11	15
TOTAL	>1608	3541	6336	8858	5835	16183	13421

ers. This diversity can be rivalled, if considered by unit of area by few places in Europe and this constitutes an asset of international importance. Additionally, Prespa is the only place in the European Union where both species of pelicans breed together, but it holds also internationally important breeding populations of rare species as the Great White Egret and the Pygmy Cormorant. Up to 1972 when both the Glossy Ibis and the Spoonbill were breeding in the area, Prespa was characterised by the highest diversity of breeding large waterbirds, rivalled only by the Danube Delta.

In Greece, Prespa National Park is one of the two most important places for birds along with Kerkini Reservoir (Crivelli et al., 1995). Besides its high biodiversity, the area hosts the only breeding populations in Greece of the Great White Pelican and the Goosander, the largest breeding populations of Dalmatian Pelican, Great White Egret and Greylag Goose and the second largest of Pygmy Cormorants and Cormorants. The importance of the area is not least for the breeding of very rare species in Greece as the Partridge, the Hazel-grouse *Bonasa bonasia*, several species of herons and raptors.

Lakes Megali Prespa and Mikri Prespa are close to one another (they formed one lake in the past) and many bird populations are so dependent on both lakes that these lakes must always be treated as one ecosystem. Lake Mikri Prespa is very important as a nesting site for many species but Lake Megali Prespa is important as a feeding site and has a limited – though considerable for Greece – importance for the wintering of some species, especially the Great Crested Grebe and the Tufted Duck. Lake Mikri Prespa has virtually no importance for wintering birds.

Efforts to conserve this important biological asset must not only be directed within Greece but also towards trans-national cooperation with Albania and FYROM in which parts of these two lakes lie.

References

Athanassiou, H., 1987. Past and present importance of the Greek wetlands for wintering waterfowl International Waterfowl Research Bureau. Slimbridge. Mimeo

Bauer, W., H. J. Bohr, U. Mattern & G. Muller, 1973. 2. 'Nachtrag zum Catalogus Faunae Graeciae; Pars Aves'. Vogelwelt 94: 1–21.

Bauer, W. & M. Hodge, 1970. 1. Nachtrag zum 'Catalogus Faunae Graeciae; Pars Aves'. Vogelwelt 52: 96–106.

Berlic, G., G. Catsadorakis & J. Matsakis, 1983. Avifaune migratrice et problemes de gestion du lac du Petit Prespa. XV Congr. Int. Fauna Cinegetica y Silvestre. Trujillo 1981: 505–509.

Biber, O. & A. J. Crivelli, 1978. Mission en Grèce, du 2 au 23 Juin 1978. Mimeo.

Brosselin, M. & A. Molinier, 1968. Visite au lac de Mikra Prespa. Mimeo.

Broussalis, P., 1974. Prespa National Park. Hellenic Society for the Protection of Nature. Athens, 39 pp.

Catsadorakis, G., 1986. Notes on the avifauna of Lake Prespa National Park. Biologia Gallo–Hellenica 12: 485–496.

Catsadorakis, G., 1989. Bird communities in Prespa (Florina, Greece). Elements of ecology and biogeography. Ph.D. Dissertation. University of Athens. Department of Biology, 238 pp. (in Greek with English summary).

Catsadorakis, G., 1991. Bird species that have become extinct and others threatened with extinction from Prespa National Park. Nature. Bull. Hellenic Soc. Protect. Nature 55: 21–25 (in Greek with English summary).

Catsadorakis, G., 1994. Recent siuation of diurnal and nocturnal birds of prey in Prespa National Park (Greece). Biologia Gallo–Hellenica 22: 143–150.

Catsadorakis, G., M. Malakou & A. J. Crivelli, 1996. The effects of the 1989/1990 drought on the colonial waterbirds nesting at Lake Mikri Prespa with special emphasis on pelicans. Colonial Waterbirds 19 (Special Publication 1): 207–218.

Catsadorakis, G. & M. Malakou, M., 1997. Conservation and management issues of Prespa National Park. Hydrobiologia 351: 175–196.

Commission of the European Communities, 1985. Commission Directive of 25 July 1985 amending Council Directive 79/409/EEc on the conservation of wild birds (85/411/EEC).

Cramp, S. & K. E. L. Simmons, 1977. The Birds of the Western Palearctic, Vol. I. Oxford Univ. Press. Oxford.

Crivelli, A. J., 1987. The ecology and behaviour of the Dalmatian Pelican, *Pelecanus crispus* Bruch, a world-endangered species. Final report to the Commission of the European communities, 94 pp + Appendices.

Crivelli, A. J. & O. Vizi, 1981.The Dalmatian Pelican *Pelecanus crispus* Bruch 1832, a recently world endangered bird species. Biol.Conserv. 20: 297–310.

Crivelli, A. J., G. Catsadorakis, H. Jerrentrup, D. Hatzilacou & T. Mitchev, 1991. In Salathe, T. (ed.), Conserving migratory birds. International Council for Bird Preservation, Technical Publication 12, Cambridge, UK: 137–152.

Crivelli, A. J., V. G. Krivenko & V. G. Vinogradov, 1994. Pelicans in the former USSR. IWRB Publication 27, Slimbridge, UK, 151 pp.

Crivelli, A. J., H. Jerrentrup, T. Nazirides & P. Grillas, 1995. Effects on fisheries and waterbirds of raising water levels at Kerkini Reservoir, a Ramsar site in Northern Greece. Envir. Mgmt 19: 431–443.

Geroudet, P., 1973. Mikra Prespa, le lac des Pelicans. Nos Oiseaux 32: 145–167.

Groombridge, B., 1994. IUCN Red List of Threatened Animals. International Union for Conservation of Nature and Natural Resources, Switzerland, 286 pp.

Handrinos, G., 1989. The I.W.R.B. 'Midwinter Waterfowl Counts' in Greece, 1967–1987: A preliminary analysis of the populations of Anatidae. Biologia Gallo-Hellenica 15: 219–236.

Handrinos, G., 1992. Birds. In Karandinos, M. & A. Legakis (eds), 1992. The Red Data Book of threatened Vertebrates of Greece. Hell. Zool. Soc. & Hell. Ornithol. Soc., Athens.

Hatzilacou, D., 1992. The breeding biology and feeding ecology of the great white pelican, *Pelecanus onocrotalus* L. 1758, at Lake Mikri Prespa (Northwestern Greece). PhD dissertation, National and Kapodestrian University of Athens: 180 pp (in Greek, English summary).

Kempf, Ch. & M. Wersinger, 1974. Notes ornithologiques sur le Deransko Jezero,le lac Scutari (Yougoslavie) les lacs Prespa et la vallée d'Avas (Grèce). L' Oiseau et R.F.O. 44: 56–61.

Makatsch, W., 1950. Die Vogelwelt Macedoniens. Akad. Verlagsges. Geest and Portig. Leipzig.

Mendel, H., 1988. Mittelsager-Brut *(Mergus serrator)* am Prespa-See. Ornith. Mitt. 40: 236.

Mikuska, J., 1992. Midwinter Waterfowl Counts in Yugoslavia. A poster presented in IWRB Int. Symposium, Grado, Feb. 1991. Managing Mediterranean Wetlands and their Birds.

Pyrovetsi, M. & A. J. Crivelli, 1988. Habitat Use by Water-birds in Prespa National Park, Greece. Biol. Conserv. 45: 135–153.

Pyrovetsi, M. & M. Karteris, 1986. Forty years of land/cover use changes in Prespa National Park,Greece. J. envir. Mgmt 23: 173–83.

Schneider-Jacoby, M. & V. Vasic, 1989. The Red–crested Pochard *Netta rufina* breeding and wintering in Yugoslavia. Wildfowl 40: 39–40.

Terrasse, J. F., M. Terrasse & M. Brosselin, 1969. Avifaune d' un lac des Balkans: Mikra Prespa (Grèce). L' Oiseau et R.F.O. 39: 185–201.

Thorpe, W. H., P. T. Cotton & P. F. Holmes, 1936. Notes on the Birds of Lakes Ohrid, Malik and Prespa and adjacent parts of Jugoslavia, Albania and Greece. Ibis XIII-vol.VI: 557–580.

Visser, R. & F. Koning, 1968. List of birds seen in Greece during the winter 1967–1968. Mimeo.

Voous, K. H., 1977. List of recent Holarctic bird species. British Ornithologists' Union. London, 85 pp.

Appendix 1. List of bird species observed in Prespa National Park. Sequence of species as in Voous (1977) B=Breeding, (FB)=Fommerly Bred, (B)=very likely breeding, (B?)=Possibly breeding, Bo=occasional breeder. R=Resident, M=Migratory, (M)=few individuals observed, migratory, (FM)=Formerly migrated W=Winter visitor, (W=) few individuals observed, A=Accidental (usually only observed once, Aw=Accidental in winter, Am=Accidental in migration period. ?= the seasonal occurence of the species is uncertain reg=regular presence, irr=irregular presence. A period between status of seasonal occurence and breeding, e.g M,W, means that some individuals migrate and some are observed in winter. No period, e.g, MB, means the species is migratory and breeding. It is not always easy to distinguish individuals that only migrate through the area. Red Data Book symbols: E=Endangered, V=Vulnerable, R=Rare, I=indeterminate, IK=Insufficiently known

	English name	Latin name	Seasonal occurence Breeding	79/409 EEC	Red Data Book of Greece
1	Red-throated Diver	*Gavia stellata*	Aw	O	
2	Black-throated Diver	*G. arctica*	(W)reg	O	
3	Little Grebe	*Tachybaptus ruficollis*	MB		
4	Great Crested Grebe	*Podiceps cristatus*	R.B.W		
5	Red-necked Grebe	*P. grisegena*	A		
6	Slavonian Grebe	*P. auritus*	A	O	
7	Black-necked Grebe	*P. nigricollis*	M,Bo,W		IK
8	Cormorant	*Phalacrocorax carbo*	M,B,(W)reg	O	
9	Pygmy Cormorant	*P. pygmaeus*	R,B	O	E
10	Great White Pelican	*Pelecanus onocrotalus*	MB	O	E
11	Dalmatian Pelican	*P. crispus*	MB,(W)irr	O	E
12	Bittern	*Botaurus stellaris*	(M)irr,(W)	O	
13	Little Bittern	*Ixobrychus minutus*	MB	O	
14	Night Heron	*Nycticorax nycticorax*	MB	O	IK
15	Squacco Heron	*Ardeola ralloides*	MB	O	
16	Little Egret	*Egretta garzetta*	MB	O	
17	Great White Egret	*E. alba*	RB	O	E
18	Grey Heron	*Ardea cinerea*	MB,(W)		
19	Purple Heron	*A. purpurea*	MB	O	V
20	Black Stork	*Ciconia nigra*	(M)	O	
21	White Stork	*C. ciconia*	MB	O	
22	Glossy Ibis	*Plegadis falcinellus*	M(FB)	O	
23	Spoonbill	*Platalea leucorodia*	M(FB)	O	
24	Mute Swan	*Cygnus olor*	(FB),(M)irr,(W)irr		
25	White-fronted Goose	*Anser albifrons*	Wirr	O	
26	Greylag Goose	*A. anser*	RB,WIrr		E
27	Ruddy Shelduck	*Tadorna ferruginea*	A	O	
28	Shelduck	*T. tadorna*	Mirr		
29	Wigeon	*Anas penelope*	M,W		
30	Gadwall	*A. strepera*	MB(W)		IK
31	Ieal	*A. crecca*	M(W)		
32	Mallard	*A. platyrhynchos*	R,B,M		
33	Pintall	*A. acuta*	M		
34	Garganey	*A. querquedula*	M,(B?)		
35	Shoveler	*A. clypeata*	M (FB)		
36	Red-crested Pochard	*Netta rufina*	Aw		
37	Pochard	*Aythya ferina*	M,W,(FB?)		
38	Ferruginous Duck	*A. nyroca*	(FB),(W)	O	V
39	Tufted Duck	*A. fuligula*	M,W,(FB)		
40	Scaup	*A. marila*	Aw		

Appendix 1. Continued

	English name	Latin name	Seasonal occurence Breeding	79/409 EEC	Red Data Book of Greece
41	Goldeneye	*Bucephala clangula*	(W)irr		
42	Smew	*Mergus albellus*	A		
43	Goosander	*M. merganser*	RB		E
44	Honey Buzzard	*Pernis apivorous*	MB	O	
45	Black Kite	*Milvus migrans*	A	O	
46	White-tailed Eagle	*Haliaeetus albicilla*	(W)irr (FB)	O	
47	Egyptian Vulture	*Neophron percnopterus*	(M)irr,(FB)	O	V
48	Griffon Vulture	*Gyps fulvus*	A	O	
49	Short-toed Eagle	*Circaetus gallicus*	MB	O	
50	Marsh Harrier	*Circus aeruginosus*	M,MB	O	V
51	Hen Harrier	*C. cyaneus*	W	O	
52	Pallid Harrier	*C. macrourus*	A	O	
53	Montagu's Harrier	*C. pygargus*	(M)reg	O	
54	Goshawk	*Accipiter gentilis*	RB		
55	Sparrowhawk	*A. nisus*	RB		
56	Levant Sparrowhawk	*A. brevipes*	A	O	
57	Buzzard	*Buteo buteo*	RB,W		
58	Long-legged Buzzard	*B. rufinus*	A(FB)	O	
59	Lesser Spotted Eagle	*Aquilla pomarina*	A(FB)	O	
60	Imperial Eagle	*A. heliaca*	(FB)	O	
61	Spotted Eagle	*A. clanga*	A	O	
62	Golden Eagle	*A. chrysaetos*	RB	O	V
63	Booted Eagle	*Hieraaetus pennatus*	A(FB)	O	
64	Osprey	*Pandion haliaetus*	(M)reg	O	
65	Lesser Kestrel	*Falco naumanni*	(FB?)	O	
66	Kestrel	*F. tinnunculus*	RB		
67	Red-footed Falcon	*F. vespertinus*	Mreg		
68	Merlin	*F. columbarius*	(W)irr	O	
69	Hobby	*F. subbuteo*	MB	O	
70	Lanner	*Falco biarmicus*	A	O	
71	Peregrine	*F. peregrinus*	A	O	
72	Hazel Grouse	*Tetrastes bonasia*	RB	O	IK
73	Rock Partridge	*Alectoris graeca*	RB		
74	Partridge	*Perdix perdix*	RB	O	V
75	Quail	*Coturnix coturnix*	MB,M?		IK
76	Pheasant	*Phasianus colchicus*	(FB)introduced		
77	Water Rail	*Rallus aquaticus*	RB		
78	Spotted Crake	*Porzana porzana*	A	O	
79	Little Crake	*P. parva*	Mirr?,(FB)	O	
80	Moorhen	*Gallinula chloropus*	RB		

Appendix 1. Continued

	English name	Latin name	Seasonal occurence Breeding	79/409 EEC	Red Data Book of Greece
81	Coot	*Fulica atra*	RB,W		
82	Crane	*Grus grus*	A	O	
83	Houbara Bustard	*Chlamydotis undulata*	(FM)		
84	Oystercatcher	*Haematopus ostralegus*	A		
85	Black-winged Stilt	*Himantopus himantopus*	M	O	
86	Avocet	*Recurvirostra avosetta*	Mirr	O	
87	Collared Pratincole	*Glareola pratincola*	A	O	
88	Little Ringed Plover	*Charadrius dubius*	M,B		
89	Ringed Plover	*C. hiaticula*	(M)irr		
90	Kentish Plover	*C. alexandrinus*	Am		
91	Grey Plover	*Pluvialis squatarola*	Am		
92	Lapwing	*Vanellus vanellus*	M,(Bo),W		
93	Little Stint	*Calidris minuta*	M		
94	Temminck's Stint	*C. teminckii*	(M)		
95	Curlew Sandpiper	*C. ferruginea*	Mirr		
96	Dunlin	*C. alpina*	M		
97	Ruff	*Philomachus pugnax*	M	O	
98	Snipe	*Gallinago gallinago*	W		
99	Great Snipe	*G. media*	Am	O	
100	Woodcock	*Scolopax rusticola*	W		
101	Black-tailed Godwit	*Limosa limosa*	M		
102	Whimbrel	*Numenius phaeopus*	Am		
103	Curlew	*N. arquata*	Mirr		
104	Spotted Redshank	*Tringa erythropus*	Mirr		
105	Redshank	*T. totanus*	M		
106	Greenshank	*T. nebularia*	Mirr		
107	Green Sandpiper	*T. ochropus*	M		
108	Wood Sandpiper	*T. glareola*	M	O	
109	Common Sandpiper	*Actitis hypoleucos*	MB?		
110	Turnstone	*Arenaria interpres*	Am		
111	Arctic Skua	*Stercorarius parasiticus*	A		
112	Mediterranean Gull	*Larus melanocephalus*	Mirr	O	
113	Little Gull	*L. minutus*	A		
114	Black-headed Gull	*L. ridibundus*	M,W		
115	Slender-billed Gull	*L. genei*	A		
116	Common Gull	*L. canus*	Am		
117	Yellow-legged Gull	*L. argentatus*	MB,(W)		
118	Gull-billed Tern	*Gelochelidon nilotica*	Am	O	
119	Caspian Tern	*Sterna caspia*	Am	O	
120	Sandwich Tern	*S. sandvicensis*	A	O	

Appendix 1. Continued

	English name	Latin name	Seasonal occurence Breeding	79/409 EEC	Red Data Book of Greece
121	Common Tern	*S. hirundo*	MB	O	
122	Little Tern	*S. albifrons*	Mirr	O	
123	Whiskered Tern	*Chlidonias hybridus*	M(FB)	O	V
124	Black Tern	*C. niger*	M	O	
125	White-winged Tern	*C. leucopterus*	Mirr		
126	Collared Dove	*Streptopelia decaocto*	MB		
127	Turtle Dove	*S. turtur*	MB		
128	Rock Dove	*Columba livia*	FB		
129	Stock Dove	*C. oenas*	(FB)	R	
130	Woodpigeon	*C. palumbus*	RB,W?		
131	Cuckoo	*Cuculus canorus*	MB		
132	Scops Owl	*Otus scops*	M(Bo)		
133	Eagle Owl	*Bubo bubo*	RB	O	
134	Tawny Owl	*Strix aluco*	RB		
135	Long-eared Owl	*Asio otus*	B		
136	Little Owl	*Athene noctua*	RB		
137	Nightjar	*Caprimulgus europaeus*	MB	O	
138	Swift	*Apus apus*	Mirr		
139	Alpine Swift	*A. melba*	MB		
140	Bee-eater	*Merops apiaster*	MB		
141	Roller	*Coracias garrulus*	(M)irr	O	
142	Hoopoe	*Upupa epops*	MB		
143	Kingfisher	*Alcedo atthis*	MB?,(W)	O	
144	Wryneck	*Jynx torquilla*	M,B?		
145	Green Woodpecker	*Picus viridis*	RB		
146	Black Woodpecker	*Dryocopus martius*	RB	O	
147	Great Spotted Woodpecker	*Dendrocopos major*	RB		
148	Syrian Woodpecker	*D. syriacus*	RB	O	
149	Middle Spotted Woodpeck	*D. medius*	RB	O	
150	White-backed Woodpecker	*D. leucotos*	RB?	O	R
151	Lesser Spotted Woodpeck	*D. minor*	RB		
152	Short-toed Lark	*Calandrella cinerea*	MB	O	
153	Crested lark	*Galerida cristata*	RB		
154	Woodlark	*Lullula arborea*	B	O	
155	Skylark	*Alauda arvensis*	MB		
156	Shore Lark	*Eremophila alpestris*	RB		
157	Sand Martin	*Riparia riparia*	MB		
158	Swallow	*Hirundo rustica*	MB		
159	Red-rumped Swallow	*H. daurica*	MB		
160	Crag Martin	*Ptyonoprogne rupestris*	MB		

Appendix 1. Continued

	English name	Latin name	Seasonal occurence Breeding	79/409 EEC	Red Data Book of Greece
161	House Martin	*Delichon urbica*	MB		
162	Water Pipit	*Anthus spinoletta*	Mb,W		
163	Tree Pipit	*A. trivialis*	MB		
164	Meadow Pipit	*A. pratensis*	W		
165	Tawny Pipit	*A. campestris*	MB	O	
166	Yellow Wagtail	*Motacilla f.feldegg +spp.*	MB,M		
167	Grey Wagtail	*M. cinerea*	MB		
168	Pied Wagtail	*M. alba*	MB		
169	Dipper	*Cinclus cinclus*	RB		
170	Wren	*Troglodytes troglodytes*	RB		
171	Dunnock	*Prunella modularis*	W,B?		
172	Alpine Accentor	*P. collaris*	MB		
173	Robin	*Erithacus rubecula*	RB		
174	Nightingale	*Luscinia megarhynchos*	MB		
175	Black Redstart	*Phoenicurus ochruros*	MB,M		
176	Redstart	*P. phoenicurus*	(M)irr		
177	Whinchat	*Saxicola rubetra*	MB		
178	Stonechat	*S. torquata*	MB		
179	Wheatear	*Oenanthe oenanthe*	MB		
180	Pied Wheatear	*O. pleschanka*	A		
181	Black-eared Wheatear	*O. hispanica*	MB		
182	Rock Thrush	*Monticola saxatilis*	MB		
183	Fieldfare	*Turdus pilaris*	W		
184	Ring Ouzel	*T. torquatus*	(W)irr		
185	Blackbird	*T. merula*	RB		
186	Redwing	*T. iliacus*	Aw		
187	Song Thrush	*T. philomelos*	MB		
188	Mistle Thrush	*T. viscivorous*	RB		
189	Cetti's Warbler	*Cettia cetti*	RB		
190	Savi's Warbler	*Locustella luscinioides*	(FB)		IK
191	Moustached Warbler	*Acrocephalus melanopogon**	(FB)	O	R
192	Sedge Warbler	*A. schoenobaenus*	MB		
193	Marsh Warbler	*A. palustris*	MB		
194	Reed Warbler	*A. scirpaceus*	MB		
195	Great Reed Warbler	*A. arundinaceus*	MB,M		
196	Icterine Warbler	*Hippolais icterina*	M		
197	Olivaceous Warbler	*H. pallida*	MB		
198	Subalpine Warbler	*Sylvia cantillans*	MB		
199	Orphean Warbler	*S. hortensis*	Mirr		
200	Barred Warbler	*S. nisoria*	MB	O	

Appendix 1. Continued

	English name	Latin name	Seasonal occurence Breeding	79/409 EEC	Red Data Book of Greece
201	Lesser Whitethroat	*S. curruca*	MB		
202	Whitethroat	*S. communis*	MB		
203	Blackcap	*S. atricapilla*	MB		
204	Bonneli's Warbler	*Phylloscopus bonelli*	M		
205	Wood Warbler	*P. sibilatrix*	M		
206	Chiffchaff	*P. collybita*	MB		
207	Willow Warbler	*P. trochilus*	M		
208	Goldcrest	*Regulus regulus*	W		
209	Firecrest	*R. ignicapillus*	W		
210	Spotted Flycatcher	*Muscicapa striata*	(M)		
211	Pied Flycatcher	*Ficedula hypoleuca*	(M)		
212	Semi-collared Flycatcher	*Ficedula semitorquata*	(M)	O	
213	Bearded Tit	*Panurus biarmicus*	RB		
214	Marsh Tit	*Parus palustris*	RB		
215	Willow Tit	*P. montanus*	A		
216	Sombre Tit	*P. lugubris*	RB		
217	Coal Tit	*P. ater*	RB		
218	Blue Tit	*P. caeruleus*	RB		
219	Great Tit	*P. major*	RB		
220	Long-tailed Tit	*Aegithalos caudatus*	RB		
221	Penduline Tit	*Remiz pendulinus*	MB		
222	Wallcreeper	*Tichtodroma muraria*	Aw		
223	Nuthatch	*Sitta europaea*	RB		
224	Rock Nuthatch	*S. neumayer*	RB		
225	Treecreeper	*Certhia familiaris*	RB		
226	Short-toed Treecreeper	*C. brachydactyla*	RB		
227	Golden Oriole	*Oriolus oriolus*	MB		
228	Red-backed Shrike	*Lanius collurio*	MB	O	
229	Lesser Grey Shrike	*L. minor*	MB	O	IK
230	Great Grey Shrike	*L. excubitor*	W		
231	Woodchat Shrike	*L. senator*	M		
232	Jay	*Garrulus glandarius*	RB		
233	Magpie	*Pica pica*	RB		
234	Alpine Chough	*Pyrrhocorax graculus*	RB		
235	Chough	*P. pyrrhocorax*	W	O	
236	Jackdaw	*Corvus monedula*	RB		
237	Rook	*C. frugilegus*	Wirr		
238	Hooded Crow	*C. corone*	RB		
239	Raven	*C. corax*	RB		
240	Rose-coloured Starling	*Sturnus roseus*	Am		

Appendix 1. Continued

	English name	Latin name	Seasonal occurence Breeding	79/409 EEC	Red Data Book of Greece
241	Starling	*S. vulgaris*	RB		
242	House Sparrow	*Passer domesticus*	RB		
243	Spanish Sparrow	*P. hispaniolensis*	MBo		
244	Tree Sparrow	*P. montanus*	RB		
245	Rock Sparrow	*Petronia petronia*	(FB)		
246	Chaffinch	*Fringilla coelebs*	RB		
247	Brambling	*F. montifringilla*	W		
248	Serin	*Serinus serinus*	M		
249	Greenfinch	*Carduelis chloris*	RB		
250	Siskin	*C. spinus*	(W)irr		
251	Goldfinch	*C. carduelis*	RB		
252	Linnet	*C. cannabina*	RB		
253	Bullfinch	*Pyrrhula pyrrhula*	M.W		
254	Hawfinch	*Coccothraustes coccothraustes*	RB		
255	Yellowhammer	*Emberiza citrinella*	RB		
256	Cirl Bunting	*E. cirlus*	RB		
257	Rock Bunting	*E. cia*	RB		
258	Ortolan Bunting	*E. hortulana*	MB	O	
259	Reed Bunting	*E. schoeniclus*	MB,W		IK
260	Black-headed Bunting	*E. melanocephala*	MB		
261	Corn Bunting	*Miliaria calandra*	RB		

Hydrobiologia **351:** 175–196, 1997.
A. J. Crivelli & G. Catsadorakis (eds), *Lake Prespa, Northwestern Greece.*

Conservation and management issues of Prespa National Park

G. Catsadorakis & M. Malakou
Society for the Protection of Prespa, 530 77 Agios Germanos, Greece

Key words: biodiversity conservation, cultural landscapes, ecological integrity, ecosystem management, Greece, human ecology, integrated management approach, National Park management, Prespa, protected area management, wetland conservation, wetland management

Abstract

The present article describes the human activities in the area and their changes and trends, given that the contemporary nature conservation key issues are closely related to past and present socio-economic and cultural conditions, both within Prespa itself and outside of it. The liberation of Prespa, a remote and rather socially isolated borderline area, from Ottoman occupation took place in 1912–13. Recent events in its environmental history have been forest clearings and human depopulation in the Civil War (1944–49), irrigation system construction in the 1960s, its 'discovery' by ornithologists in the late 1960s, the high emigration rates of 1960–1980, its designation as a National Park in 1974, the destructive development works of 1984–86 after Greece's entry to the EEC and the conversion to intensive bean cultivation in the mid 1980s.

The human population of ca 7000 at the start of this century suffered an almost 80% decrease after the Civil War. A resettlement in the mid 1950s increased the population by 40% but in the 1970s emigration reduced it to its previous size of around 1500, still its present size. The extensive and diverse farming systems of previous centuries changed gradually after the construction of an irrigation network in the 1960s, which twenty years later and along with international changes in production and economy patterns, made possible the present day domination of a bean monoculture within irrigated croplands. This was accompanied by increased energy inputs, mechanisation and the use of chemicals. Intensification of agriculture brought increased incomes which attracted also livestock keepers, resulting in a 56% decrease in numbers of livestock in 1964–1993. Cattle are presently kept for meat production only. They graze on the mountains, and no longer in the meadows around the lakeshore. A local shorthorn breed accounting for 93% of all cattle in 1963 declined to less than 18% in 1993. Fishing aimed mainly at carp and Prespa bleak, continues today as in the past to be a source of supplementary income. Eighty-six per cent of the – mainly oak and beech – forests are State owned and the rest belong to local municipalities. The forest area has not changed essentially in the last 40 years, but forest roads have expanded. Forests provide mainly fuelwood for local needs and small quantities of industrial timber. No specific management is conducted to safeguard the conservation values of the forest, the majority of which consists of even aged coppiced stands with limited value for wildlife. Manufacturing never played an important role for the economy of the area. Tourism has increased in the last 20 years, concentrated mainly in summer and spring. Accommodation in the villages is not yet satisfactory though it has improved enormously in recent years. The National Park's infrastructure for visitors remains poor. Tourist pressure generally is not yet that high to impact drastically the social and environmental fabric. In places, hunting, angling and poaching can negatively affect the target species. The construction of the irrigation network in the eastern part of Lake Mikri Prespa was the most important change to the landscape and the habitats of Prespa the last 30 years. It resulted in the embankment of free-flowing streams, drainage of wet meadows and dramatic reduction of trees and hedges. In the last 15 years, the reedbeds, no longer burnt or mowed and assisted by increased nutrient loadings from agricultural runoff, have expanded landward to occupy formerly open, periodically flooded, areas. The above changes in habitats and land use have brought an impoverishment of biodiversity which can be seen in the extermination or the decrease in the numbers of certain plant and bird species.

The problems preventing a better management of the National Park and the perpetuation of its values are inappropriate legislation, the poor degree of coordination of the public services, the lack of both a specific National Park Service in Greece and a management authority for the Park and the hostile attitude of local people due to their limited information and involvement. The scenic beauty, the wetlands and a number of rare habitats, the diversity of habitats, flora and fauna, the endemic life forms, the rare mammals and the colonial waterbirds, the local architecture and the cultural expressions such as the Byzantine monuments, are the values of Prespa that must be preserved as a whole. The initial conservation efforts aimed at preserving biodiversity but later it became apparent that not only are humans an inseparable part of the whole, but additionally, several of their extensive activities have contributed substantially to the rich biodiversity of the area. Through the prerequisites of keeping a balance between primary and tertiary sector activities and of revitalizing the social fabric, the following are the key management and conservation issues of today: an unequivocal agreement on the values to be preserved and their limits; the formation of a management authority; a new realistic protection and land use zoning; the diversification and extensification of all farming activities; the restoration of the wet meadow habitats; the ensuring of a high water level and a good water quality for both lakes through tri-lateral cooperation between Greece, Albania and FYROM; and the establishment of a permanent environmental monitoring scheme.

Introduction

Though designated as a National Park by Greek legislation, the Prespa area (hereafter the name Prespa will denote only the Greek part of the Prespa lakes and their catchment area) does not fulfil the IUCN National Park (Category II) definition, since it is small, only 28,000 ha, inhabited by 1,200 people in twelve villages with activities mainly in the primary sector taking place within its territory. However, for the time being this is the only existing operational legislation to protect the natural values of such an area in Greece. In addition, the policy of the Greek State strongly encourages and supports the continuous presence of people in these remote frontier areas. Thus, the conservation of habitats and wildlife is based not upon the exclusion of man but upon coexistence with him.

The key conservation and management issues of today which are closely associated with the past and present socio-economic and cultural conditions of Prespa are better understood if it is recognised that the small, isolated agricultural communities of Prespa were created and flourished upon a subsistence economy basis. The present-day internationalised socio-economic systems in the developed and developing countries are completely different and have led to problems for the survival of such communities, if not to a complete degradation worldwide.

The aim of the present article is to describe the human presence in the area, its history and trends. To assess the effects these activities had and still have upon the ecosystems of Prespa, giving at the same time their social, political and economic dimensions so that they can be understood in depth. In the light of the previous articles of this book (Crivelli & Catsadorakis, in press) the present management and conservation issues of the Prespa National Park are discussed and reference is made to those that might emerge in the future.

An outline of the history of the area in the 20th century

The following brief passage through the history of Prespa in the current century serves to place Prespa within a historic context necessary for understanding its present-day social situation:

1900: The turn of the century finds Prespa in a remote corner of the Ottoman empire, away from large urban centres. The closest towns were Koritsa (today Korce in Albania) and Monastir (today Bitola in FYROM; Figure 1). Prespiots participate in 'the Macedonian Struggle' for their liberation from the Turks and their incorporation into Greece and not into Bulgaria.

1912–1913: The end of the two Balkan wars finds Prespa embodied in the Greek State.

1913: Bucharest Treaty. Prespa is ascribed to Greece. The agreed boundaries conceded to Greece the present Albanian part of Lake Megali Prespa and many villages around its periphery.

1914–1918: During World War I Prespa hosts French troops as do the rest of Central and Western Macedonia.

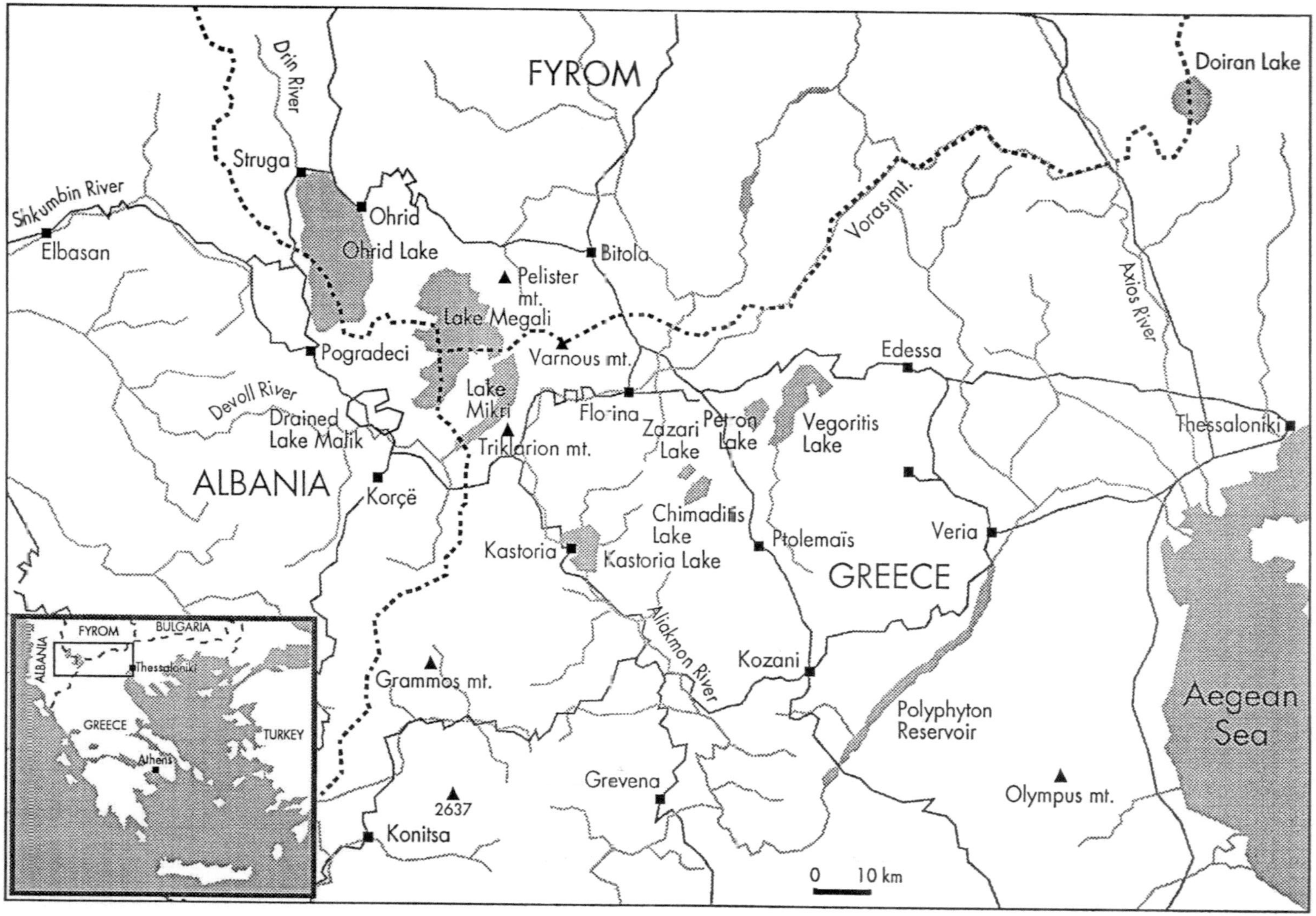

Figure 1. Geographic location of the Prespa region, and its immediate adjacent areas in Greece, Albania and the Former Yugoslavian Republic of Macedonia (FYROM).

1915–1930: A large emigration wave of Prespiots takes place mainly to the U.S.A. and Canada.

1919: Neuilly's Treaty. Prespa is again ascribed to Greece.

1923: A total of 325 Greek refugees from the Black Sea coast of Turkey settle in three villages (Lemos, Lefkonas and Pili) (Figure 2) and add to a population of ca. 5630.

1920–1930: Opaia, a lakeshore village, is abandoned due to malaria.

1920–1940: Intense building occurs with funds coming from emigrants abroad. Most present-day houses date from that era (Karadedos and Tsolakis 1989).

1935–1945: The Agios Germanos stream is partly diverted to discharge into Lake Megali Prespa instead of Lake Mikri Prespa, to avert flooding and reclaim agricultural land.

1940–1944: Italian troops occupy the area during the war.

1944–1949: Greek Civil war. The area is among those that suffered the most. Many people flee to the Eastern Bloc countries after the end of the war. Seven villages are deserted.

1953: Three of the abandoned villages (Vrondero, Agios Germanos, Kalithea) are resettled by Vlachs, nomadic livestock herders from Epirus and Thessaly; four villages fall to ruins.

1961–1972: Further emigration takes away ca. 45% of the 1961 population (Planning Service of W. and C. Macedonia 1979).

1965: The land reclamation works start for the installation of a surface irrigation system. The farmland zone around the lake is changed dramatically by the works.

1968: Foreign naturalists discover in Prespa an ornithological paradise with pelican colonies and a rich wildlife and flora.

1971: First protection measures are taken by prefectoral decrees for five different sub-areas with waterbird colonies. A paved road is constructed, connecting most of Prespa villages to the town of Florina, the prefectoral centre.

1973–1977: The second phase of works for the irrigation scheme starts. Farming is still extensive.

1974: Designation of the National Park.

1975–1981: A second wave of emigration removes almost 25% of the 1973 population, mainly to work in the fur industry in the nearby town of Kastoria and in Western Germany (Karamani 1983).

1981: Greece becomes a member of European Union (then EEC) and the impact upon the development of the area is soon felt.

1984–1987: The area is chosen for a pilot Integrated Mediterranean Programme (IMP) action financed by EEC. A third phase to accomplish the irrigation scheme is embodied in the works which have an adverse environmental impact. An unnecessary fish breeding station is built on invaluable wet meadows.

1987–1990: A quick turn towards intensive agriculture of beans is observed, which eventually occupies all the cultivable area within the irrigation scheme.

1991: The Society for the Protection of Prespa is established.

Population and employment

The increase of population between 1920 and 1940 (Figure 3) must be attributed to the settlement of refugees from the Black Sea and natality due to a robust age structure and the predominant family and production organisation patterns and social values. World War II and especially the civil war, caused a population decrease of nearly 79% for the period 1940–1950. For comparison, during the same period there was a reduction of 15% for the whole of Florina prefecture and of 3.1% for the whole of Macedonia, while Greece as a whole increased by 2.8% (Tziafetas, 1994). The resettlement by 1030 Vlachs in 1953–54 increased the population, as is shown in the 1961 census, but the following 20 years of emigration claimed more than 53% of the 1961 population, while for the whole of the Florina prefecture this decrease was 42% (Tziafetas, 1994). There were 16 settlements, but these were reduced to 12 with the abandonment of Daseri, Pixos, Angathoto and Kranies after the Civil War (Figure 2).

Up to the early 1970s the majority of people made their living from arable farming with a supplementary income from livestock and fishing. A few kept shops and small manufacturing units. The overall subsistence economy system started to change rapidly after the civil war.

In the 1990s, Milionas village became practically deserted. The overall population seems to have been stabilised at around 1500 inhabitants (National Statistical Service of Greece 1991, population survey) but the present resident population is ca 1200 people. The average population density of the Prefecture of Florina is ca 30 ind km^{-2} while that of Prespa is 6 ind km^{-2}.

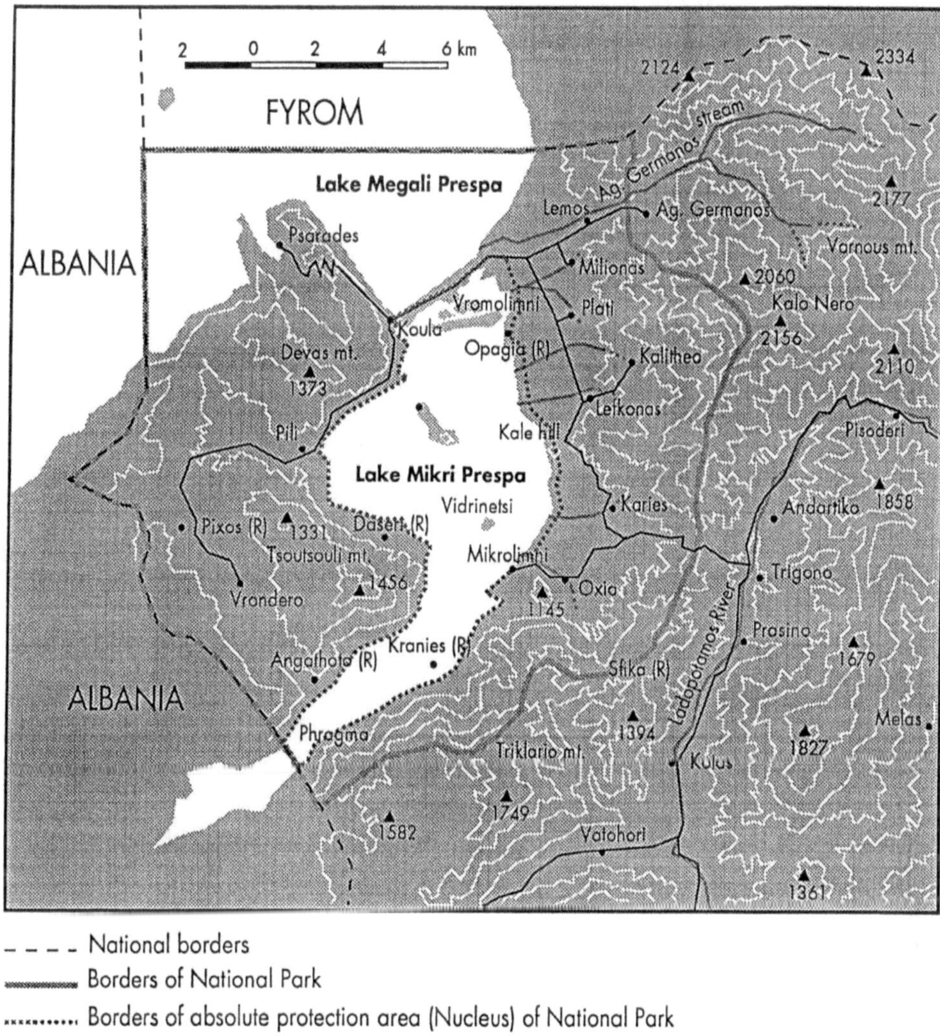

Figure 2. Geomorphologic map of the Greek part of Prespa, with names of villages, localities and other physical features mentioned in the text. Contours per 200 m.

The stabilisation of the population must be attributed both to the return of some emigrants, but also to the halting of emigration thanks to the increased incomes provided by bean cultivation, which has predominated in recent years. The economic activity of the last 30 years is illustrated in Figure 4 with the evolution of numbers of building permits in the area. Bean cultivation has occupied a ca 90% of the economically active population in recent years, while livestock raising is decreasing, but the tertiary sector including tourist services is increasing substantially.

The age pyramid deteriorated much between 1973–1983 due to emigration, but a recovery is clear between 1983 and 1989, though still remaining heavily shifted towards older ages (Figure 5).

Agriculture

Up to the end of the 1940s the predominant subsistence economy system combined traditional crop farming with livestock farming, fishing and collecting from nature.

Almost half of croplands were in the lowlands close to the lakeshores but a large part of the mountain slopes up to 1800 m asl was cultivated with cereals on small terraces. Crop holdings were small and scattered. In

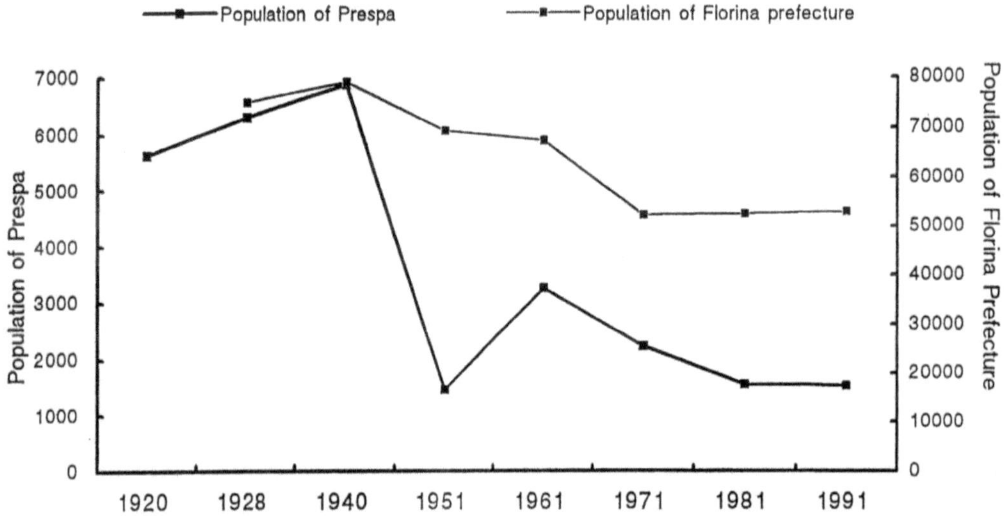

Figure 3. The fluctuation of the population of Prespa over the last 70 years in comparison to tht of the entire Florina Prefecture. For 1940 the population of Prespa was ca 1000–1500 people higher than 6880, because the populations of the later abandoned villages of Angathoto, Kranies, Daseri and Pixos is not included in this number.

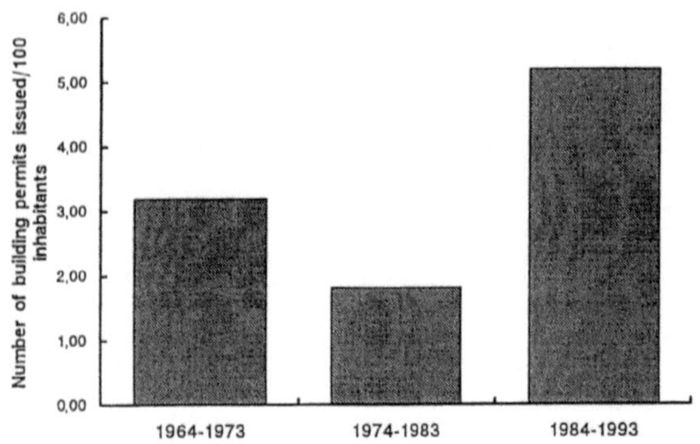

Figure 4. Number of building permits per 100 inhabitants issued for the villages of Prespa between 1964 and 1993 (Data of the Housing Service of the Prefecture of Florina). The drop in the second period reflects the economic inactivity caused by the high emigration.

the 1940s, 178 ha were irrigated by traditional methods (Vafiadis, 1940). There was almost no subsidiary energy and all transportation and draught power was provided by cattle and equids. Fertilizing was with animal manure. Cereals, mainly wheat, covered almost half of the croplands, while 37% were hay pastures (Table 1). Pulses included beans, lentils and chickpeas. Cultivation on mountain slope terraces was abandoned after the war.

In the early 1960s a surface irrigation network scheme was designed for the croplands bordering the eastern side of the lakes, using water both from Lake Mikri Prespa and the Agios Germanos stream. Its aim was to convert rainfed crops to irrigated ones and thus increase production and farmers' incomes. In so doing, policy makers hoped to retain the population at this border area and reduce emigration. At that time the region was one of the least developed in Greece, mainly due its border location and the civil war sufferings and depopulation (Psychoudakis et al., 1993). The cultivated area with meadows and fallows was 1650–1830 ha (Broikos, 1962; Moutaftsis, 1960; National Statistical Service of Greece). The average farm size was 2.3 ha. (Moutaftsis, 1960). In 1964, just before the initiation of the irrigation network construction works, the situation had remained almost unaltered except for an increase in wheat (Table 1). Only at the eastern side of the lake at least 294 ha were irrigated by traditional methods (Psychoudakis et al., 1993). The works started in 1965 but are not yet completed after 30 years, because of several unforeseen problems and bad planning (Psychoudakis et al., 1993). The plan was to construct six main canals and a distribution network of secondary and tertiary concrete canals, accompanied by a system of drainage ditches and roadways giving access to each field.

The present day network was accomplished in three phases. In 1977, 1221 ha of irrigated land were provided, but there had been no essential change in crops (Table 1). Farmers still used the majority of the irrigated croplands to cultivate rainfed cereals. Only 350 ha of croplands were actually irrigated in 1983.

Table 1. Percent cover of different crops in the whole cropland area of Prespa. (Florina Agricultural Service).

	1940*	1964**	1973**	1983	1986	1991	1993
Wheat	23	42.8	16.6	15.3	12.7	9.5	10.7
Barley	3	7.4	33.5	48.4	23.9	7.1	10.1
Maize	2	2.1	1.6	2.0	0.7	0.1	0.5
Other cereals (rye, oats)	23	3.1	1.7	3.3	14.2	19.1	9.7
Pulses	7	8.3	12.3	8.7	22.4	43.1	45.6
Lucerne	0	2.3	10.3	9.6	9.2	8.2	0.8
Hay pastures	37	25.1	13.0	5.2	4.8	3.0	11.0
Other arable (orchards, vegetables, vineyards, fallows)	6	10.8	13.3	7.5	6.9	3.1	11.5
Irrigated land (in ha)	178.4	294.0		ca 350	ca 520	ca 950	ca 1100

* Vafiadis, 1940
** Planning Service of C. & W. Macedonia, 1975

However, development started to affect Prespa and an increasing intensification of agriculture gradually displaced extensive methods. In 1973 there were already 38 tractors and 1 combine harvester while in 1994 there were 215 tractors and 28 combine harvesters, an increase of 565% and 2800% respectively (Florina Agricultural Service). Land redistribution took place three times between 1959 and 1979 mainly to serve the irrigation system (Psychoudakis et al., 1993). The average farm size in 1973 was 2.5 ha of which 0.58 was irrigated, but the average number of crop holdings which an average farm was shared was 7 (Planning Service of W. and C. Macedonia, 1975). Even in 1983 very few farms were as large as 2.0 ha (Pyrovetsi & Gerakis, 1987). In 1993, there were about 320 farms within the irrigated land and the average farm size was 4.4 ha but because almost 40% of land belongs to absentee landowners (Psychoudakis et al., 1993) a part of income was being transferred to other regions in the form of rents.

After 1985 an impressive trend was observed towards the capital-intensive cultivation of bean plants for the production of dry white haricot beans. Cultivation of other traditional pulses has ceased, along with unprofitable cereals. The drastic shrinkage of hay pastures and lucerne is associated with the concurrent decrease of cattle.

In the 1990s almost the whole of the irrigable part of the irrigation system was intensively cultivated with beans (ca 1000 ha) with increased use of chemicals and a gradual elimination of trees, orchards and natural hedges. Cereals are now restricted to rainfed areas.

The detailed study of the Planning Service of C. and W. Macedonia (1975) did not mention any data for fertilisers – a sign that very few were used at that time. The available data for fertiliser use in Prespa between 1980 and 1995 (Table 2) show that after an initial increase of ca 30–40% between 1980–86, the chemicals used changed and the overall quantities dropped to previous levels.

Available data on the pesticides used are not very elucidating. This is because the types and quantities of pesticides used vary greatly according to crop area, weather and crop conditions, but also according to marketing rules and social factors, since pesticide use is not solely a matter of expert advice. However, the most commonly used pesticides in bean cultivation include the insecticides and acaricides Chlorpyrifos, Propagrite and Azinphos; the herbicides Paraquat, Trifluralin, Ethalfluralin, Bentazon and Metobromuron+Metolachlor; and Carboxin fungicides.

Gerakis & Tsiouris (1992) concluded that in general, all fertilisers, except nitrates, were not overused but often misused, pesticides were also misused and not properly applied and much irrigation water was wasted. On the other hand the State did not give enough attention to promoting cultivation practices that minimise non-point pollution.

Livestock

Livestock numbers in 1993 were 56% of those in 1964 just before the construction of the irrigation system

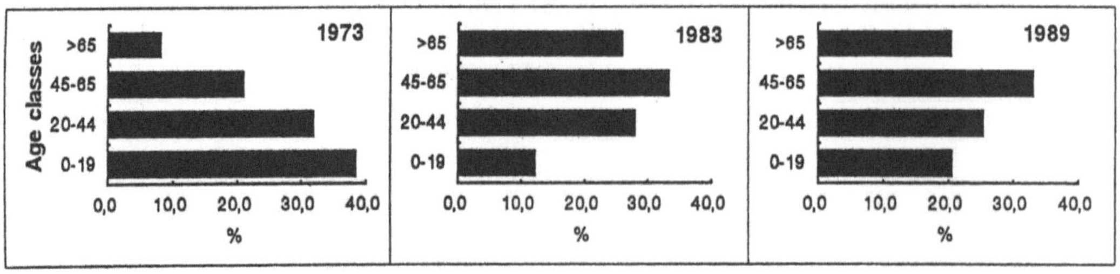

Figure 5. Changes in the age pyramids of the population of Prespa showing respectively a healthy age structure (1973), a very aged one caused by emigration (1983) and a recovery (1989).

Table 2. Fertilizers (in tons) sold to the farmers of Prespa through the Union of Agricultural Cooperatives of Florina Prefecture which was the sole seller in the area up to 1993. For 1994 and 1995 small quantities of ca 30 tons were acquired via other sources.

Fertiliser type	1980	1981	1982	1983	1984	1985	1986	1990	1991	1994	1995
Ammonium phosphate 20-10-0	357	213	251	271	359	456	339	331	232	98	ca 172
Ammonium phosphate 16-20-0	25	1	10	0	1	17	1	8	0	0	0
Composite 11-15-15	0	0	0	0	0	85	187	213	124	168	169.5
Ammonium nitrate	321	192	321	324	318	187	113	95	101	108	108
Calcium ammonium nitrate	1	1	1	1	1	125	83	2.5	0.4	25	25
Potassium sulfate	0	0	0	0	0	2	0	0	0	0	0
Hyperphosphate 0-20-0	17	17	19	19	45	114	83	42	112	110	85*
Potassium nitrate + Magnesium (12-0-48 + 2 Magnesium)	0	0	0	0	0	0	0	0	0	2.5	2.5
Total	721	424	602	615	724	987	807	692	570	542	592

* 0-46-0 for 1995

Table 3. Numbers of livestock in Prespa.

	1964	1973	1983	1993
1 Sheep	27780	16130	8528	9730
2 Goats	4658	1708	2810	4145
3 Sows	328	154	191	120
4 Cattle (local breeds)	1450	300	145	105
5 Cattle (improved and introduced breeds)	83	838	666	429
Cattle Total	1533	1138	811	534
6 Poultry	no data	8860*	6500	2630
7 Bee hives	"	no data	563	216
8 Horses	"	253	38	9
9 Mules	"	107	28	7
10 Donkeys	"	377	140	22
Total livestocks units**	7652	3892	3118	3301

* = 1978 data
** = One livestock unit is equivalent to: 1 cattle of improved or introduced breed, 1.42 cattle of local breed, 5 sheep, 5 goats or 5 sows.

(Table 3). The abrupt decrease of ca. 50% between 1964–1973 was not however solely due to changes in the landscape and land uses brought about by the irrigation system, but also to the high emigration rates. A majority among emigrants had been livestock keepers. Livestock rearing is labour intensive and changes in the age pyramid tend to decrease this activity. Changing prevalent social values in an evolving Greek society, not favouring pastoral life but promoting urban models, were also vital in diminishing livestock farming.

In the past, livestock numbers were much more evenly distributed among villages than today and grazing was spread amongst various biotopes including the lowlands and the pastures around the lake. Less than 50 Water Buffalos *(Bubalus bubalis)* were kept in the area up to the early 1960s and large flocks of domestic geese were kept before the war.

Equids were replaced by cars and tractors (94.8% decrease in 1973–1993). Local cattle breeds also showed a dramatic 92.7% decline between 1963–1993, although before the war they represented the entire cattle population in Prespa and still accounted for 95% in 1964. The local cattle breed is one of the few surviving Hellenic breeds of cattle. It belongs to the Hellenic shorthorn breed which has close similarities to the Balkan shorthorn breeds (A.Georgoudis, in litt.). The Greek type is the smallest of all these: 100–115 cm shoulder height and 100–160 kg of adult weight. They were kept for a combination of labour and meat and milk production. They are extremely rugged, tough, disease resistant and highly adapted to local conditions.

Improvement in the size and milk production of local races through hybridization with European breeds, as well as introduction of such breeds started in the 1950s and has continued thereafter. These animals make an increasing proportion of all cattle of the area: 73.6% in 1973 and 80.3% in 1993.

Before the war and the introduction of lucerne, hay and especially reeds (*Phragmites australis*) were the main winter fodder of cattle. Cattle also grazed on reeds and other aquatic macrophytes in spring and summer in the wet meadows and flooded areas along the lakeshore.

The almost stabilised numbers of livestock in 1983–1993, combined with the reduction in the proportion of livestock keepers reflects the existence of fewer but larger herds.

Nowadays Agios Germanos and Kalithea hold more than 81.5% of the total cattle in the area and Vrondero and Pili together hold 71.2% of total sheep and goats. These four villages at present are still dominated by Vlachs and all together hold 85.6% of the total livestock units in Prespa.

Livestock nowadays is concentrated in fewer villages than before, herds of cattle mainly graze the mountain pastures above the tree line and at the western side sheep and goats graze in areas of oak forests. The cattle are free ranging and are kept for meat production only. Milk producing cattle were wiped out owing to the conversion to bean agriculture and also because they were unprofitable because of the small quantities produced and remoteness from canning factories. Nowadays the remaining pure-bred local cattle number about 50–70. These last animals are kept only at the village of Psarades, grazing freely on hilly, calcareous stony substrates.

Coypu (*Myocastor coypus*) were introduced into Lake Mikri Prespa for fur farming in 1968 but the whole enterprise was abandoned in the late 1970s. In the early 1990s all the remaining animals vanished. Mink farming for fur was carried out in Agios Germanos village from 1988 to 1991. Some animals escaped and an unknown few still survive in Lake Mikri Prespa (Catsadorakis, unpublished data).

Fisheries

Up to the mid 1960s the fishing rights of the lakes were leased by the State to private individuals who subsequently leased fishing rights to fishermen for 10% of the value of their catch. Afterwards the State conceded the fishing rights directly to the individual fishermen while taking 10% of their catch value.

Seven of the 13 traditional fishing methods used in Prespa lakes (Catsadorakis, 1996) up to the introduction of cotton and later nylon nets, were practised at the shallows of the lake, up to depths of one to two metres. All of these methods required a clean bottom, so the flooded meadows and the shallows around the lake were necessarily kept clean of reeds and other vegetation. Additionally, up to the 1950s the main fishing gear of fishermen was constructed from reeds woven together to form fences and traps. In late winter when the water level was low, fishermen also burnt all the remaining reeds so that in the next spring floods there would be no old stems, and new shoots would thrive and enhance carp spawning habitats. These activities in combination with reed grazing and mowing, removed a large amount of organic plant material from the lake water. The ban on reed cutting and burning in 1971 'to avoid destruction of birds' nests', only increased the abandonment of these lakeshore methods which had started some years ago with the introduction of nylon gill nets. This method gradually displaced all other labour intensive fishing methods.

In 1984–1986 a fish breeding station was built on Mikri Prespa lakeshore with EU funds with the intention of rearing carp and releasing them into the lake to enhance fish stocks and hence fisheries (Crivelli, 1984). This station never operated.

There are four fishing methods currently applied: trammel and gill nets, pezovola, pelaizia and long lines. The two traditional unique methods, namely 'pelaizia' and 'pezovola' will probably completely die out in a few years.

At present there are a few exclusively fishermen only in the village of Psarades and some elderly retired people in Mikrolimni village. The main markets for the fish are the towns of Florina and Kastoria and the villages in the prefecture of Florina. Some fish are sold by the fishermen themselves but most are sold to merchants who drive around and sell their fish. The number of professional fishermen is decreasing each year, because old people retire and fewer young join the profession, and are absorbed into agriculture and tourist professions. Fishing still continues to be a supplementary source of income with almost the same importance as in the past. There has certainly been an increase of the footage of nets per fisherman but its magnitude remains unknown. Many fishermen from the village of Psarades actually supplement their income by taking visitors on their boats for trips in the lake and some also keep fish taverns where they cook part of their catch.

In 1964 the gross product from fisheries accounted for 14.8% of the total local gross product (Planning Service of W.& C. Macedonia, 1975), while in 1994 it was estimated not to be more than 3–4%, (Catsadorakis, unpublished data), although under the current conditions and the nature of the profession, an accurate estimate is almost impossible. As the total number of full-time fishermen diminishes and the footage of nets increases, a higher income for those remaining in the profession is ensured.

Forestry

The main forests of the area are oak *Quercus*, beech *Fagus*, mixed oak-beech, and mixed hardwoods (*Quercus, Juniperus, Ostrya, Fraxinus, Acer, Buxus, Corylus*, etc). On the NE parts of Mt. Varnous there are also fir and mixed beech-fir woodlands. There are at least seven species of oak forming the most extensive forests.

The majority of forest lands are State owned (80%) and the remaining are municipal forests belonging to the communities of Vrondero, Agios Germanos and Karies.

For centuries the forests of the area were used for collecting fuelwood and timber for buildings, constructions and bean sticks. Other important uses included grazing, cutting branches for fodder, hunting and collecting fruits.

The first intensive large scale logging occurred during World War I by the Entend troops occupying the area, for fortifications and other works (Pavlides, 1985). During the Civil War, extensive forested areas around military camps were cleared with fire for better visibility and logged for timber for construction works. Fires ignited by shelling also decimated large tracts of forests (Pavlides, 1985; Catsadorakis, unpublished data).

Pyrovetsi & Karteris (1986), comparing aerial photos of 1945 and 1984 found a total net decrease of only 0.2% in forestland, but they noted that some forest areas in 1945 appear to be less dense. In 1945 there were very few forest roads and the transportation was done with animals. Fuelwood collection was more anarchic and accordingly the forest areas close to the villages were much more degraded from both logging and grazing. Large trees were by far more abundant. The collection of bean sticks was important, since no canes (*Arundo donax*) were used at that time to support bean plants as is done today. Exploitation of forests for industrial wood was non-existent and the forests covered the local needs for fuelwood only. However, especially at the southern parts of Prespa local people cultivated terraces on high-slope openings in the forests due to the scarcity of cultivable land. Since the war, with the abandonment of terrace cultivation, these patches have been recolonised by forest vegetation. The farmland zone around the middle and NE part of the lake was however covered partly with small woodlots which were not considered as forest coverage in Pyrovetsi & Karteris (1986).

In the 1970s, the length of forest roads increased as did exploitation of timber. Clearcutting was the commonest management method. Two forest cooperatives were formed at Pili and Vrondero in 1968 and 1972 and another at Agios Germanos in 1984.

After years of uncontrolled and unsuccessful management by the locals and the forest service, ca 45% of forestland is currently classified as partly forested, since the existing vegetation does not fulfil minimum requirements in terms of stocking and productivity (Matheas, 1975; Karidas, 1979).

The entire oak woodland consists of an even-aged and young, coppiced woodland with trees of a breast height diameter less than 20 cm. The same is true for mixed hardwoods. The woodland is degraded and the trees are mostly malformed and can be used only for fuelwood. Only the beech forests still contain old trees, capable of providing industrial timber. Except for the fuelwood used by local people, the exploitation rights of forests are rented to private entrepreneurs, because no active cooperative exists any more in the area. Gross income from forest exploitation in direct terms (forest products) decreased from 1.9% of the total gross income of the area in the period 1964–1973 (Planning Service of C. and W. Macedonia) to less than 0.5% in the 1990s (Catsadorakis, unpublished data).

Presently, grazing is only allowed in a 50% of the whole area of forests but over-grazing by goats exists only in the W and SE parts only. Hunting is also allowed and practised in almost all the forestlands except those of the south-eastern part. No special game management measures are taken and poaching is extensive. The notion of conserving the other values of the forest (fauna, flora, rare species, recreational values, etc.) besides the timber producing capacity, is totally absent from all management planning.

Manufacturing

No manufacturing at all in the profit making, professional sense of the term, ever took place in the area up to the late 1930s. In 1936 a small cheese dairy unit started on a cooperative basis in Vrondero but worked for only three years. A limited number of loom handicrafts/products, embroidery, baskets and glass bead ornaments were made and sold on a family basis, but this also slowly died out. A large number of everyday articles were made by the people themselves. After the war another cheese dairy operated in Agios Germanos from 1955 to 1975.

A marble quarry and a 10-worker marble processing plant operated between 1966 and 1975. A canning factory for Prespa bleak and locally produced vegetables was established in 1969, employing seasonally 11–60 workers but was closed as unprofitable in 1975. Both plants were situated at the fringe of Lake Mikri Prespa.

In the mid 1970s, young, economically active people emigrated to the nearby town of Kastoria and later mainly to Western Germany to work as craftsmen in the fur compilation/processing industry there. Many returned after 1980, but along with farming activities they continued working on fur at home to supplement their income.

The contribution of the secondary sector to the total gross product of the area was 9.7% in 1973 and 2.4% in 1979 (Koroni, 1979) and was estimated as being certainly even lower in 1994–5.

In 1986 a bean packing unit was built in Lefkonas village but never operated. Presently, a few people still work, home based, in fur compilation. No other secondary sector activities exist in the area, and there are few prospects for the creation of any.

Tourism

The area remained secluded and unapproachable for decades and even after the war, access to visitors was kept under strict military control up to the late 1960s.

In the decades before the war, Prespa attracted mainly day visitors from the nearby towns of Kastoria and Florina, who came principally to swim in Lake Megali Prespa, hike or picnic and visit the Byzantine monuments (Vafiadis, 1940). This continued in the 1960s (Catsadorakis, unpublished data). The scenic and natural values of the area started to become known to a wider public especially after its designation as a National Park in 1974. Initially the area attracted mainly foreign birdwatchers and naturalists and was able to accommodate them in a poor infrastructure in the old family-hospitality way. Gradually the number of tourists increased and led to the building of a few rooms and tavernas in the most frequented villages of Psarades and Mikrolimni. An organised beach was established by the Hellenic Tourist Organisation in the early 1970s on the Megali Prespa coast. The 1980s saw an impressive increase of publicity about Prespa. The destruction of the natural environment in the area caused by the Integrated Mediterranean Programme pilot project brought publicity to Prespa. Slowly, rooms, guest houses and tavernas started to grow. On the other hand, the National Park infrastructure for visitors never developed adequately. Unfortunately there was never any collection of statistical data at all for tourist numbers, lengths of stay, etc in Prespa. In 1973 there were 27 beds in the area, distributed in 14 small rooms in family houses and two small hostels, dispersed among three villages. The quality of services remained low to moderate (Planning Service of C. and W. Macedonia, 1975).

Presently, the National Park infrastructure for visitors is still almost non-existent. Three guest houses and a hotel were built, and the number of available rooms increased in 1994 to 57 (307% increase since 1973) increasing the total number of beds to 124 (360% increase), which is still far lower than the demand in the peak seasons (Catsadorakis, unpublished data). The number of tavernas and cafes has increased. The quality of services is improving. The number of Greek visitors is showing an increasing trend and is more equally distributed around the year than that of foreigners (Figure 6) but almost all are confined to seasons other than winter. Many people, especially professional fishermen from the villages of Psarades and Agios Achillios turn to touristic professions (tavernas and boat trips). The contribution of tourism to the gross local income has probably increased, but no recent data exist.

The development of ecotourism, a mild impact kind of tourism was proposed by many, especially after the 1990s, as one possible alternative and rather complementary source of income for the area. A few timid efforts have already started. Although Prespa presents endless opportunities for the development of

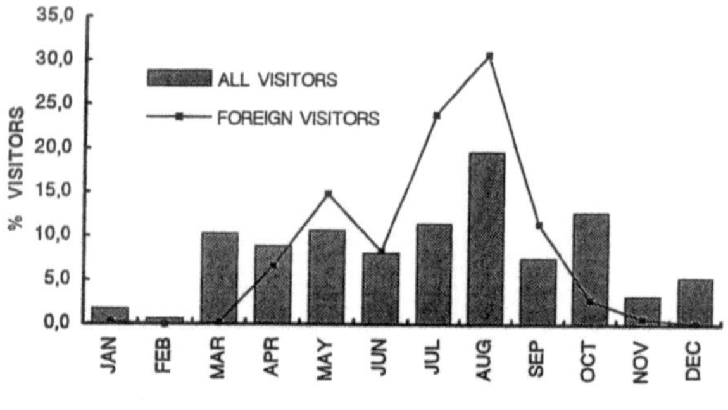

Figure 6. Temporal distribution of number of visitors within an annual cycle. Data of years 1993–1995 based on numbers of people visiting the Information Centre of the Society for the Protection of Prespa in Agios Germanos village.

such tourism, its development in the future will depend upon the initiative of local entrepreneurs, since it is rather improbable that outside investors will be attracted. A kind of balance between primary sector activities and tourism must be kept, so that the area retains its authentic physiognomy and its high biodiversity.

Hunting, angling and collecting from nature

In the 1960s limited numbers of hunters from all over Greece came to Prespa to hunt mainly winged game. Later, the increasing fame of Prespa as a protected area rather discouraged any increase of hunter numbers. Before the designation of the protected areas and the National Park in 1974, hunting was permitted everywhere in the area but after was banned in Lake Mikri Prespa and an area roughly 3 km around its eastern periphery. It is still permitted everywhere else in the National Park, including Lake Megali Prespa, an important wetland for wintering aquatic birds in Greece (Catsadorakis, this volume).

Hunting by local people was never very intensive. The most important mammalian game for locals has been and still is the wild boar (*Sus scrofa*). The continuously increasing kilometres of forest roads providing access to many remote and formerly inaccessible places in combination with the increased use of 4x4 wheel drive vehicles, led to an increasing number of hunters since boar hunting became much 'easier'. The above, combined with the wider use of advanced automatic rifles and often various portable wireless communication systems exert a pressure to the population of wild boar, the extent of which remains unassessed.

Since the wardening system for hunting is meagre, poaching can be widespread. Roe deer (*Capreolus capreolus*) a non-game species in Greece, is frequently shot. Rare and protected species such as brown bears (*Ursus arctos*) and wolves (*Canis lupus*) are also sometimes shot.

Illegal methods such as night-hunting with car lights have been in the recent past partly responsible for the decimation of hares (*Lepus europaeus*).

Angling is permitted all the year round and in both lakes but is practised by only a few people. Some also fish for the endemic subspecies of trout in the Agios Germanos stream, in which there is however a closed season of 105 days for spawning. There is no control and often bag limits are exceeded as is the case with the hand catching of the endemic Prespa barbel spawning in the Agios Germanos stream (Catsadorakis et al., 1996).

Collection of mushrooms for self consumption is insignificant. Collection for commercial reasons lasted only for a few years in the 1970s. Collected species included mainly *Agaricus* spp. and *Macrolepiota procera*.

Land snails of the species *Helix lucorum* are collected mainly for commercial reasons. Collection of *Cornus mas*, and *Rubus* spp. fruits for preparation of liquors is limited.

Plants currently collected for food are mainly *Taraxacum officinale*, *Chenopodium bonus-henricus* and *Urtica dioica*. The species *Sideritis syriaca* which grows on the middle and high altitudes of the Mts Triklarion, Devas and Tsoutsouli is intensively collected for the preparation of a local hot 'tea'. Large quantities are sold outside Prespa. The species is considered by Pavlides (1985) as in need of protection, with undefined status of threat. In 1956, overcollection of *Gentiana lutea* by the local people to sell to German industries led the species to extinction (Makris, 1979; Pavlides, 1985).

The landscape and land cover/use

The land belonging to the communities of Prespa covers an area of 249.8 km^2, representing 14.5% of the Prefecture of Florina to which Prespa belongs. The average altitude of all settlements in the Florina Prefecture is 706 m while the average altitude of the settlements of Prespa is 957 m (855–1100 m asl).

The most important change to the landscape of Prespa was brought by the construction of the irrigation network which started in the mid 1960s and resulted in: clearcutting of all small forest tracts and natural hedges within the scheme area, felling of high numbers of trees and scrub; embankments and straightening of natural streams; ground levelling which removed huge quantities of soil; drainage of marshes and wet meadows within the network; opening of many farm roads and the redesigning and asphalting of the main road in 1970. In general, after the 1960s there was an ever increasing opening of new forest and mountain roads.

In 1984–1986 within the IMP works the last remnant tract of old forest within the agricultural area was cut and the last inland marshes were drained; 60 000 trees growing within streambeds and draining ditches were cut; a fish-breeding station with open ponds was built on 5 ha of wet meadows and pastures.

In 1988 a fire burned 2000 ha of hardwoods on Mt. Triklarion.

Pyrovetsi & Karteris (1986) comparing aerial photos of 1945 and 1984 found that forestland, open water and urban areas did not differ much. Abandoned agricultural land was non-existent in 1945 but comprised 215 ha in 1984. Rangeland and wet meadows suffered the biggest losses, while barren-eroded land dramatically increased by 338%. Reedbeds also increased by 25%, but as with wet meadows this is not reliable since the area of reedbeds and wet meadows depends upon many parameters such as the water level of the lake, the intensity of reedbed management, etc. Karteris & Pyrovetsi (1986) acknowledged the problem of differentiating reedbeds from wet meadows, but they finally gave numbers. Nevertheless, a reduction of wet meadows was surely observed.

After the mid 1970s a continuous increase of reedbeds was observed to the detriment of wet meadows around the lakes. After 1990 a steady lowering of Lake Megali Prespa's water level resulted in the exposure of extensive alluvial flats with temporary marshes forming and drying near the coast line. The emergence of new areas with open water and disappearance of others in the marshy areas around Vromolimni happened just after the peak drought year of 1990. A decrease of rainfed winter cereals took place and a dramatic increase of beans was accompanied by a decimation of trees within croplands. Forests became denser and expanded, encroaching on abandoned fields.

Biodiversity

Many previous publications have given an indicative picture of the high biodiversity of the area. The reasons for this wealth are of two origins: natural and human. The natural reasons only will be explained hereafter. The reasons of human origin will be discussed in a subsequent paragraph.

The natural reasons for the high biodiversity of Prespa are biogeographical, geomorphological, climatic and geological:

- the location of the area in the perimediterranean zone which has intermediate climatic characteristics between continental and Mediterranean types and which was only partly affected by glacial periods while in parts retaining some of its – few though – Mediterranean features.
- the very old age of the lakes.
- the high altitudinal differences within a small area (1500 m in a straight distance of 11 kilometres).
- the intense relief as well as the amphitheatric arrangement of mountains around the lakes which creates various microenvironments with variable expositions.
- the existence of the large water bodies at the bottom of the drainage basin, which create a microclimate with less extremities, within a wider mountainous area with very harsh winter conditions (actually the harshest in Greece).
- in paleogeographic terms the Psarades-Vrondero peninsula between the two lakes had been an island or an isolated peninsula surrounded by waters for some thousand years and it likely acted as a refugium during glacial periods of the Pleistocene.
- the existence of two main types of quite different geological substrates: limestones and granites.

The first comprehensive lists of species of birds and mammals in the area appeared in the 1980s (Catsadorakis, 1986a,b; Catsadorakis & Kollaros, 1986). The first comprehensive work on the flora and vegetation of Prespa area appeared in 1985 (Pavlides, 1985). It is noteworthy that the wildlife described by the first naturalists in the late 1960s had not yet been severely affected by the dramatic habitat changes brought by the irrigation system construction in the lowlands of Prespa but the first extinctions and strong declines of many species started in the early 1970s (Catsadorakis b, this volume) after a certain time lag. Between 1970–1980, six species of birds ceased breeding or became completely extinct and in the period 1981–1994 this

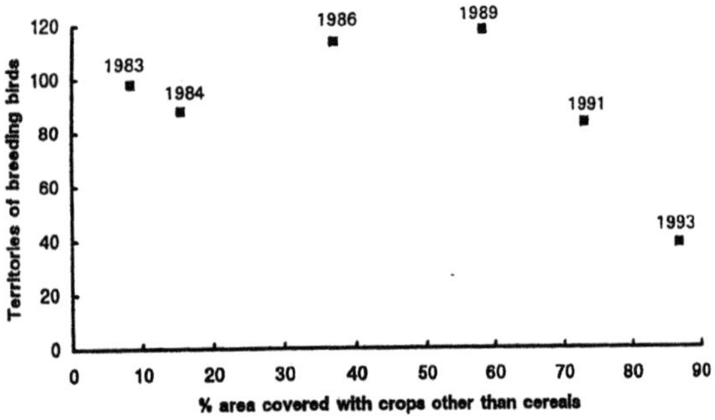

Figure 7. Densities of the five commoner bird species of the farmland zone of Prespa (Corn Bunting, Black-headed Bunting, Red-backed Shrike, Yellow Wagtail and Great Reed Warbler) expressed as number of territories per 50 ha in relation to percent coverage of croplands with crops other than cereals, mainly beans (Catsadorakis, unpublished data).

happened to eight species of birds (Catsadorakis b, 1997).

Pavlides (1985) mentioned that the following plant species became extinct or probably so: *Salix silesiaca, Tamarix parviflora, Salix alba* subsp. *micans, Pistacia terebinthus, Pinus heldreichii* and *Pinus peuce.* The first three disappeared after the works of the IMP projects in 1984–1985. At the end of 1980s a rare, for Greece, vegetational formation, the meadows with *Narcissus poeticus poeticus var. hellenicum,* which covered extensive areas between the villages of Mikrolimni and Karies, were converted to cultivation despite having a strong cultural value for the local populations (Pavlides, 1985).

Presently, plant species which are found in very low numbers and are actually threatened with extinction are: *Salvinia natans, Narcissus poeticus, Vallisneria spiralis* and *Phelypea boissieri* (Pavlides, 1985; Pavlides, a,b, 1997; Catsadorakis, unpublished data).

In the past, reed management practices had several direct and indirect effects on wading birds. The two most obvious were: (a) there were no extensive unmanaged reedbeds anywhere and the birds had to nest in the trees of the only unvisited area on the 'no man's land' strip of the lake on the Greek-Albanian border; (b) extensive shallows free of emergent vegetation were available for these species as feeding grounds.

The intensification of agriculture, the complete predominance of bean croplands and the ousting of cereals especially in the irrigated zone threatens many bird species that inhabit this zone and depend upon cereals. Figure 7 shows the dramatic decline of densities of breeding birds in the cropland areas in relation to the recession of cereals. Species such as the Partridge *Perdix perdix*, Skylark *Alauda arvensis*, Black-headed Bunting *Emberiza melanocephala* and above all Corn bunting *Miliaria calandra* and Red-backed Shrike *Lanius collurio* have declined (Catsadorakis a, this volume). In addition to removal of hedges, bush and trees, and heavy use of pesticides, bean fields remain without any vegetation cover up to mid-April and throughout the whole season bean fields offer no cover and host farmer workers daily.

Other bird species which are at the verge of extinction are referred to elsewhere (Catsadorakis, b, this volume).

No data exist to show that any mammal species are threatened, but poaching of Brown Bears, Chamois *Rupicapra rupicapra*, Wolf and Roe Deer is certainly a threat. Some species such as the Otter *Lutra lutra* might have well benefited from increase of reedbeds as cover and a lower hunting pressure. No data exist on population trends of reptiles and amphibians.

The future of wading birds feeding in the increasingly scarcer wet meadows seems unsafe (Catsadorakis, b, this volume), since no measures are taken to increase this type of habitat.

Worsening of water quality in the lake might affect the density of benthic organisms but no detailed monitoring exist of their abundance. Some small remnants of the gallery forests which existed in the past in many parts around the lake, still remain at the north-eastern part of the lake and on Agios Achillios island.

The remaining number of 50–70 specimens of the local cattle breed is really at the lowest edge of a minimum viable population size and at maximum danger of irreversible genetic deterioration from inbreeding. Unless strict measures are taken to safeguard this breed it will soon vanish.

Legislation, protection status and the National Park

In 1971 the first measures towards the protection of the area were taken when by prefectoral decrees, five areas with nesting aquatic birds were declared protected and their patrol was assigned to the military units of the area.

With presidential Decree 46/1974 the area was declared a National Park. With Law 191/1974 the Parliament accepted the Ramsar Convention which the Greek government had signed some years ago. Lake Mikri Prespa was among the 11 wetlands of international importance declared by the Greek State.

With Ministerial Decree 23211/1747/1975 the area of the lakes and around them, that is the National Park area, was declared protected as a Site of Exceptional Natural Beauty. Also with Ministerial Decree 165594/3007/1985 the Operational Regulations of Prespa National Park were established.

Prespa National Park is an Important Bird Area in Greece and the European Union and it is fully eligible for protection under the EEC Directives 79/409 for the protection of birds, the Bern Convention, the Biodiversity Convention, Council Directive 92/43/EC of 21/5/92 on the conservation of wild habitats and of wild fauna and flora, Council Directive 8/659/EC of 18/7/78 on the quality of freshwaters needing protection or improvement in order to support fish life and of course all other EU Directives relating to environment protection. It is also a constituent of the Natura 2000 network of valuable nature interest areas. In 1986 two wardens were employed by the Florina Forestry Service which is legally responsible for the National Park. There is neither a National Park Headquarters, nor any permanent installation, accommodation or bureau. A forester situated in Florina is nominally responsible among many other duties for the management of the National Park, but in fact he/she scarcely visits the area. Three wooden birdwatching towers constructed in the 1970s had fallen apart in the late 1980s and have not been replaced since. The only visible installation in the National Park area are a few insufficient signs. There is no information centre for visitors, no public awareness programs, no relevant publications and no kind of scientific monitoring executed by the Florina Forest Service. Since 1992, the Society for the Protection of Prespa, a local non-profit organisation has established and runs two visitor centres.

A number of matters must be clarified before a sufficient understanding of the situation in Prespa National Park can be gained: with Greek legislation all National Parks in Greece fall under the jurisdiction of the Ministry of Agriculture and specifically the Section of National Parks and Aesthetic Forests, an Athens based service manned by no more than 4 non-specialised foresters.

The management of the National Parks in Greece is the responsibility of the Forestry Department of the prefecture to which each one belongs. On the other hand, Ramsar wetlands in Greece fall under the jurisdiction of the Ministry of Environment, Housing and Public Works. The two Ministries should be closely cooperating in the case of Prespa which is the only area to be both a National Park and a Ramsar site, but unfortunately they rather compete. This results in a swamping of progress but there are two more reasons for that: the improper legislation and the negative reaction of local people towards any kind of management or control.

The legislation on National Parks

The legislation on National Parks, Aesthetic Forests, and Monuments of Nature in Greece come under the Forest Code (Law 998/1971). The whole body of law refers to uninhabited forested areas such as all the National Parks in Greece except Prespa, which is inhabited and in which farming and fishing take place on a considerable scale and whose zone of absolute protection is a wetland.

Thus, the legislation for National Parks in Greece is irrelevant to the real-life situation in Prespa. In many cases the reality has obliged the state itself not to apply the law, as for example when tacitly permitting the continuation of fishing and grazing in the nucleus of the Park. In other issues the state has not fulfilled its own obligations stemming out of the respective law, e.g. to compulsorily expropriate and compensate all private lands within the nucleus of the park and cease all human activities there.

In addition to the above, even if the local Forestry Service had sufficient personnel, infrastructure and competence on biodiversity conservation, it would still be incapable of tackling the complex and interrelated problems of agriculture, housing, zonation of activities, pollution and conservation in an inhabited area.

Thus, improper legislation and lack of a competent administrative authority is an important drawback to any progress towards a better preservation of Prespa's values.

Unfortunately, the recent law 1650/1986 on environment protection, which is much more progressive and includes solutions for the problem of Prespa, is still inactive due to lack of political will and the reluctance of the Ministry of Agriculture to lose power over the National Parks since this law promotes the authority of the Ministry of Environment.

The hostile attitude of local people

A serious matter to consider in order to understand the drawbacks of further promoting conservation in Prespa, is the limited information provided to local people and the limited opportunities of participation in decision-making they have had, up to the recent past.

The problematic legislation and lack of competent personnel in the Forest Service are the reasons behind this omission. Another cause could be the belief of many narrow-minded bureaucrats that preservation of natural and cultural values can be achieved without the consensus and active participation of local people. The only management measures which sprung out of such a posture have been prohibitions for the locals which in combination with the feeling of lack of participation in decision-making, the lack of understanding and the lack of any compensation measures, led to a very hostile attitude of the locals to anything associated with the National Park. Eventually, the underdevelopment of the area, the misfortunes of the Civil War, the miseries brought about by emigration and the changes brought about by the sudden passage of the rural Greek societies into the industrial era, were attributed to the National Park's existence, thus creating an extremely dull and negative perspective.

The integrity of the area – values to be preserved

The ideas prevailing universally about conservation policies at the time when Prespa was 'discovered' and first protected and the degree of understanding of the ecosystem's structure and functioning at that time, prevented the scientific community, the activists and a wider public from realising a simple thing: the invaluable biodiversity of Prespa was not only due to natural factors but was also due in part to the human activities in the area and the methods of land use. And even more, only a few seemed to understand that if this biodiversity was to be protected, then the best long term method for doing this would be through the perpetuation of this kind of human primary sector activities.

Wardening and prohibitions would protect perhaps a few individuals from being exterminated, but the real danger would come from a shift to land use patterns different from those that had led to the creation of the biodiversity. Nevertheless, it was an irony that Prespa's wealth had been discovered just before the impact brought about by the land reclamation works upon the populations of several species started to become tangible.

There was also a second fact that few decision-makers realize even today. Biodiversity was only a part of the wealth of Prespa, the other part being cultural: scenic beauty, monuments, local civilisation and architecture. The uniqueness of Prespa was mainly that one place brought together so many values, and these should be preserved as a whole. Often, the interdependency of ecological and social situations was such and the scale of the area so small, that in order to preserve one value, many others must be affected.

Catsadorakis (1995) tried to provide a list of the main values of the area which should be protected as a whole:

- the scenic beauty
- the multiple values of the two lakes as wetlands (see Maltby, 1986; Mitch & Gosselink, 1986, Tsiouris & Gerakis, 1991; Pearce & Crivelli, 1994).
- a number of habitats increasingly rare in the national and European context, namely reedbeds, wet meadows, habitats with floating and emergent aquatic macrovegetation, extensively managed farmlands, natural hedges, mixed hardwood woodlands little affected by humans, beech and fir forests, centuries-old juniper stands.
- the considerable number of habitats per unit area; the particularly rich flora, made up of more than 1326 plant species; the extremely diverse fauna including 23 species of fish, 11 amphibian, 21 reptile, at least 42 mammal and over 260 avian species.
- the endemic life forms such as: the plant *Centaurea prespana*, the fishes *Chalcalburnus belvica, Barbus prespensis, Rutilus ohridanus prespensis, Chondrostoma prespensis, Alburnoides bipunctatus prespensis, Cobitis meridionalis* and *Salmo trutta peristericus*. Also species having a very restricted distribution in Greece or/and in the Balkans such as: *Phelypaea boissieri* and birch *Betula pendula.*
- the internationally important breeding populations of rare waterbirds such as the Dalmatian Pelican and the Pygmy Cormorant. Additionally the nationally important populations of Cormorant, Great White Egret, Goosander, Great White Pelican, Grey-lag Goose, Hazel Grouse, etc.
- the presence of mammal species endangered throughout Europe such as the Brown Bear, Wolf and Otter. The Chamois as an endangered mammal in Greece.
- the local architecture of old houses found in some villages, two of which have a protected status.
- the unique 'pelaizia' and 'pezovolo' fishing methods, the local food preservation methods, the legends and folk tales developed in the region.
- the Byzantine and post-Byzantine monuments, such as temples, chapels, monasteries, hermitages and rock frescoes dating back to the 10th century A.D.

- the local shorthorn cattle breed.
- the close interactions of natural processes and human activities which along with the natural diversity of the area provide excellent opportunities for environmental interpretation, research and environmental education.
- the chance to organise and implement an innovative integrated management policy in the area thanks to – in addition to all the above – its small size and its relative isolation, which could be used as a pilot model for all similar Greek and Balkan mountain areas.

Of course, in the time that lapsed since the designation of the park some of the assets that existed at that time have deteriorated, others are still intact and a few have vanished. It is however extremely useful that by understanding that human activities are partly responsible for the high biodiversity and by knowing the extent and patterns of these activities the manager of today can work with an invaluable compass towards the restoration of certain procedures, populations and communities.

The conservation efforts

The French naturalist brothers Terrasse first discovered the wildlife paradise of Prespa and made it more widely known. After their initial visit to Prespa in 1967 many other visits of European naturalists to the area were organised and/or facilitated by the Hellenic Society for the Protection of Nature (HSPN). In 1970 three outstanding supporters of Greek nature reported on behalf of the International Conservation Union (IUCN) to the Greek government for the protection of areas in Northern Greece with Prespa among them (Hoffmann et al., 1971). Two years after the designation of the National Park in 1974 a Field Biological Station was built by the Hellenic Society for the Protection of the Environment and the Cultural Heritage (Elliniki Eteria). Scientific missions to the area continued with the particular contribution of the Station Biologique de La Tour du Valat of France. In the 1980s the presence of Greek scientists working in the area became more regular. In 1984 the destruction caused by the IMP works were not halted by the proposals of an integrated environmental study made by a multidisciplinary group of scientists, but the dissemination of the facts to the media was made through actions of locally working scientists. The Royal Society for the Protection of Birds of the United Kingdom (RSPB) and the Hellenic Ornithological Society (HOS) sued Greece through the European Commission and Greece was condemned by the European Parliament. The whole affair was on the other hand an opportunity for a wider public to hear about the assets of Prespa. The primary role in the above movements was played by a small group of active citizens 'The Friends of Prespa'.

Soon after, it was realised that better public education and awareness was a must and WWF-International funded a project which generated various publications, a series of public presentations and lectures were made locally and in the schools of the wider area, a portable exhibition was prepared and travelled around Greece and a Management Plan proposal was prepared.

In December 1987 a workshop was organised in Prespa by a group of Greek NGOs, with participants from local and regional services, the local people and the environmental NGOs, to discuss 'A future for Prespa'. The main conclusion was a plea for the compilation of a Management Plan (Master Plan, M.P.). And indeed in 1988 an interministerial group of scientists was formed to work on this. The group asked the help of WWF's experience and a multidisciplinary multinational expert team was immediately formed (Hollis et al., 1989). Unfortunately, the WWF team's final proposals were not taken into consideration and additionally the M.P. has since been buried due to lack of political will to forward it. In 1988 several environmental NGOs assisted the local people in stopping the construction of a small hydroelectric power plant on the Agios Germanos stream. Among the activities of a short-lived local NGO (Prespa Centre for Man and Nature, PCMN) founded by scientists working locally was to provide scientific advice for ecologically sounder agricultural practices.

In 1990 the RSPB, the Tour du Valat and the Danish Ornithological Society joined HSPN, HOS, Elliniki Eteria, Goulandris Natural History Museum, the Friends of Prespa, the PCMN and later WWF-Greece, to form the Society for the Protection of Prespa (SPP), the aim of which was to coordinate all efforts towards the protection of Prespa's natural and cultural values. A locally working biologist was appointed as its managing director.

One of the first aims of SPP had been to push the M.P. forward, but this has proved difficult. In 1992 SPP established an Information/Visitor Centre in Agios Germanos village, the first ever in a Greek National Park. The centre, which is visited annually by more than 6000 people, is run by five young local people who were trained as ecoguides and work in an environmental education program which attracts more than 5000 students annually. SPP produced and distributed various environmental education publications on Prespa for primary and secondary schools. Organic agriculture was supported through provision of technical advice to interested farmers and the promotion of products. The marketing of labelled local products such as organic beans and nuts, woollen socks and herbs was assisted. Another project used EU funds to save the last specimens of the local cattle breed. Two booklets were published and distributed free to local people with instructions how to restore or mend their traditional houses without spoiling their character.

In parallel, lectures were given, articles were written and a collaboration was built with the prefectoral services. Public campaigns were organised, awareness material was printed and volunteers were used to warden pelican colonies in order to avert disturbance. In collaboration with the International Pelican Research and Conservation Programme many management measures were applied to safeguard pelicans and all the effort led to the increase of pelican numbers, so that the colony of Dalmatian pelicans in Prespa became one of the largest on earth.

With its daily actions and presence, the trust and respect of the locals is slowly being gained towards SPP and collaboration is being sought. In 1995 the prospects for the future were optimistic. An alliance of environmentalists and local people was for the first time in Greece very close to start mobilising the State to participate in the effort to safeguard and enhance Prespa's values.

Key conservation and management issues

The previous articles of this volume and the chapters of this article have provided most of the necessary background knowledge that is needed to appreciate in depth the main present management and conservation issues in the area. So that these rather complex issues can be more easily discussed, three main subject categories are shown in Figure 8: (a) The principal management and conservation objectives, (b) the basic means needed to achieve them and (c) the main existing or potential limitations and hindrances which might inhibit the fulfilment of the objectives. In the following text these issues are described and commented upon one by one in the light of all previous information.

– A detailed, integrated inventory of the natural and cultural values of the area which is to be preserved is lacking and should be made from the beginning so that it can form the basis of a detailed, long-term strategy and planning for a sustainable development that is flexible to changes. This development plan should be based upon the sustainable exploitation of these values.

– An administrative and executive authority is necessary in order to apply the above development measures. The participation of the local and prefectoral authorities and staffing with a multidisciplinary scientific team are essential. If all this is to be carried out, the existing legal framework for National Parks should be changed to conform to real life or a new one should be created to suit the special features of the area. The current protection zonation ('nucleus' and 'peripheral zone') has repeatedly been proved to be totally inappropriate in Prespa National Park. A new zonation must be made according to law 1650/86 and the newly defined zones must include: 'absolute nature protection areas', where no activities are permitted, 'nature protection areas', where only non-impacting primary sector activities are permitted, 'protected natural formations' and 'ecodevelopment areas', where settlements and traditional activities belong. Through the new zonation conflicts such as the occasional encroachment of agriculture onto wet meadows can be resolved and a control over residential development can be achieved so that traditional residential patterns do not change.

– In the long term the monoculture of white beans will be detrimental to biodiversity, scenic beauty and lake water quality. An effort to extensify the cultivations should include crop diversification, creation and maintenance of natural hedges and trees, a more scientific use of chemicals, promotion of biological agriculture and other traditional ways of extensification such as crop rotation, fallows, etc. There are still many funding opportunities provided for this aim through EU programs. However a major drawback to any effort for a more ecological agriculture remains the land ownership problem. Land belongs to many absentee landlords and is leased to the farmers. Additionally, every farm is split up in several parcels. Farmers, not owning the land, do not care for its sustainable use and this leads to overexploitation resulting in diminishing returns and the overall deterioration of land quality.

– Availability of extensive wet meadows and periodically flooded lands at the edge of the lake is the single most important factor for ensuring a good spawning success for fish and proper feeding habitats for wading birds. In the last twenty years, the elimination of cattle herds which grazed the vegetation of wet meadows and prevented reed invasion, the cessation of reed mowing

PRIORITY OBJECTIVES

- Extensify agriculture
- Diversify croplands
- Increase native fish stocks
- Increase waterbird populations and overall biodiversity
- Improve landscape aesthetics
- Keep water availability and improve quality
- Restore wet meadows

BASIC MEANS TO REACH OBJECTIVES

- Clear list of values to preserve
- Planning, setting priorities and strategy for development
- Scientific research, surveys ans monitoring
- Education and participation of local people
- High maximum water level and adequate fluctuation
- Zonation of land uses and residential development
- Reed control and management
- Amendment of existing legal framework

EXISTING AND/OR POTENTIAL LIMITATIONS

- Land ownership
- Adverse climatic conditions
- Lack of transnational cooperation
- Availability of financial sources
- Wetland soils within irrigation system
- Lack of political will
- Deterioration of social fabric
- Unforseeable disasters
- Degree of understanding natural systems
- Lack of specific coordination authority
- Broader scale economic, social and cultural processes

Figure 8. Schematic representation of the complex: main objectives, principal means, hindrances and limitations, for the achievement of a better management for conservation of Prespa National Park.

and burning practices, low water levels and increased nutrient seepage from croplands have promoted a thriving of reeds, and have alarmingly reduced wet meadows, negatively affecting fish stocks and wading birds. Restoration of wet meadows depends upon the combination of two factors: reed control and ensuring a high water level. Reed control can be easily done with purposeful selective mowing or grazing. Ensuring a high water level depends upon adequate precipitation on the one hand and also on the control of the sluice culvert at the outlet of the waters of lake Mikri Prespa. Enough shallow flooded areas are created when the maximum water level exceeds 854.60 m asl. With this water level however a part of croplands within the irrigation system is flooded and/or becomes uncultivable. These areas should have never been included in the irrigation system since they consist of typical wetland soils (Kosmas et al., this volume). Notwithstanding this fact, the owners of this land exert pressure to ensure that the lake does not exceed this water level. This matter can only be resolved by the State, by expropriating these lands and either compensating the owners or offering other land parcels to them.

Another serious problem for maintaining a high enough water level for both lakes, is the trans-national cooperation among Greece, Albania and FYROM, since all three countries use lake water in an uncoordinated way. The ensuring of a high maximum water level alone is not however enough to guarantee a high spawning success of fish and a satisfactory feeding site for birds. During the crucial period between March and June the recession rate of waters must be slow enough to fit the needs of the developing fish eggs and the food requirements for rearing chicks. The management of the sluice is thus a very delicate matter that has to be handled with great care.

– For all the above interventions and management issues there is one common prerequisite: a scheme for continuous scientific surveys, inventories, censuses and monitoring which must be established for a number of crucial factors and functions such as: complete meteorological data, water quality and availability, fish population sizes, population sizes of selected bird species, the densities of selected bio-indicator animals and plants, the areas of certain habitat types, etc..

– However, the human population of the area is already below the lower limits of social self-sustainability, and the opportunities for a healthy social life for young people are very few. An important social problem, common however to all remote mountain areas in Greece, is that young women do not wish to lead a farmer's life and they emigrate in a much higher proportion than males to cities and marry there. As a result many unmarried young males are forced into emigration to find a partner and eventually stay in town for life. High unemployment in the cities however, pushes many such people back to agriculture. Thus, many young people choose to stay in Prespa only for spring and summer to cultivate their fields, their families remaining in the nearby town and after the harvest they spend the winter together in the town. This creates a new category of people who see Prespa as a working place rather than a living place, an attitude which discourages any kind of active participation and the possibilities for a social revitalization fade away.

In addition to any other reasoning, the State policy has always been and still is to help these border populations to stick to their native lands, for reasons of national security. The creation of the expensive irrigation system for a relatively small cropland area and the total subsidization of irrigation water that is still continued despite financially difficult times, show that this policy is being continued. However, under modern social archetypes, financial aid is not sufficient to stop depopulation and the feeling that life is somewhere else. A further deterioration of the social fabric and a further depopulation will create unforseeable changes to the whole system of relationships between man, land use and nature conservation in Prespa and whether these changes will bring more negative or positive changes to the preservation of natural and cultural values is difficult to anticipate.

The core questions and answers

A detailed prescription of those management measures that would lead to the conservation of the closely interwoven biodiversity, the cultural assets and the landscape of Prespa is well beyond the scope of this article and only a brief synthesis of the above described is attempted below.

The definition of biodiversity is yet unclear and most proposed definitions fail to include processes, such as interspecific interactions, natural disturbances and nutrient cycles (Noss, 1990) but things become more complex in 'cultural landscapes', such as Prespa, where natural conditions have for hundreds of years been disturbed through human interventions and especially when despite the changes, the natural character of the landscape is retained, but is far from being primeval, as in the case of Prespa and many other non-wetland mountain areas in Europe (e.g. in the Alps; Broggi, 1988).

Extensive land-use practices often created conditions which favoured a high biodiversity, as is known to have happened elsewhere in Europe with ancient types of farmland or wet meadows along lakeshores (Götmark, 1992). Natural successions were disturbed and a diverse patchwork of habitats was created, influencing positively the overall biodiversity. Examples of practices that have enhanced biodiversity in Prespa have been: grazing, mowing and collection or/and burning of the reedbeds each year, cultivation of small woodland openings, cultivation practices with intercropping, crop rotations, small and intermingling fields with a variety of crops, maintaining natural hedges and trees, the non-use of chemicals, and the combination of arable farming and livestock rearing in a system of high spatial and temporal entropy.

Such land use practices cannot be revived because the whole economic and technological environment has changed. On the other hand the society as a whole requires the preservation of various species that flourished in the past under the old-fashioned land practices (e.g. birds, mammals, plants, etc.), but since these practices will no longer have a direct tangible financial revenue for the locals, cannot now be carried out under their own initiative. Thus society must always at least partly support financially specific management activities and the planning and coordination authority necessary to implement them. Additionally the planning of these practices needs to be supported by multidisciplinary experts, because no procedures can be left to chance anymore, the interventions must be devoid of faults and successful at many levels and in many aspects, economic as well as social.

Besides the problems of planning these subsidized practices carefully, ensuring the necessary funds in the long term and involving the locals in their implementation, the unavoidable partial incompatibility with the wider social, conceptual and financial environment dominated by market laws, continuously poses managers, conservationists, politicians and the whole society with the following two questions: which are the elements worth preserving? and at which levels exactly do they want to maintain each one of these elements? The preservation of natural processes is always behind these questions as is the question what is natural and what not.

Since, as mentioned above, the largest part of the primeval cultural landscape is already changed to a large extent, an effort should be made to transform it to a condition as close as possible to that before the irrigation system construction. To what degree this can proceed can not be given an easy answer. The limits will be put by the degree of local understanding and participation, the available funds and the pressure of the entire society. Knowledge of the previous generation is still available through local elders and can be well used towards that direction.

What is required is not to reach 'as much a natural state as possible' but on the contrary 'as much an ancient man-made state as possible'.

However, it must be admitted that long before managers and scientists start to tackle the associated real problems of restoration, management and conservation theory, there are many tangible things that must be resolved by the Greek State, some of which are described in the previous chapter. On the other hand there are the choices for local people, for the socio-economic dynamic processes, and the local and broader socio-economic conditions and changes. For example: for how long will the white bean monoculture be rewarding? For how long will the agroecosystems and the other ecosystems affected endure the pressure and at what cost? What can be the future of tourism and the residential development in the area? How far-reaching will the primary and tertiary sectors' impacts be within a few years, if measures are not taken for a sustainable use of water, ecosystems, the settlements? How much actual participation can be expected from a society with such a degraded social fabric? For how long will tourism enterprises be small and locally owned ? A sophisticated, long-term and carefully planned strategy, which is imperative, must be a product of a joint dynamic process by the local people, politicians and Greek society as a whole and will inevitably have to tackle tough philosophical questions such as the very definitions of 'sustainable development', 'conservation ethics', 'biodiversity' and 'cultural and natural heritage'.

References

Athanassopoulos, G., 1922. Le plateau des lacs de Macedoine occidentale. Int. Rev. Ges. Hydrobiol. 10: 31–39.

Broggi, M. F., 1988. The Alpine region in danger. Naturopa 60: 18–20.

Broikos, A., 1962. Final study for the development of the area around Prespa lakes, Florina. Ministry of Agriculture. Land Reclamation Service. Direction II. Duplicated Report (in Greek), 18 pp.

Catsadorakis, G., 1986a. Notes on the avifauna of Lake Prespa National Park. Biologia Gallo-hellenica 12: 485–496.

Catsadorakis, G., 1986b. Biotopes and vertebrates in Prespa National Park. Ministry of Agriculture and University of Athens, Department of Biology, Section of Ecology and Systematics. Programme for the Survey of the Fauna and Flora of Greece. Duplicated Report (in Greek), 161 pp.

Catsadorakis, G., 1995. The Texts of Prespa Information Centre. Society for the Protection of Prespa. Agios Germanos, 125 pp. (in Greek).

Catsadorakis, G., 1996. Fish and fisheries at Prespa lakes. Society for the Protection of Prespa, 52 pp. (in Greek).

Catsadorakis, G., 1997a. Breeding birds from reedbeds to alpine meadows. Hydrobiologia 351: 143–155.

Catsadorakis, G., 1997b. The importance of Prespa National Park for breeding and wintering birds. Hydrobiologia 351: 157–174.

Catsadorakis, G. & D. Kollaros, 1986. Preliminary notes on the mammalian fauna of Lake Prespa National Park (Greece). 3rd Int. Cong. Zoogeogr. and Ecol. Greece. Patras, 1984. Biologia Gallo-hellenica 12: 507–519.

Catsadorakis, G. & J. Petrides, 1988. The problems of reedbeds in Lake Mikri Prespa. Nature. Bulletin of the Hellenic Society for the Protection of Nature 35: 15–19 (in Greek).

Catsadorakis, G., M. Malakou, M. & A. J. Crivelli, 1996. The Prespa barbel, *Barbus prespensis*, Karaman 1924, in the Prespa lakes basin, north-western Greece. Tour du Valat, Arles, 79 pp.

Crivelli, A.J., 1984. Fisheries. In Pyrovetsi, M.D., A. J. Crivelli, P. A. Gerakis, M. A. Karteris, E. P. Kastro & N. Komninos, Integrated Environmental Study of Prespa National Park, Greece. Final Report to the C.E.C. DG XI.ix+ 205 pp. + maps.

Crivelli, A. J., G. Catsadorakis, M. Malakou & E. Rosecchi, 1997. Fish and fisheries of the Prespa lakes. Hydrobiologia 351: 107–125.

Crivelli, A. J. & G. Catsadorakis, (eds), 1997. Lake Prespa, Northwestern Greece. [A unique Balkan wetland.] Developments in Hydrobiology 122. Kluwer Academic Publishers, Dordrecht, xiv + 196 pp. Reprinted from Hydrobiologia 351.

Gerakis, P. A. & S. E. Tsiouris, 1992. Agricultural techniques for a better protection of wetlands. Final report of a research programme. Aristotelian University of Thessaloniki. School of Agronomy. Laboratory of Ecology and Environmental protection. Thessaloniki, (Mimeographed report in Greek), 104 pp.

Götmark, F., 1992. Naturalness as an evaluation criterion in nature conservation: a response to Anderson. Conserv. Biol. 6: 455–458.

Hoffmann, L., W. Bauer & G. Müller, 1971. Proposals for Nature Conservation in Northern Greece. IUCN Occasional paper No.1. Morges, Switzerland.

Hollis, G. E., H. Athanassiou, S. Bradford, G. Catsadorakis, A. J. Crivelli, R. Flower, A. P. Gerakis, B. Goldsmith, N. Rose, N. Stedman, A. C. Stevenson, D. Thomas & Y. Troumbis, 1989. A Management Plan for the Prespa National Park, Greece. Volume 1: Five Research Studies. Report to WWF (International) on Project 3535, Department of Geography, University College London, 108 pp.

Hutchins, M., 1995. Olympic mountain goat controversy continues. Conserv. Biol. 9: 1324–1326.

Karadedos, G. & P. Tsolakis, 1989. *Prespes*. In the series: Greek traditional architecture. Melissa, Athens, 46 pp.

Karamani, K., 1983. Research by sampling about the external emigration in the Prefecture of Florina. Ministry of National Economy. Service of Regional Development of Central and Western Macedonia, Thessaloniki. Duplicated Report (in Greek).

Karidas, Ch., 1979. Timber exploitation plan for the forests of the Western part of Prespa area. Florina Forestry Service (Unpublished report, in Greek).

Karteris, M. A. & M. Pyrovetsi, 1986. Land Cover/Use Analysis of Prespa National Park, Greece. Environmental Conservation 13: 319–330.

Koussouris, Th. & A. Diapoulis, 1983. Conservation and Development of the country's freshwaters. I. Lake Mikri Prespa. Institute of Oceanographic and Fisheries Research Special Publication No 6. (in Greek), 89 pp.

Koussouris, Th., A. Diapoulis & E. T. Balopoulos, 1989. Assessing the trophic status of Lake Mikri Prespa, Greece. Annls. Limnol. 25: 17–24.

Koussouris, Th. & J. Satsmadjis, 1987. Changes in plankton assemblages from spring to summer in a Greek lake. Rev. Int. Oceanogr. Med. 87/88: 51–66.

Koutsoumbidis, E., 1987. Ecological research in the rivers and lakes of the Prefecture of Florina. Prefecture of Florina. Chemical Service, Florina, (in Greek), 183 pp.

Koroni, M., 1979. Geographical region of Prespa: A Development Plan proposal. Ministry of Coordination. Regional Development Service of Central & Western Macedonia, Thessaloniki (in Greek).

Kosmas, C. S., N. G. Danalatos & N. K. Moustakas, 1997. The soils. Hydrobiologia 351: 21–33.

Lichtenberger, E., 1993. Mountainous perspectives. Lessons from history. Naturopa 72: 9–11.

Maltby, E., 1986. Waterlogged wealth. Earthscan. 200 pp.

Makris, K., 1979. The trade of medicinal and aromatic plants and their need for protection. In Proceedings of Conference on the Protection of the Flora, Fauna and Biotopes in Greece. Athens, October 1979. Hellenic Society for the Protection of Nature, Athens, 87–90.

Matheas, Z., 1975. Timber exploitation plan for the forests of the Eastern side of Lake Mikri Prespa. Florina Forestry Service (in Greek). Unpublished report.

Messerli, P., 1993. Managing this vulnerable environment. Naturopa 72: 6–7.

Mitch, W. J. & J. G. Gosselink, 1986. Wetlands. Van Nostrand Reinhold. 539 pp.

Mourkidis, G., G. Tsikritsis, S. Tsiouris & O. Menkisoglou, 1978. Lakes of Northern Greece. I. Trophic levels. Sci. Ann. Agronomy & Forestry School, the Aristotelian University of Thessaloniki 21: 94–119.

Noss, R. F., 1990. Indicators for monitoring biodiveristy: A hierarchical approach. Conserv. Biol. 4: 355–364.

Pavlides, G., 1985. Geobotanical study of the Prespa National Park. Florina. Part A'. Ecology, flora, phytogeography, vegetation. Aristotelian University of Thessaloniki 308 pp., (in Greek, with English summary).

Pavlides, G., 1997a. The flora of Prespa National Park with emphasis on species of conservation interest. Hydrobiologia 351: 35–40.

Pavlides, G., 1997b. Aquatic and terrestrial vegetation of the Prespa area. Hydrobiologia 351: 41–60.

Pearce, F. & A. J. Crivelli, 1994. Characteristics of Mediterranean Wetlands. A MedWet Publication. Tour du Valat, Arles, France, 90 pp.

Planning Service of Central & Western Macedonia, 1975. Development study for Prespa area. Vol. I & II. 250 pp. Unpublished report 250 pp., (in Greek).

Psychoudakis, A., S. Papoutsi-Psychoudaki & A. M. M. McFarquhar, 1993. An Assessment of the irrigation Project Affecting the Wetlands of Prespa, Greece. Greek Biotope/Wetland Centre and Department of Agricultural Economics, School of Agriculture, Aristotle University of Thessaloniki. 44 pp.

Pyrovetsi, M. D., 1989. Integrated Mediterranean Programmes and the Natural Environment: A Case Study in Greece. Environmentalist 9: 201–211.

Pyrovetsi, M. D. & P. A. Gerakis, 1987. Environmental problems from practising agriculture in Prespa National Park, Greece. Environmentalist 7: 35–42.

Pyrovetsi, M. D. & M. Karteris, 1986. Forty years of land cover/use changes in Prespa National Park, Greece. J. envir. Mgmt. 23: 173–183.

Scoullos, M. & J. Hatzianestis, 1989. Dissolved and particulate trace metals in a wetland of international importance:Lake Mikri Prespa, Greece. Water, Air, Soil Pollut. 44: 307–320.

Stevenson, A. C. & R. J. Flower, 1991. A Palaeoecological Evaluation of Environmental Degradation in Lake Mikri Prespa, NW Greece. Biol. Conserv. 57: 89–109.

Tryphon, E., M. Moustaka-Gouni, G. Nikolaidis & I. Tsekos, 1994. Phytoplankton and physical-chemical features of the shallow Lake Mikri Prespa, Macedonia, Greece. Arch. Hydrobiol. 131: 477–494.

Tsiouris, S. & P. A. Gerakis, 1991. Wetlands of Greece: values, alterations, conservation. WWF & Laboratory of Ecology and Environmental Protection of Faculty of Agriculture of Aristotelian University of Thessaloniki & IUCN, Thessaloniki, 96 pp.

Tziafetas, G. N., 1994. Population changes in W.Macedonia. Panorama 3: 10–16.

Vafiadis, L., 1940. Prespa and its beauties. Athens, 93 pp. (in Greek).

MIX
Papier aus verantwortungsvollen Quellen
Paper from responsible sources
FSC® C105338

If you have any concerns about our products,
you can contact us on
ProductSafety@springernature.com

In case Publisher is established outside the EU,
the EU authorized representative is:
Springer Nature Customer Service Center GmbH
Europaplatz 3, 69115 Heidelberg, Germany

Printed by Libri Plureos GmbH
in Hamburg, Germany